Beginning Data Science in R 4

Data Analysis, Visualization, and Modelling for the Data Scientist

Second Edition

Thomas Mailund

Apress®

Beginning Data Science in R 4: Data Analysis, Visualization, and Modelling for the Data Scientist

Thomas Mailund
Aarhus, Denmark

ISBN-13 (pbk): 978-1-4842-8154-3
https://doi.org/10.1007/978-1-4842-8155-0

ISBN-13 (electronic): 978-1-4842-8155-0

Managing Director, Apress Media LLC: Welmoed Spahr
Acquisitions Editor: Steve Anglin
Development Editor: James Markham
Coordinating Editor: Mark Powers

Cover designed by eStudioCalamar

Cover image by Pixabay (www.pixabay.com)

Distributed to the book trade worldwide by Apress Media, LLC, 1 New York Plaza, New York, NY 10004, U.S.A. Phone 1-800-SPRINGER, fax (201) 348-4505, e-mail orders-ny@springer-sbm.com, or visit www. springeronline.com. Apress Media, LLC is a California LLC and the sole member (owner) is Springer Science + Business Media Finance Inc (SSBM Finance Inc). SSBM Finance Inc is a Delaware corporation.

For information on translations, please e-mail booktranslations@springernature.com; for reprint, paperback, or audio rights, please e-mail bookpermissions@springernature.com.

Apress titles may be purchased in bulk for academic, corporate, or promotional use. eBook versions and licenses are also available for most titles. For more information, reference our Print and eBook Bulk Sales web page at http://www.apress.com/bulk-sales.

Any source code or other supplementary material referenced by the author in this book is available to readers on GitHub (https://github.com/Apress). For more detailed information, please visit http://www. apress.com/source-code.

Printed on acid-free paper

Table of Contents

About the Author

Thomas Mailund is an associate professor in bioinformatics at Aarhus University, Denmark. His background is in math and computer science, but for the last decade, his main focus has been on genetics and evolutionary studies, particularly comparative genomics, speciation, and gene flow between emerging species.

About the Technical Reviewer

Jon Westfall is an associate professor of psychology at Delta State University. He has authored *Set Up and Manage Your Virtual Private Server, Practical R 4, Beginning Android Web Apps Development, Windows Phone 7 Made Simple,* and several works of fiction including *One in the Same, Mandate,* and *Franklin: The Ghost Who Successfully Evicted Hipsters from His Home and Other Short Stories.* He lives in Cleveland, Mississippi, with his wife.

Acknowledgments

I would like to thank Asger Hobolth for many valuable comments on earlier versions of this manuscript that helped me improve the writing and the presentation of the material.

Introduction

Welcome to *Beginning Data Science in R 4*. I wrote this book from a set of lecture notes for two classes I taught a few years back, "Data Science: Visualization and Analysis" and "Data Science: Software Development and Testing." The book is written to fit the structure of these classes, where each class consists of seven weeks of lectures followed by project work. This means that the book's first half consists of eight chapters with core material, where the first seven focus on data analysis and the eighth is an example of a data analysis project. The data analysis chapters are followed by seven chapters on developing reusable software for data science and then a second project that ties the software development chapters together. At the end of the book, you should have a good sense of what data science can be, both as a field covering analysis and developing new methods and reusable software products.

What Is Data Science?

That is a difficult question. I don't know if it is easy to find someone who is entirely sure what data science is, but I am pretty sure that it would be difficult to find two people without having three opinions about it. It is undoubtedly a popular buzzword, and everyone wants to hire data scientists these days, so data science skills are helpful to have on the CV. But what is it?

Since I can't give you an agreed-upon definition, I will just give you my own: data science is the science of learning from data.

This definition is very broad—almost too broad to be useful. I realize this. But then, I think data science is an incredibly general field. I don't have a problem with that. Of course, you could argue that any science is all about getting information out of data, and you might be right. However, I would say that there is more to science than just transforming raw data into useful information. The sciences focus on answering specific questions about the world, while data science focuses on how to manipulate data efficiently and effectively. The primary focus is not which questions to ask of the data but how we can answer them, whatever they may be. It is more like computer science and mathematics than it is like natural sciences, in this way. It isn't so much about studying the natural world as it is about computing efficiently on data and learning patterns from the data.

Included in data science is also the design of experiments. With the right data, we can address the questions in which we are interested. This can be difficult with a poor design of experiments or a poor choice of which data we gather. Study design might be the most critical aspect of data science but is not the topic of this book. In this book, I focus on the analysis of data, once gathered.

Computer science is mainly the study of computations, hinted at in the name, but is a bit broader. It is also about representing and manipulating data. The name "computer science" focuses on computation, while "data science" emphasizes data. But of course, the fields overlap. If you are writing a sorting algorithm, are you then focusing on the computation or the data? Is that even a meaningful question to ask?

There is considerable overlap between computer science and data science, and, naturally, the skill sets you need overlap as well. To efficiently manipulate data, you need the tools for doing that, so computer programming skills are a must, and some knowledge about algorithms and data structures usually is as well. For data science, though, the focus is always on the data. A data analysis project focuses on how the data flows from its raw form through various manipulations until it is summarized in some helpful way. Although the difference can be subtle, the focus is not on what operations a program does during the analysis but how the data flows and is transformed. It is also focused on why we do certain data transformations, what purpose those changes serve, and how they help us gain knowledge about the data. It is as much about deciding what to do with the data as it is about how to do it efficiently.

Statistics is, of course, also closely related to data science. So closely linked that many consider data science as nothing more than a fancy word for statistics that looks slightly more modern and sexy. I can't say that I strongly disagree with this—data science does sound hotter than statistics—but just as data science is slightly different from computer science, data science is also somewhat different from statistics. Only, perhaps, somewhat less so than computer science is.

A large part of doing statistics is building mathematical models for your data and fitting the models to the data to learn about the data in this way. That is also what we do in data science. As long as the focus is on the data, I am happy to call statistics data science. But suppose the focus changes to the models and the mathematics. In that case, we are drifting away from data science into something else—just as if the focus shifts from the data to computations, we are straying from data science to computer science.

Data science is also related to machine learning and artificial intelligence—and again, there are huge overlaps. Perhaps not surprising since something like machine learning has its home both in computer science and statistics; if it focuses on data analysis, it is also at home in data science. To be honest, it has never been clear to me when a mathematical model changes from being a plain old statistical model to becoming machine learning anyway.

For this book, we are just going to go with my definition, and, as long as we are focusing on analyzing data, we will call it data science.

Prerequisites for Reading This Book

For the first eight chapters in this book, the focus is on data analysis and not programming. For those eight chapters, I do not assume a detailed familiarity with software design, algorithms, data structures, etc. I do not expect you to have any experience with the R programming language either. However, I assume that you have had some experience with programming, mathematical modelling, and statistics.

Programming R can be quite tricky at times if you are familiar with scripting languages or object-oriented languages. R is a functional language that does not allow you to modify data. While it does have systems for object-oriented programming, it handles this programming paradigm very differently from languages you are likely to have seen, such as Java or Python.

For the data analysis part of this book, the first eight chapters, we will only use R for very straightforward programming tasks, so none of this should pose a problem. We will have to write simple scripts for manipulating and summarizing data, so you should be familiar with how to write basic expressions like function calls, if statements, loops, and such—these things you will have to be comfortable with. I will introduce every such construction in the book when we need them to let you see how they are written in R, but I will not spend much time explaining them. Mostly, I will expect you to be able to pick it up from examples.

Similarly, I do not expect you to already know how to fit data and compare models in R. I do assume that you have had enough introduction to statistics to be comfortable with basic terms like parameter estimation, model fitting, explanatory and response variables, and model comparison. If not, I expect you to at least be able to pick up what we are talking about when you need to.

I won't expect you to know a lot about statistics and programming, but this isn't "Data Science for Dummies," so I expect you to figure out examples without me explaining everything in detail.

After the first seven chapters is a short description of a data analysis project that one of my students did for my class the first time I held it. It shows how such a project could look, but I suggest that you do not wait until you have finished the first seven chapters to start doing such analysis yourself. To get the most benefit out of reading this book, you should continuously apply what you learn. Already when you begin reading, I suggest that you find a data set that you would be interested in finding out more about and then apply what you learn in each chapter to that data.

For the following eight chapters of the book, the focus is on programming. To read this part, you should be familiar with object-oriented programming—I will explain how we handle it in R and how it differs from languages such as Python, Java, or C++. Still, I will expect you to be familiar with terms such as class hierarchies, inheritance, and polymorphic methods. I will not expect you to be already familiar with functional programming (but if you are, there should still be plenty to learn in those chapters if you are not already familiar with R programming). The final chapter is yet another project description.

Plan for the Book

In the book, we will cover basic data manipulation:

- Filtering and selecting relevant data

- Transforming data into shapes readily analyzable

- Summarizing data

- Visualization data in informative ways both for exploring data and presenting results

- Model building

These are the critical aspects of doing analysis in data science. After this, we will cover how to develop R code that is reusable and works well with existing packages and that is easy to extend, and we will see how to build new R packages that other people will be able to use in their projects. These are the essential skills you will need to develop your own methods and share them with the world.

R is one of the most popular (and open source) data analysis programming languages around at the moment. Of course, popularity doesn't imply quality. Still, because R is so popular, it has a rich ecosystem of extensions (called "packages" in R) for just about any kind of analysis you could be interested in. People who develop statistical methods often implement them as R packages, so you can usually get the state-of-the-art techniques very easily in R. The popularity also means that there is a large community of people who can help if you have problems. Most problems you run into can be solved with a few minutes on Google or Stack Overflow because you are unlikely to be the first to run into any particular issue. There are also plenty of online tutorials for learning more about R and specialized packages. And there are plenty of books you can buy if you want to learn more.

Data Analysis and Visualization

The topics focusing on data analysis and visualization I cover in the first eight chapters:

1. Introduction to R Programming: In this chapter, we learn how to work with data and write data pipelines.

2. Reproducible Analysis: In this chapter, we find out how to integrate documentation and analysis in a single document and how to use such documents to produce reproducible research.

3. Data Manipulation: In this chapter, we learn how to import data, tidy up data, transform, and compute summaries from data.

4. Visualizing Data: In this chapter, we learn how to make plots for exploring data features and presenting data features and analysis results.

5. Working with Large Data Sets: In this chapter, we see how to deal with data where the number of observations makes our usual approaches too slow.

6. Supervised Learning: In this chapter, we learn how to train models when we have data sets with known classes or regression values.

7. Unsupervised Learning: In this chapter, we learn how to search for patterns we are not aware of in data.

8. Project 1: Hitting the Bottle: Following these chapters is the first project, an analysis of physicochemical features of wine, where we see the various techniques in use.

Software Development

The next nine chapters cover software and package development:

1. Deeper into R Programming: In this chapter, we explore more advanced features of the R programming language.

2. Working with Vectors and Lists: In this chapter, we explore two essential data structures, namely, vectors and lists.

3. Functional Programming: In this chapter, we explore an advanced feature of the R programming language, namely, functional programming.

4. Object-Oriented Programming: In this chapter, we learn how R handles object orientation and how we can use it to write more generic code.

5. Building an R Package: In this chapter, we learn the necessary components of an R package and how we can program our own.

6. Testing and Package Checking: In this chapter, we learn techniques for testing our R code and checking our R packages' consistency.

7. Version Control: In this chapter, we learn how to manage code under version control and how to collaborate using GitHub.

8. Profiling and Optimizing: In this chapter, we learn how to identify code hotspots where inefficient solutions are slowing us down and techniques for alleviating this.

9. Project 2: Bayesian Linear Regression: In the final chapter, we get to the second project, where we build a package for Bayesian linear regression.

Getting R and RStudio

You will need to install R on your computer to do the exercises in this book. I suggest that you get an integrated environment since it can be slightly easier to keep track of a project when you have your plots, documentation, code, etc., all in the same program.

I use RStudio (`www.rstudio.com/products/RStudio`), which I warmly recommend. You can get it for free—just follow the link—and I will assume that you have it when I need to refer to the software environment you are using in the following chapters. There won't be much RStudio specific, though, and most tools for working with R have mostly the same features, so if you want to use something else, you can probably follow the notes without any difficulties.

Projects

You cannot learn how to analyze data without analyzing data, and you cannot understand how to develop software without developing software either. Typing in examples from the book is nothing like writing code on your own. Even doing exercises from the book—which you really ought to do—is not the same as working on your own projects. Exercises, after all, cover minor isolated aspects of problems you have just been introduced to. There is not a chapter of material presented before every task you have to deal with in the real world. You need to work out by yourself what needs to be done and how. If you only do the exercises in this book, you will miss the most crucial lesson in analyzing data:

- How to explore the data and get a feeling for it

- How to do the detective work necessary to pull out some understanding from the data

- How to deal with all the noise and weirdness found in any data set

And for developing a package, you need to think through how to design and implement its functionality such that the various functions and data structures fit well together.

I will go through a data analysis project to show you what that can look like in this book. To learn how to analyze data on your own, you need to do it yourself as well—and you need to do it with a data set that I haven't explored for you. You might have a data set lying around you have worked on before, a data set from something you are just interested in, or you can probably find something interesting at a public data repository, for example, one of these:

- RDataMining.com: `www.rdatamining.com/resources/data`

- UCI Machine Learning Repository: `http://archive.ics.uci.edu/ml/`

- KDNuggets: `www.kdnuggets.com/datasets/index.html`

- Reddit R Data sets: `www.reddit.com/r/datasets`

- GitHub Awesome Public Data sets: `https://github.com/caesar0301/awesome-public-datasets`

I suggest that you find yourself a data set and that you, after each lesson, use the skills you have learned to explore this data set. Pick data structured as a table with observations as rows and variables as columns since that is the form of the data we will consider in this book. At the end of the first eight chapters, you will have analyzed this data. You can write a report about your analysis that others can evaluate to follow and maybe modify it: you will be doing reproducible science.

For the programming topics, I will describe another project illustrating the design and implementation issues involved in making an R package. There, you should be able to learn from implementing your own version of the project I use, but you will, of course, be more challenged by working on a project without any of my help at all. Whatever you do, to get the full benefit of this book, you really ought to make your own package while reading the programming chapters.

CHAPTER 1

Introduction to R Programming

We will use R for our data analysis, so we need to know the basics of programming in the R language. R is a full programming language with both functional programming and object-oriented programming features, and learning the complete language is far beyond the scope of this chapter. We return to it later, when we have a little more experience using R. The good news is, though, that to use R for data analysis, we rarely need to do much programming. At least, if you do the right kind of programming, you won't need much.

For manipulating data—how to do this is the topic of the next chapter—you mainly have to string together a couple of operations, such as "group the data by this feature" followed by "calculate the mean value of these features within each group" and then "plot these means." Doing this used to be more complicated to do in R, but a couple of new ideas on how to structure data flow—and some clever implementations of these in packages such as `magrittr` and `dplyr`—have significantly simplified it. We will see some of this at the end of this chapter and more in the next chapter. First, though, we need to get a taste of R.

Basic Interaction with R

Start by downloading RStudio if you haven't done so already. If you open it, you should get a window similar to Figure 1-1. Well, except that you will be in an empty project while the figure shows (on the top right) that this RStudio is opened in a project called "Data Science." You always want to be working on a project. Projects keep track of the state of your analysis by remembering variables and functions you have written and keep track of which files you have opened and such. Go to File and then New Project to create a

© Thomas Mailund 2022
T. Mailund, *Beginning Data Science in R 4*, https://doi.org/10.1007/978-1-4842-8155-0_1

project. You can create a project from an existing directory, but if this is the first time you are working with R, you probably just want to create an empty project in a new directory, so do that.

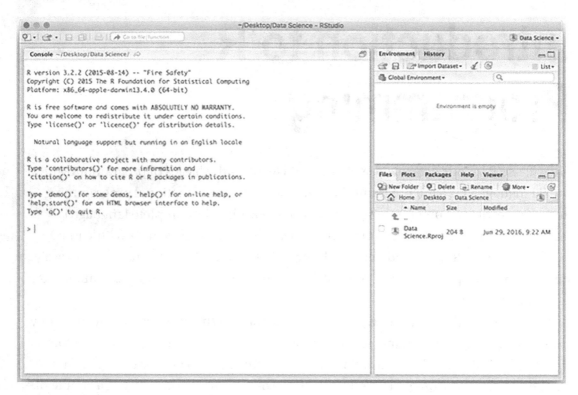

Figure 1-1. *RStudio*

Once you have RStudio opened, you can type R expressions into the console, which is the frame on the left of the RStudio window. When you write an expression there, R will read it, evaluate it, and print the result. When you assign values to variables, and we will see how to do this shortly, they will appear in the Environment frame on the top right. At the bottom right, you have the directory where the project lives, and files you create will go there.

To create a new file, you go to File and then New File.... There you can select several different file types. Those we are interested in are the R Script, R Notebook, and R Markdown types. The former is the file type for pure R code, while the latter two we use for creating reports where documentation text is mixed with R code. For data analysis projects, I would recommend using either Notebook or Markdown files. Writing documentation for what you are doing is helpful when you need to go back to a project several months down the line.

For most of this chapter, you can just write R code in the console, or you can create an R Script file. If you create an R Script file, it will show up on the top left; see Figure 1-2. You can evaluate single expressions using the Run button on the top right of this frame or evaluate the entire file using the Source button. For writing longer expressions, you might want to write them in an R Script file for now. In the next chapter, we will talk about R Markdown, which is the better solution for data science projects.

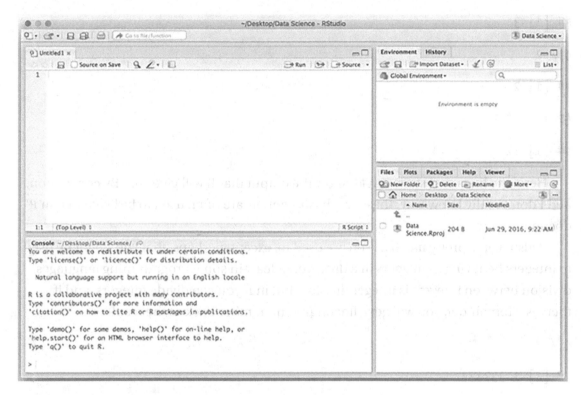

Figure 1-2. *RStudio with a new R Script file open*

Using R As a Calculator

You can use the R console as a calculator where you type in an expression you want to calculate, hit "enter," and R gives you the result. You can play around with that a little bit to get familiar with how to write expressions in R—there is some explanation for how to write them in the following—and then moving from using R as a calculator to writing

3

more sophisticated analysis programs is only a matter of degree. A data analysis program is little more than a sequence of calculations, after all.

Simple Expressions

Simple arithmetic expressions are written, as in most other programming languages, in the typical mathematical notation that you are used to:

```
1 + 2
```

```
## [1] 3
```

```
4 / 2
```

```
## [1] 2
```

```
(2 + 2) * 3
```

```
## [1] 12
```

Here, the lines that start with ## show the output that R will give you. By convention, and I don't really know why, these two hash symbols are often used to indicate that in R documentation.

It also works pretty much as you are used to, except, perhaps, that you might be used to integers behaving as integers in a division. At least in some programming languages, division between integers is integer division, but in R you can divide integers, and if there is a remainder, you will get a floating-point number back as the result:

```
4 / 3
```

```
## [1] 1.333333
```

When you write numbers like 4 and 3, they are always interpreted as floating-point numbers, even if they print as integers, that is, without a decimal point. To explicitly get an integer, you must write 4L and 3L:

```
class(4)
```

```
## [1] "numeric"
```

```
class(4L)
```

```
## [1] "integer"
```

It usually doesn't matter if you have an integer or a floating-point number, and everywhere you see numbers in R, they are likely to be floats.

You will still get a floating-point if you divide two integers, and there is no need to tell R explicitly that you want floating-point division. If you do want integer division, on the other hand, you need a different operator, %/%:

```
4 %/% 3
```

```
## [1] 1
```

In many languages, % is used for getting the remainder of a division, but this doesn't quite work with R where % is used for something else (creating new infix operators), so in R the operator for this is %%:

```
4 %% 3
```

```
## [1] 1
```

In addition to the basic arithmetic operators—addition, subtraction, multiplication, division, and the modulus operator we just saw—you also have an exponentiation operator for taking powers. For this, you can use either ^ or ** as infix operators:

```
2 ^ 2
```

```
## [1] 4
```

```
2 ** 2
```

```
## [1] 4
```

```
2 ^ 3
```

```
## [1] 8
```

```
2 ** 3
```

```
## [1] 8
```

There are some other data types besides numbers, but we won't go into an exhaustive list here. There are two types you do need to know about early, though, since they are frequently used and since not knowing about how they work can lead to all kinds of grief. Those are strings and "factors."

Strings work as you would expect. You write them in quotes, either double quotes or single quotes, and that is about it:

```
"Hello,"
## [1] "Hello,"
'world!'
## [1] "world!"
```

Strings are not particularly tricky, but I mention them because they look a lot like factors, but factors are not like strings, they just look sufficiently like them to cause some confusion. I will explain the factors a little later in this chapter when we have seen how functions and vectors work.

Assignments

To assign a value to a variable, you use the arrow operators. So to assign the value 2 to the variable x, you would write

```
x <- 2
```

and you can test that x now holds the value 2 by evaluating x:

```
x
## [1] 2
```

and of course, you can now use x in expressions:

```
2 * x
## [1] 4
```

You can assign with arrows in both directions, so you could also write

```
2 -> x
```

An assignment won't print anything if you write it into the R terminal, but you can get R to print it by putting the assignment in parentheses:

```
x <- "invisible"
(y <- "visible")

## [1] "visible"
```

Actually, all of the above are vectors of values...

If you were wondering why all the values printed earlier had a [1] in front of them, it is because we are usually not working with single values anywhere in R. We are working with vectors of values (and you will hear more about vectors in the next section). The vectors we have seen have length one—they consist of a single value—so there is nothing wrong about thinking about them as individual values. But they are vectors and what we can do with a single number we can do with multiple in the same way.

The [1] does not indicate that we are looking at a vector of length one. The [1] tells you that the first value after [1] is the first value in the vector. With longer vectors, you get the index each time R moves to the next line of output. This output makes it easier to count your way into a particular index.

You will see this if you make a longer vector, for example, we can make one of length 50 using the : operator:

```
1:50

##  [1]   1   2   3   4   5   6   7   8   9  10  11  12  13  14  15
## [16]  16  17  18  19  20  21  22  23  24  25  26  27  28  29  30
## [31]  31  32  33  34  35  36  37  38  39  40  41  42  43  44  45
## [46]  46  47  48  49  50
```

The : operator creates a sequence of numbers, starting at the number to the left of the colon and increasing by one until it reaches the number to the right of the colon, or just before if an increment of one would move past the last number:

```
-1:1

## [1] -1  0  1

0.1:2.9

## [1] 0.1 1.1 2.1
```

If you want other increments than 1, you can use the seq function instead:

```
seq(.1, .9, .2)
```

```
## [1] 0.1 0.3 0.5 0.7 0.9
```

Here, the first number is where we start, the second where we should stop, as with :, but the third number gives us the increment to use.

Because we are practically always working on vectors, there is one caveat I want to warn you about. If you want to know the length of a string, you might—reasonably enough—think you can get that using the length function. You would be wrong. That function gives you the length of a vector, so if you give it a single string, it will always return 1:

```
length("qax")
```

```
## [1] 1
```

```
length("quux")
```

```
## [1] 1
```

```
length(c("foo", "bar"))
```

```
## [1] 2
```

In the last expression, we used the function c() to concatenate two vectors of strings. Concatenating "foo" and "bar"

```
c("foo", "bar")
```

```
## [1] "foo" "bar"
```

creates a vector of two strings, and thus the result of calling length on that is 2. To get the length of the actual string, you want nchar instead:

```
nchar("qax")
```

```
## [1] 3
```

```
nchar("quux")
```

```
## [1] 4
```

```
nchar(c("foo", "bar"))
```

```
## [1] 3 3
```

If you wanted to concatenate the strings "foo" and "bar", to get a vector with the single string "foobar", you need to use paste:

```
paste("foo", "bar", sep = "")
```

```
## [1] "foobar"
```

The argument sep = "" tells paste not to put anything between the two strings. By default, it would put a space between them:

```
paste("foo", "bar")
```

```
## [1] "foo bar"
```

Indexing Vectors

If you have a vector and want the i'th element of that vector, you can index the vector to get it like this:

```
(v <- 1:5)
```

```
## [1] 1 2 3 4 5
```

```
v[1]
```

```
## [1] 1
```

```
v[3]
```

```
## [1] 3
```

We have parentheses around the first expression to see the output of the operation. An assignment is usually silent in R, but by putting the expression in parentheses, we make sure that R prints the result, which is the vector of integers from 1 to 5. Notice here that the first element is at index 1. Many programming languages start indexing at zero, but R starts indexing at one. A vector of length n is thus indexed from 1 to n, unlike in zero-indexed languages where the indices go from 0 to $n - 1$.

If you want to extract a subvector, you can also do this with indexing. You just use a vector of the indices you want inside the square brackets. We can use the : operator for this or the concatenate function, c():

```
v[1:3]
```

```
## [1] 1 2 3
```

```
v[c(1,3,5)]
```

```
## [1] 1 3 5
```

You can use a vector of boolean values to pick out those values that are "true":

```
v[c(TRUE, FALSE, TRUE, FALSE, TRUE)]
```

```
## [1] 1 3 5
```

This indexing is particularly useful when you combine it with expressions. We can, for example, get a vector of boolean values telling us which values of a vector are even numbers and then use that vector to pick them out:

```
v %% 2 == 0
```

```
## [1] FALSE  TRUE FALSE  TRUE FALSE
```

```
v[v %% 2 == 0]
```

```
## [1] 2 4
```

You can get the complement of a vector of indices if you change the sign of them:

```
v[-(1:3)]
```

```
## [1] 4 5
```

It is also possible to give vector indices names, and if you do, you can use those to index into the vector. You can set the names of a vector when constructing it or use the names function:

```
v <- c("A" = 1, "B" = 2, "C" = 3)
v
```

```
## A B C
```

```
## 1 2 3

v["A"]

## A
## 1

names(v) <- c("x", "y", "z")
v

## x y z
## 1 2 3

v["x"]

## x
## 1
```

Names can be handy for making tables where you can look up a value by a key.

Vectorized Expressions

Now, the reason that the expressions we saw earlier worked with vector values instead of single values is that in R, arithmetic expressions all work component-wise on vectors. When you write an expression such as

```
x <- 1:3 ; y <- 4:6
x ** 2 - y

## [1] -3 -1  3
```

you are telling R to take each element in the vector x, squaring it, and subtracting element-wise by y:

```
(x <- 1:3)

## [1] 1 2 3

x ** 2

## [1] 1 4 9

y <- 6:8
```

```
x ** 2 - y
```

```
## [1] -5 -3  1
```

This also works if the vectors have different lengths, as they do in the preceding example. The vector 2 is a vector of length 1 containing the number 2. The way expressions work, when vectors do not have the same length, is you repeat the shorter vector as many times as you need to:

```
(x <- 1:4)
```

```
## [1] 1 2 3 4
```

```
(y <- 1:2)
```

```
## [1] 1 2
```

```
x - y
```

```
## [1] 0 0 2 2
```

If the length of the longer vector is not a multiple of the length of the shorter, you get a warning. The expression still repeats the shorter vector a number of times, just not an integer number of times:

```
(x <- 1:4)
```

```
## [1] 1 2 3 4
```

```
(y <- 1:3)
```

```
## [1] 1 2 3
```

```
x - y
```

```
## Warning in x - y: longer object length is not a
## multiple of shorter object length
```

```
## [1] 0 0 0 3
```

Here, y is used once against the 1:3 part of x, and the first element of y is then used for the 4 in x.

Comments

You probably don't want to write comments when you are just interacting with the R terminal, but in your code, you do. Comments let you describe what your code is intended to do, and how it is achieving it, so you don't have to work that out again when you return to it at a later point, having forgotten all the great thoughts you thought when you wrote it.

R interprets as comments everything that follows the # character. From a # to the end of the line, the R parser skips the text:

```
# This is a comment.
```

If you write your analysis code in R Markdown documents, which we will cover in the next chapter, you won't have much need for comments. In those kinds of files, you mix text and R code differently. But if you develop R code, you will likely need it, and now you know how to write comments.

Functions

You have already seen the use of functions, although you probably didn't think much about it when we saw expressions such as

```
length("qax")
```

You didn't think about it because there wasn't anything surprising about it. We just use the usual mathematical notation for functions: $f(x)$. If you want to call a function, you simply use this notation and give the function its parameters in parentheses.

In R, you can also use the names of the parameters when calling a function, in addition to the positions; we saw an example with sep = "" when we used paste to concatenate two strings.

If you have a function $f(x, y)$ of two parameters, x and y, calling $f(5, 10)$ means calling f with parameter x set to 5 and parameter y set to 10. In R, you can specify this explicitly, and these two function calls are equivalent:

```
f(5, 10)
f(x = 5, y = 10)
```

(Don't try to run this code; we haven't defined the function f, so calling it will fail. But if we had a function f, then the two calls would be equivalent.)

If you specify the names of the parameters, the order doesn't matter anymore, so another equivalent function call would be

```
f(y = 10, x = 5)
```

You can combine the two ways of passing parameters to functions as long as you put all the positional parameters before the named ones:

```
f(5, y = 10)
```

Except for maybe making the code slightly more readable—it is usually easier to remember what parameters do than which order they come in—there is not much need for this in itself. Where it becomes useful is when combined with default parameters.

A lot of functions in R take many parameters. More than you really can remember the use for and certainly the order of. They are a lot like programs that take a lot of options but where you usually just use the defaults unless you need to tweak something. These functions take a lot of parameters, but most of them have useful default values, and you typically do not have to specify the values to set them to. When you do need it, though, you can specify it with a named parameter.

Getting Documentation for Functions

Since it can be hard to remember the details of what a function does, and especially what all the parameters to a function do, you often have to look up the documentation for functions. Luckily, this is very easy to do in R and RStudio. Whenever you want to know what a function does, you can just ask R, and it will tell you (assuming that the author of the function has written the documentation).

Take the function length from the example we saw earlier. If you want to know what the function does, just write ?length in the R terminal. If you do this in RStudio, it will show you the documentation in the frame on the right; see Figure 1-3.

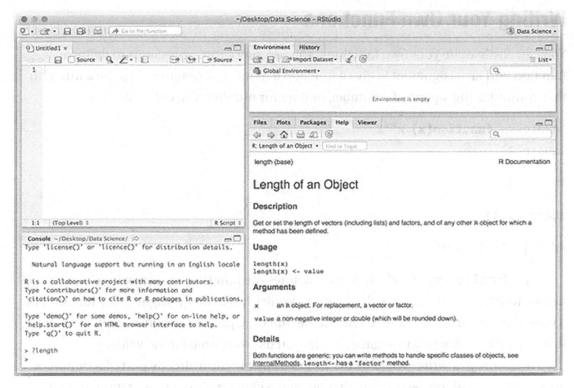

Figure 1-3. *RStudio's help frame*

Try looking up the documentation for a few functions, for example, the `nchar` function we also saw earlier.

All infix operators, like + or %%, are also functions in R, and you can read the documentation for them as well. But you cannot write ?+ in the R terminal and get the information. The R parser doesn't know how to deal with that. If you want help on an infix operator, you need to quote it, and you do that using back quotes. So to read the documentation for +, you would need to write

```
?`+`
```

You probably do not need help to figure out what addition does, but people can write new infix operators, so this is useful to know when you need help with those.

Writing Your Own Functions

You can easily write your own functions. You use `function` expressions to define a function and an assignment to give a function a name. For example, to write a function that computes the square of a number, or a vector number, you can write

```
square <- function(x) x**2
square(2)
```

```
## [1] 4
```

```
square(1:4)
```

```
## [1]  1  4  9 16
```

The "`function(x) x**2`" expression defines the function, and anywhere you would need a function, you can write the function explicitly like this. Assigning the function to a name lets you use the name to refer to the function, just like assigning any other value, like a number or a string to a name, will let you use the name for the value.

Functions you write yourself work just like any function already part of R or part of an R package, with one exception, though: you will not have documentation for your functions unless you write it, and that is beyond the scope of this chapter (but covered in the chapter on building packages).

The `square` function just does a simple arithmetic operation on its input. Sometimes, you want the function to do more than a single thing. If you want the function to do several operations on its input, you need several statements for the function. In that case, you need to give it a "body" of several statements, and such a body has to go in curly brackets:

```
square_and_subtract <- function(x, y) {
    squared <- x ** 2
    squared - y
}
square_and_subtract(1:5, rev(1:5))
```

```
## [1] -4 0 6 14 24
```

(Check the documentation for `rev` to see what is going on here. Make sure you understand what this example is doing.)

In this simple example, we didn't really need several statements. We could just have written the function as

```
square_and_subtract <- function(x, y) x ** 2 - y
```

As long as there is only a single expression in the function, we don't need the curly brackets. For more complex functions, you will need it, though.

The result of a function—what it returns as its value when you call it—is the last statement or expression (there actually isn't any difference between statements and expressions in R; they are the same thing). You can make the return value explicit, though, using the return() expression:

```
square_and_subtract <- function(x, y) return(x ** 2 - y)
```

Explicit returning is usually only used when you want to return a value before the end of the function. To see examples of this, we need control structures, so we will have to wait a little bit to see an example. It isn't used as much as in many other programming languages.

One crucial point here, though, if you are used to programming in other languages: The return() expression needs to include the parentheses. In most programming languages, you could just write

```
square_and_subtract <- function(x, y) return x ** 2 - y
```

Such an expression doesn't work for R. Try it, and you will get an error.

Summarizing and Vector Functions

As we have already seen, when we write arithmetic expressions such as x**2 - y, we have an expression that will work for both single numbers for x and y, but also element-wise for vectors x and y. If you write functions where the body consists of such expressions, the function will work element-wise as well. The square and square_and_ subtract functions we wrote earlier work like that.

Now all functions work like this, however. While we often can treat data one element at a time, we also often need to extract some summary of a collection of data, and functions handle this as well.

Take, for example, the function sum which adds together all the values in a vector you give it as an argument (check ?sum now to see the documentation):

```
sum(1:4)
```

```
## [1] 10
```

This function summarizes its input into a single value. There are many similar functions, and, naturally, these cannot be used element-wise on vectors; rather, they reduce an entire vector into some smaller summary statistics, here the sum of all elements.

Whether a function works on vector expressions or not depends on how it is defined. While there are exceptions, most functions in R either work on vectors or summarize vectors like sum. When you write your own functions, whether the function works element-wise on vectors or not depends on what you put in the body of the function. If you write a function that just does arithmetic on the input, like square, it will work in vectorized expressions. If you write a function that does some summary of the data, it will not. For example, if we write a function to compute the average of its input like this:

```
average <- function(x) {
    n <- length(x)
    sum(x) / n
}
average(1:5)
```

```
## [1] 3
```

This function will not give you values element-wise. Pretty obviously. It gets a little more complicated when the function you write contains control structures, which we will get to in the next section. In any case, this would be a nicer implementation since it only involves one expression:

```
average <- function(x) sum(x) / length(x)
```

Oh, and by the way, don't use this average function to compute the mean value of a vector. R already has a function for that, mean, that deals much better with special cases like missing data and vectors of length zero. Check out ?mean.

Just because you are summarizing doesn't mean that you have to return a single value. In this function, we return both the mean and the standard deviation of the values in a vector:

```
mean_and_sd <- function(x) c(mean = mean(x), sd = sd(x))
mean_and_sd(1:10)

##      mean        sd
##   5.50000   3.02765
```

We use the functions mean and sd to compute the two summary statistics, and then we combine them into a vector (with named elements) that contains the two summaries. This isn't a vectorized function, because we do not process the values in the input element-wise. It doesn't compute a single summary, but returns something (ever so slightly) more complex. Complicated functions often return data more complex than vectors or single values, and we shall see examples in later chapters. If you can avoid it, though, do so. Simple functions, with simple input and output, are easier to use, and when we write functions, we want to make things as simple for us as we can. With this mean_and_sd function, we do not gain anything that we do not already have with the mean and sd function, and combining both operations in a single function only complicates things needlessly.

The rough classification of functions into the vectorized, which operate element-wise on data, and the summarizing functions, is only a classification of how we can use them. If you compute a value for each element in one or more vectors, you have the former, and if you summarize all the data in one or more vectors, you have the latter. The implementation of a function can easily combine both.

Imagine, for example, that we wish to normalize data by subtracting the mean from each element and then dividing by the standard deviation. We could implement it like this:

```
normalise <- function(x) (x - mean(x)) / sd(x)
normalise(1:10)

##  [1]  -1.4863011  -1.1560120 -0.8257228 -0.4954337
##  [5]  -0.1651446   0.1651446  0.4954337  0.8257228
##  [9]   1.1560120   1.4863011
```

We compute a value for each element in the input, so we have a vectorized function, but in the implementation, we use two summarizing functions, mean and sd. The expression (x - mean(x)) / sd(x) is a vector expression because mean(x) and sd(x) become vectors of length one, and we can use those in the expression involving x to get a value for each element.

A Quick Look at Control Flow

While you get very far just using expressions, for many computations, you need more complex programming. Not that it is particularly complex, but you do need to be able to select a choice of what to do based on data—selection or if statements—and ways of iterating through data, looping or for statements.

If statements work like this:

```
if (<boolean expression>) <expression>
```

If the boolean expression evaluates to true, the expression is evaluated; if not, it will not:

```
# this won't do anything
if (2 > 3) "false"
```

```
# this will
if (3 > 2) "true"
```

```
## [1] "true"
```

For expressions like these, where we do not alter the program state by evaluating the expression, there isn't much of an effect in evaluating the if expression. If we, for example, are assigning to a variable, there will be an effect:

```
x <- "foo"
if (2 > 3) x <- "bar"
x
```

```
## [1] "foo"
```

```
if (3 > 2) x <- "baz"
x
```

```
## [1] "baz"
```

20

If you want to have effects for both true and false expressions, you have this:

```
if (<boolean expression>) <true expression> else <false expression>
```

```
if (2 > 3) "bar" else "baz"
```

```
## [1] "baz"
```

If you want newlines in if statements, whether you have an else part or not, you should use curly brackets.

You don't always have to. If you have a single expression in the if part, you can leave them out:

```
if (3 > 2)
    x <- "bar"
x
```

```
## [1] "bar"
```

or if you have a single statement in the else part, you can leave out the brackets:

```
if (2 > 3) {
    x <- "bar"
} else
    x <- "qux"
x
```

```
## [1] "qux"
```

but we did need the brackets in the preceding if part for R to recognize that an else bit was following. Without it, we would get an error:

```
if (2 > 3)
    x <- "bar"
else
    x <- "qux"
```

```
## Error: <text>:3:1: unexpected 'else'
## 2:     x <- "bar"
## 3: else
##    ^
```

If you always use brackets, you don't have to worry about when you strictly need them or when you do not, and a part can have multiple statements without you having to worry about it. If you put a newline in an `if` or `if-else` expression, I recommend that you always use brackets as well.

An `if` statement works like an expression:

```
if (2 > 3) "bar" else "baz"
```

```
## [1] "baz"
```

This evaluates to the result of the expression in the "if" or the "else" part, depending on the truth value of the condition:

```
x <- if (2 > 3) "bar" else "baz"
x
```

```
## [1] "baz"
```

It works just as well with braces:

```
x <- if (2 > 3) { "bar" } else { "baz" }
x
```

```
## [1] "baz"
```

but when the entire statement is on a single line, and the two parts are both a single expression, I usually do not bother with that.

You cannot use it for vectorized expressions, though, since the boolean expression, if you give it a vector, will evaluate the first element in the vector:

```
x <- 1:5
if (x > 3) "bar" else "baz"
```

```
## Warning in if (x > 3) "bar" else "baz": the
## condition has length > 1 and only the first
## element will be used
```

```
## [1] "baz"
```

If you want a vectorized version of `if` statements, you can instead use the `ifelse()` function:

```
x <- 1:5
ifelse(x > 3, "bar", "baz")

## [1] "baz" "baz" "baz" "bar" "bar"
```

(read the ?ifelse documentation to get the details of this function).

This, of course, also has consequences for writing functions that use if statements. If your function contains a body that isn't vectorized, your function won't be either. So, if you have an if statement that depends on your input—and if it doesn't depend on the input, it is rather useless—then that input shouldn't be a vector:

```
maybe_square <- function(x) {
    if (x %% 2 == 0) x ** 2 else x
}
maybe_square(1:5)

## Warning in if (x%%2 == 0) x^2 else x: the
## condition has length > 1 and only the first
## element will be used

## [1] 1 2 3 4 5
```

This function was supposed to square even numbers, and it will if we give it a single number, but we gave it a vector. Since the first value in this vector, the only one that the if statement looked at, was 1, it decided that x %% 2 == 0 was false—it is if x[1] is 1—and then none of the values were squared. Clearly not what we wanted, and the warning was warranted.

If you want a vectorized function, you need to use ifelse():

```
maybe_square <- function(x) {
    ifelse(x %% 2 == 0, x ** 2, x)
}
maybe_square(1:5)

## [1] 1 4 3 16 5
```

or you can use the `Vectorize()` function to translate a function that isn't vectorized into one that is:

```r
maybe_square <- function(x) {
    if (x %% 2 == 0) x ** 2 else x
}
maybe_square <- Vectorize(maybe_square)
maybe_square(1:5)
```

```
## [1]  1  4  3 16  5
```

The `Vectorize` function is what is known as a "functor"—a function that takes a function as input and returns a new function. It is beyond the scope of this chapter to cover how we can manipulate functions like other data, but it is a very powerful feature of R that we return to in later chapters.

For now, it suffices to know that `Vectorize` will take your function that can only take single values as input and then create a function that handles an entire vector by calling your function with each element. You only see one element at a time, and `Vectorize`'s function makes sure that you can handle an entire vector, one element at a time.

To loop over elements in a vector, you use `for` statements:

```r
x <- 1:5
total <- 0
for (element in x) total <- total + element
total
```

```
## [1] 15
```

As with `if` statements, if you want the body to contain more than one expression, you need to put it in curly brackets.

The `for` statement runs through the elements of a vector. If you want the indices instead, you can use the `seq_along()` function, which given a vector as input returns a vector of indices:

```r
x <- 1:5
total <- 0

for (index in seq_along(x)) {
    element <- x[index]
```

```
    total <- total + element
}
total

## [1] 15
```

There are also while statements for looping. These repeat as long as an expression is true:

```
x <- 1:5
total <- 0
index <- 1
while (index <= length(x)) {
    element <- x[index]
    index <- index + 1
    total <- total + element
}
total

## [1] 15
```

If you are used to zero-indexed vectors, pay attention to the index <= length(x) here. You would normally write index < length(x) in zero-indexed languages. Here, that would miss the last element.

There is also a repeat statement that loops until you explicitly exit using the break statement:

```
x <- 1:5
total <- 0
index <- 1
repeat {
    element <- x[index]
    total <- total + element
    index <- index + 1
    if (index > length(x)) break
}
total

## [1] 15
```

There is also a `next` statement that makes the loop jump to the next iteration.

Now that I have told you about loops, I feel I should also say that they generally are not used as much in R as in many other programming languages. Many actively discourage using loops, and they have a reputation for leading to slow code. The latter is not justified in itself, but it is easier to write slow code using loops than the alternatives. Instead, you use functions to take over the looping functionality. There is usually a function for doing whatever you want to accomplish using a loop, and when there is not, you can generally get what you want by combining the three functions `Map`, `Filter`, and `Reduce`.

But that is beyond the scope of this chapter; we return to it later in the book.

Factors

Now let us return to data types and the factors I hinted at a while ago. Factors are mostly just vectors but of categorical values. That just means that the elements of a factor should be considered as categories or classes and not as numbers or strings. For example, categories such as "small," "medium," and "large" could be encoded as numbers, but there aren't any natural numbers to assign to them. We could encode soft drink sizes like 1, 2, and 3 for "small," "medium," and "large." By doing this, we are implicitly saying that the difference between "small" and "medium" is half of the difference between "small" and "large" which may not be the case. Data with sizes "small," "medium," and "large" should be encoded as categorical data, not numbers, and in R that means encoding them as factors.

A factor is usually constructed by giving the `factor()` function a list of strings. The function translates these into the different categories, and the factor becomes a vector of the categories:

```
f <- factor(c("small", "small", "medium",
            "large", "small", "large"))
f

## [1] small  small  medium large  small  large
## Levels: large medium small
```

The categories are called "levels":

```
levels(f)
```

```
## [1] "large"   "medium" "small"
```

By default, these are ordered alphabetical, which in this example gives us the order "large," "medium," "small." You can change this order by specifying the levels when you create the factor:

```
ff <- factor(c("small", "small", "medium",
               "large", "small", "large"),
             levels = c("small", "medium", "large"))
ff
```

```
## [1] small  small  medium large  small  large
## Levels: small medium large
```

Changing the order of the levels like this changes how many functions handle the factor. Mostly it affects the order that summary statistics or plotting functions present results in.

```
summary(f)
```

```
##  large medium  small
##      2      1      3
```

```
summary(ff)
```

```
##  small medium  large
##      3      1      2
```

The summary function, when used on factors, just counts how many of each kind we see, and here we have three "small," one "medium," and two "large." The only thing the order of the levels does is determine in which order summary prints the categories.

The order in which the levels are given shouldn't be thought of as "ordering" the categories, though. It is just used for displaying results; there is not an order semantics given to the levels unless you explicitly specify this.

Some categorical data has a natural order, like "small," "medium," and "large." Other categories are not naturally ordered. There is no natural way of ordering "red," "green," and "blue." When we print data, it will always come out ordered since text always comes out ordered. When we plot data, it is usually also ordered. But in many mathematical models, we would treat ordered categorical data different from unordered categorical data, so the distinction is sometimes important.

By default, factors do not treat the levels as ordered, so they assume that categorical data is like "red," "green," and "blue," rather than ordered like "small," "medium," and "large." If you want to specify that the levels are ordered, you can do that using the ordered argument to the factor() function:

```
of <- factor(c("small", "small", "medium",
                "large", "small", "large"),
            levels = c("small", "medium", "large"),
            ordered = TRUE)
of
```

```
## [1] small   small   medium large   small   large
## Levels: small < medium < large
```

You can also use the ordered() function:

```
ordered(ff)
```

```
## [1] small   small   medium large   small   large
## Levels: small < medium < large
```

```
ordered(f, levels = c("small", "medium", "large"))
```

```
## [1] small   small   medium large   small   large
## Levels: small < medium < large
```

In many ways, you can work with a combination of strings and factors. For example, you can check if a factor value is from a certain level by comparing it with the string of that label:

```
f
```

```
## [1] small   small   medium large   small   large
## Levels: large medium small
```

```
f == "small"
```

```
## [1]  TRUE  TRUE FALSE FALSE  TRUE FALSE
```

28

Here, we test each of the elements in the factor f against the string "small," and we get TRUE for those that have the level small. However, factors are not strings, and in some places they behave fundamentally different. The fact that they so often look like strings makes this extra tricky, when something that looks perfectly innocent can hide a fatal error.

The case where I have seen this the most is when R users try to use factors to index into vectors. While this is a little more advanced than most of what we see in this chapter, I want to show it early so you are aware of the dangers.

When we create a vector, we can give the indices names. We can do this in the same expression as we create the vector:

```
v <- c(a = 1, b = 2, c = 3, d = 4)
v
```

```
## a b c d
## 1 2 3 4
```

or we can add the names later:

```
v <- 1:4
names(v) <- letters[1:4]
v
```

```
## a b c d
## 1 2 3 4
```

(the letters vector contains all the lowercase letters, so letters[1:4] are a, b, c, and d).

If we have named the elements in the vector, we can use them to index, just as we can use numbers. If we want indices 2 and 3, we can index with 2:3, but we could also index with c("b", "c"):

```
v[2:3]
```

```
## b c
## 2 3
```

```
v[c("b", "c")]
```

```
## b c
## 2 3
```

The indexing does not have to be in the same order as the elements are in vector, so we could, for example, extract indices 3 and 2, in that order with

```
v[c(3, 2)]
```

```
## c b
## 3 2
```

or using their names

```
v[c("c", "b")]
```

```
## c b
## 3 2
```

and if we repeat an index, we get the corresponding value more than once:

```
v[c("c", "b", "c")]
```

```
## c b c
## 3 2 3
```

Here, we are using a vector of strings to index, but what would happen if we used a factor?

A factor is not stored as strings, even though we create it from a vector of strings. It is stored as a vector of integers where the integers are indices into the levels. This representation can bite you if you try to use a factor to index into a vector.

Read the following code carefully. We have the vector v that can be indexed with the letters A, B, C, and D (LETTERS is a vector that contains the uppercase letters). We create a factor, ff, that consists of these four letters in that order. When we index with it, we get what we would expect. Since ff is the letters A to D, we pick out the values from v with those labels and in that order:

```
v <- 1:4
names(v) <- LETTERS[1:4]
v
```

```
## A B C D
## 1 2 3 4
```

```
(ff <- factor(LETTERS[1:4]))
```

```
## [1] A B C D
## Levels: A B C D
```

```
v[ff]
```

```
## A B C D
## 1 2 3 4
```

We are lucky to get the expected result, and it is only luck though, because this expression is not indexing using the names we might expect it to use. Read the following even more carefully!

```
(ff <- factor(LETTERS[1:4], levels = rev(LETTERS[1:4])))
```

```
## [1] A B C D
## Levels: D C B A
```

```
v[ff]
```

```
## D C B A
## 4 3 2 1
```

This time, ff is still a vector with the categories A to D in that order, but we have specified that the levels are D, C, B, and A, in that order. So the numerical values that the categories are stored as are actually these:

```
as.numeric(ff)
```

```
## [1] 4 3 2 1
```

What we get when we use it to index into v are those numerical indices—so we get the values pulled out of v in the reversed order from what we would expect if we didn't know this (which you now know).

The easiest way to deal with a factor as if it contained strings is to translate it into a vector of strings. You can use such a vector to index:

```
as.vector(ff)
```

```
## [1] "A" "B" "C" "D"
```

```
v[as.vector(ff)]
```

```
## A B C D
## 1 2 3 4
```

If you ever find yourself using a factor to index something—or in any other way treat a factor as if it was a vector of strings—you should stop and make sure that you explicitly convert it into a vector of strings. Treating a factor as if it was a vector of strings—when, in fact, it is a vector of integers—only leads to tears and suffering in the long run.

Data Frames

The vectors we have seen, whatever their type, are just sequences of data. There is no structure to them except for the sequence order, which may or may not be relevant for how to interpret the data. That is not how data we want to analyze look like. What we usually have is several related variables from some collection of observations. For each observed data point, you have a value for each of these variables (or missing data indications if some variables were not observed). Essentially, what you have is a table with a row per observation and a column per variable. The data type for such tables in R is the `data.frame`.

A data frame is a collection of vectors, where all must be of the same length, and you treat it as a two-dimensional table. We usually think of data frames as having each row correspond to some observation and each column correspond to some property of the observations. Treating data frames that way makes them extremely useful for statistical modelling and fitting.

You can create a data frame explicitly using the `data.frame` function:

```
df <- data.frame(a = 1:4, b = letters[1:4])
df
```

```
##   a b
## 1 1 a
## 2 2 b
## 3 3 c
## 4 4 d
```

but usually you will read in the data frame from files.

To get to the individual elements in a data frame, you must index it. Since it is a two-dimensional data structure, you should give it two indices:

```
df[1,1]
```

```
## [1] 1
```

You can, however, leave one of these empty, in which case you get an entire column or an entire row:

```
df[1,]
```

```
##   a b
## 1 1 a
```

```
df[,1]
```

```
## [1] 1 2 3 4
```

If the rows or columns are named, you can also use the names to index. This is mostly used for column names since it is the columns that correspond to the observed variables in a data sets. There are two ways to get to a column, but explicitly indexing

```
df[,"a"]
```

```
## [1] 1 2 3 4
```

or using the $column_name notation that does the same thing but lets you get at a column without having to use the [ ] operation and quote the name of a column:

```
df$b
```

```
## [1] "a" "b" "c" "d"
```

Before R version 4, a data frame would consider a character vector as a factor and implicitly convert it. It saves a little space, but was a source of errors as the one I described in the section on factors, so with R4 the default is now to keep string vectors as string vectors. If df$b was a factor when you run the preceding code, you are using an older version of R, and I suggest you update it.

Turning string vectors into factors, or keeping them as they are, is just the default behavior, though. You can control it with the `stringsAsFactors` parameter. If you set this to TRUE, you will get the old behavior that turns strings into factors:

```
data.frame(a = 1:4, b = letters[1:4],
           stringsAsFactors = TRUE)
```

```
##   a b
## 1 1 a
## 2 2 b
## 3 3 c
## 4 4 d
```

If you used `stringsAsFactors = FALSE`, you would get the now default behavior of keeping string vectors as strings.

You can combine two data frames row-wise or column-wise by using the `rbind` and `cbind` functions:

```
df2 <- data.frame(a = 5:7, b = letters[5:7])
rbind(df, df2)
```

```
##   a b
## 1 1 a
## 2 2 b
## 3 3 c
## 4 4 d
## 5 5 e
## 6 6 f
## 7 7 g
```

```
df3 <- data.frame(c = 5:8, d = letters[5:8])
cbind(df, df3)
```

```
##   a b c d
## 1 1 a 5 e
## 2 2 b 6 f
## 3 3 c 7 g
## 4 4 d 8 h
```

These data frames are built into R, but there are various alternatives implemented in packages. They differ from the built-in data frames by being optimized for certain use patterns or just based on programmer taste.

The most popular variant, which you are practically guaranteed to run into sooner rather than later, is the so-called "tibble." You can get access to it by loading the package `tibble`:

```
library(tibble)
```

or by loading the large collection of packages known as the "tidyverse":

```
library(tidyverse)
```

The tidyverse is a large framework for working with data in a structured way, implemented in numerous packages, but you can load all the common ones in a single instructing by loading `tidyverse`.

If these two commands did not work when you tried them, it is because you haven't installed them yet. We return to working with packages shortly, but for now, you can just do

```
install.packages("tidyverse")
```

After that, both of the preceding `library(...)` commands should work.

Then, to create a tibble instead of a built-in data frame, you can use

```
tibble(a = 1:4, b = letters[1:4])

## # A tibble: 4 × 2
##       a b
##   <int> <chr>
## 1     1 a
## 2     2 b
## 3     3 c
## 4     4 d
```

As you can see, the syntax is much the same as when you create a data frame with the `data.frame` function, and the result is similar as well. Generally, you can use tibbles as drop-in replacements for data frames. The operations you can do on data frames you can also do on tibbles.

In day-to-day programming, there is not a big difference between data frames and tibbles, but the latter prints a little better by giving a nicer summary of large data collections, and as already mentioned, they are heavily used by the tidyverse framework, so you are more likely to use them than the classical data frames if you start using the packages there, which I strongly suggest that you do.

For more sophisticated manipulation of data frames, you really should use the `dplyr` package, also part of the tidyverse, but we will return to this in Chapter 3.

Using R Packages

Out of the box, R has a lot of functionality, but where the real power comes in is through its package mechanism and the large collection of packages available for download and use.

When you install RStudio, you also install a set of default packages. You can see which packages are installed by clicking the Packages tab in the lower-right frame; see Figure 1-4.

Files	Plots	Packages	Help	Viewer			
Install	Update	Packrat					
	Name		Description			Version	
System Library							
	admixturegraph		Admixture Graph Manipulation and Fitting			1.0.0.9000	
	arules		Mining Association Rules and Frequent Itemsets			1.2–1	
	assertthat		Easy pre and post assertions.			0.1	
	BH		Boost C++ Header Files			1.58.0–1	
	bitops		Bitwise Operations			1.0–6	
	blm		A Package For Implementing Bayesian Linear Regression			0.0.0.9002	
	blmPackage		Bayesian Linear Regression			0.1	
	boot		Bootstrap Functions (Originally by Angelo Canty for S)			1.3–17	
	brew		Templating Framework for Report Generation			1.0–6	
	caTools		Tools: moving window statistics, GIF, Base64, ROC AUC, etc.			1.17.1	
	class		Functions for Classification			7.3–13	
	cluster		"Finding Groups in Data": Cluster Analysis Extended Rousseeuw et al.			2.0.3	
	coda		Output Analysis and Diagnostics for MCMC			0.18–1	
	codetools		Code Analysis Tools for R			0.2–14	
	coin		Conditional Inference Procedures in a Permutation Test Framework			1.1–0	
	colorspace		Color Space Manipulation			1.2–6	

Figure 1-4. *RStudio packages*

From here, you can update packages—new versions of essential packages are released regularly—and you can install new packages. You might have already done this when we talked about tibbles, but try another one. Try installing the package `magrittr`. We are going to use it shortly.

You can also install packages from the R console. Just write

```
install.packages("magrittr")
```

Once you have installed a package, you have access to the functionality in it. You can get function `f` in `package` by writing `package::f()`, or you can load all functions from a package into your global namespace to have access to them without using the `package::` prefix.

Loading the functionality from the `magrittr` package is done like this:

```
library(magrittr)
```

Dealing with Missing Values

Most data sets have missing values—parameters that weren't observed or that were incorrectly recorded and had to be masked out. How you deal with missing data in an analysis depends on the data and the analysis, but it must be addressed, even if all you do is remove all observations with missing data.

Missing data is represented in R by the special value NA (not available). Values of any type can be missing and represented as NA, and importantly R knows that NA means missing values and treats NAs accordingly. You should always represent missing data as NA instead of some particular number (like -1 or 999 or whatever). R knows how to work with NA but has no way of knowing that -1 means anything besides minus one.

Operations that involve NA are themselves NA—you cannot operate on missing data and get anything but more missing values in return. This also means that if you compare two NAs, you get NA. Because NA is missing information, it is not even equal to itself:

```
NA + 5
```

```
## [1] NA
```

```
NA == NA
```

```
## [1] NA
```

```
NA != NA
```

```
## [1] NA
```

If you want to check if a value is missing, you must use the function is.na:

```
is.na(NA)
```

```
## [1] TRUE
```

```
is.na(4)
```

```
## [1] FALSE
```

Functions such as sum() will by default return NA if its input contains NA:

```
v <- c(1,NA,2)
sum(v)
```

```
## [1] NA
```

If you want just to ignore the NA values, there is often a parameter for specifying this:

```
sum(v, na.rm = TRUE)
```

```
## [1] 3
```

Data Pipelines

Most data analysis consists of reading in some data, performing various operations on that data and, in the process, transforming it from its raw form into something we can start to extract meaning out of, and then doing some summarizing or visualization toward the end.

These steps in an analysis are typically expressed as a sequence of function calls that each change the data from one form to another. It could look like the following pseudocode:

```
my_data <- read_data("/some/path/some_file.data")
clean_data <- remove_dodgy_data(my_data)
data_summaries <- summarize(clean_data)
plot_important_things(data_summaries)
```

There isn't anything wrong with writing a data analysis in this way. But there are typically many more steps involved than listed here. When there is, you either have to get very inventive in naming the variables you are saving the data in or you have to overwrite variable names by reassigning to a variable after modifying the data. Both having many variable names and reassigning to variables can be problematic.

If you have many variables, it is easier accidentally to call a function on the wrong variable. For example, you might summarize the `my_data` variable instead of the `clean_data`. While you would get an error if you called a function with a variable name that doesn't exist, there is nothing to catch when you call a function with the wrong data. You will likely get the wrong result, and the error will not be easy to find. It would not be an error easy to debug later.

There is slightly less of a problem with reassigning to a variable. It is mostly an issue when you work with R interactively. There, if you want to go back and change part of the program you are writing, you have to go back to the start, where the data is imported. You cannot just start somewhere in the middle of the function calls with a variable that doesn't refer to the same data it did when you ran the program from scratch. It is less of a problem if you always run your R scripts from the beginning, but the typical use of R is to work with it in an interactive console or Markdown document, and there this can be a problem.

A solution, then, is not to call the functions one at a time and assign each temporary result to a variable. Instead of having four statements in the preceding example, one per function call, you would just feed the result of the first function call into the next:

```
plot_important_things(
    summarize(
        remove_dodgy_data(
            read_data("/some/path/some_file.data"))))
```

You get rid of all the variables, but the readability suffers, to put it mildly. You have to read the code from right to left and inside out.

Writing Pipelines of Function Calls

The `magrittr` package introduced a trick to alleviate this problem, which was later followed by a built-in solution in R 4.1. The solution is to introduce a "pipe operator," %>% in `magrittr` and |> in R 4.1, that lets you write the functions you want to combine from left to right but get the same effect as if you were calling one after the other and sending the result from one function to the input of the next function.

The operator works such that writing

```
x %>% f()
```

or

```
x |> f()
```

is equivalent to writing

```
f(x)
```

With the magrittr operator, you can leave out the parentheses, writing x %>% f instead, but with the built-in operator, x |> f is considered a syntax error.

How the two pipe operators work and how you can use them overlap, but they are not equivalent. The built-in operator is a little faster; when you write x |> f(), it is just syntactic sugar for f(x), meaning that the two are completely equivalent and that there is no overhead in using the operator rather than a function call. With the magrittr operator, x %>% f(), you are calling a function, %>%, every time you use the operator, and there is some overhead to that. But there is also more flexibility to this, and the %>% operator is more flexible and can handle use cases that the |> operator cannot.

You can combine sequences of such operators such that writing

```
x |> f() |> g() |> h()
```

or

```
x %>% f() %>% g() %>% h()
```

or

```
x %>% f %>% g %>% h
```

is equivalent to writing

```
h(g(f(x)))
```

The preceding example would become

```
read_data("/some/path/some_file.data") %>%
    remove_dodgy_data %>%
```

```
    summarize %>%
    plot_important_things
```

with the `magrittr` operator, or

```
read_data("/some/path/some_file.data") |>
    remove_dodgy_data() |>
    summarize() |>
    plot_important_things()
```

with the built-in operator.

Reading code like this might still take some getting used to, but it is much easier to read than combining functions from the inside and out.

If you have ever used pipelines in UNIX shells, you should immediately see the similarities. It is the same approach for combining functions/programs. By combining several functions, which each do something relatively simple, you can create very powerful pipelines.

Writing pipelines using the %>% or |> operator is a relatively new idiom introduced to R programming, but one that is very powerful and is being used more and more in different R packages. We will use pipelines extensively in the coming chapters.

Incidentally, if you are wondering why the package that implements pipes in R is called `magrittr`, it refers to Belgian artist René Magritte who famously painted a pipe and wrote "Ceci n'est pas une pipe" ("This is not a pipe") below it. But enough about Belgian surrealists.

Writing Functions That Work with Pipelines

The pipeline operator actually does something very simple, which in turn makes it simple to write new functions that work well with it. It just takes whatever is computed on the left-hand side of it and inserts it as the first argument to the function given on the right-hand side, and it does this left to right. So x %>% f becomes f(x), x %>% f %>% g becomes f(x) %>% g and then g(f(x)), and x %>% f(y) becomes f(x,y). If you are providing additional parameters to a function in the pipeline, the left-hand side of %>% or |> is inserted before the explicit parameters passed to it.

If you want to write functions that work well with pipelines, you should, therefore, make sure that the most likely parameter to come through a pipeline is the first

41

parameter of your function. Write your functions such that the first parameter is the data it operates on, and you have done most of the work.

For example, if you wanted a function that would sample n random rows of a data frame, you could write it such that it takes the data frame as the first argument and the parameter n as its second argument:

```
subsample_rows <- function(d, n) {
  rows <- sample(nrow(d), n)
  d[rows,]
}
```

and then you could simply pop it right into a pipeline:

```
d <- data.frame(x = rnorm(100), y = rnorm(100))
d %>% subsample_rows(n = 3)
```

```
##              x         y
## 31  0.3159150 1.3485491
## 76 -0.1553485 0.3320349
## 54 -0.5918348 0.8083360
```

or

```
d |> subsample_rows(n = 3)
```

```
##              x          y
## 69 -1.2723834 -0.01965686
## 25  0.8190089  1.05925039
## 87  2.3872326 -0.18939869
```

Since we are simulating random data here, your output will differ from mine, but you should see something similar.

The Magical "." Argument

Now, you cannot always be so lucky that all the functions you want to call in a pipeline take the left-hand side of the pipe operator as its first parameter. If this is the case, you can still use the function, though, but here the two operators differ in how easy they make it.

The operator from `magrittr`, but not the built-in operator, interprets the symbol "." in a special way. If you use "." in a function call in a pipeline, then that is where the left-hand side of the `%>%` operation goes instead of as the default first parameter of the right-hand side. So if you need the data to go as the second parameter, you put a "." there, since x `%>%` f(y, .) is equivalent to f(y, x). The same goes when you need to provide the left-hand side as a named parameter since x `%>%` f(y, z = .) is equivalent to f(y, z = x), something that is particularly useful when the left-hand side should be given to a model fitting function. Functions fitting a model to data are usually taking a model specification as their first parameter and the data they are fitted to as a named parameter called `data`:

```
d <- data.frame(x = rnorm(10), y = rnorm(10))
d %>% lm(y ~ x, data = .)
```

```
##
## Call:
## lm(formula = y ~ x, data = .)
##
## Coefficients:
## (Intercept)         x
##      0.2866    0.3833
```

We will return to model fitting, and what an expression such as y ~ x means, in a later chapter, so don't worry if it looks a little strange for now. If you are interested, you can always check the documentation for the lm() function.

The built-in operator does not interpret "." this way, and d |> lm(y ~ x, data = .) will give you an error (unless you have defined "." somewhere, which you probably shouldn't). The |> operator always puts the left-hand side as the first argument to the right-hand side. If that doesn't fit your function, you have to adapt the function.

With lm, the data is not the first argument, but we can make a function where it is:

```
my_lm <- function(d) lm(y ~ x, data = d)
d |> my_lm()
```

```
##
## Call:
## lm(formula = y ~ x, data = d)
##
```

```
## Coefficients:
## (Intercept)           x
##      0.2866      0.3833
```

We usually don't like having such specialized functions lying around, and we don't have any use for it outside of the pipeline, so this is not ideal. However, we don't have to first define the function, give it a name, and then use it. We could just use the function definition as it is:

```
d |> (function(d) lm(y ~ x, data = d))()
```

```
##
## Call:
## lm(formula = y ~ x, data = d)
##
## Coefficients:
## (Intercept)           x
##      0.2866      0.3833
```

The syntax here might look a little odd at first glance, with the function definition in parentheses and then the extra () after that, but it is really the same syntax as what we have been using so far. We have written pipes such as d |> f() where f refers to a function. It is the same now, but instead of a function name, we have an expression, (function(d) lm(...)), that gives us a function. It needs to be in parentheses so the () that comes after the function are not considered part of the function body. In other words, take d |> f() and put in (function(d) lm(...)) instead of f, and you get the preceding expression.

Such functions that we do not give a name are called anonymous functions, or with a reference to theoretical computer science, lambda expressions. From R 4.1, perhaps to alleviate that using |> without the "." is cumbersome, there is a slightly shorter way to write them. Instead of writing function(...), you can use \(...) and get

```
d |> (\(d) lm(y ~ x, data = d))()
```

```
##
## Call:
## lm(formula = y ~ x, data = d)
##
```

```
## Coefficients:
## (Intercept)          x
##        0.2866     0.3833
```

The notation \() is supposed to look like the Greek letter lambda, λ.

Of course, even with the shorter syntax for anonymous functions, writing d |>
(\(d) lm(y ~ x, data = d))() instead of just lm(y ~ x, data = d) doesn't give us
much, and it is more cumbersome to use the |> operator when you try to pipe together
functions where the data doesn't flow from the first argument to the first argument from
function to function. If you are in a situation like that, you will enjoy using the magrittr
pipe more.

Anonymous functions do have their uses, though, both for the built-in and
magrittr's pipe operator. Pipelines are great when you can call existing functions
one after another, but what happens if you need a step in the pipeline where there is
no function doing what you want? Here, anonymous functions usually are the right
solution.

As an example, consider a function that plots the variable y against the variable x and
fits and plots a linear model of y against x. We can define and name such a function to
get the following code:

```
plot_and_fit <- function(d) {
  plot(y ~ x, data = d)
  abline(lm(y ~ x, data = d))
}

x <- rnorm(20)
y <- x + rnorm(20)
data.frame(x, y) |> plot_and_fit()
```

Since giving the function a name doesn't affect how the function works, it isn't
necessary to do so; we can just put the code that defined the function where the name of
the function goes to get this:

```
data.frame(x, y) |> (\(d) {
    plot(y ~ x, data = d)
    abline(lm(y ~ x, data = d))
})()
```

45

with the built-in operator, or like this

```
data.frame(x, y) %>% (\(d) {
    plot(y ~ x, data = d)
    abline(lm(y ~ x, data = d))
})()
```

with the %>% operator.

With the `magrittr` pipe operator, we could also leave out the final `()` and do simply:

```
data.frame(x, y) %>% (\(d) {
    plot(y ~ x, data = d)
    abline(lm(y ~ x, data = d))
})
```

This is because the %>% operator takes both a function call and a function on the right-hand side, so we can write x %>% f() or x %>% f, and similarly we can write x %>% (\(x) ...)() or x %>% (\(x) ...). You cannot leave out the parentheses around the function definition, though. A function definition is also a function call, and the %>% operator would try to put the left-hand side of the operator into that function call, which would give you an error. The |> operator explicitly checks if you are trying to define a function as the right-hand side and tells you that this is not allowed.

With the `magrittr` operator, though, you do not need to explicitly define an anonymous function this way; you can use "." to simulate the same effect:

```
data.frame(x, y) %>% {
  plot(y ~ x, data = .)
  abline(lm(y ~ x, data = .))
}
```

By putting the two operations in curly braces, we effectively make a function, and the first argument of the function goes where we put the "."

The `magrittr` operator does more with "." than just changing the order of parameters. You can use "." more than once when calling a function, and you can use it in expressions or in function calls:

```
rnorm(4) %>% data.frame(x = ., is_negative = . < 0)

##            x is_negative
```

```
## 1 -1.5782344          TRUE
## 2 -0.1215720          TRUE
## 3 -1.7966768          TRUE
## 4 -0.4755915          TRUE
```

```
rnorm(4) %>% data.frame(x = ., y = abs(.))
```

```
##              x          y
## 1 -0.8841023 0.8841023
## 2 -3.4980590 3.4980590
## 3 -0.3819834 0.3819834
## 4  0.9776881 0.9776881
```

There is one caveat: if "." only appears in function calls, it is still given as the first argument to the function on the right-hand side of %>%:

```
rnorm(4) %>% data.frame(x = sin(.), y = cos(.))
```

```
##            .          x          y
## 1 -0.5580409 -0.5295254  0.8482941
## 2 -0.6264551 -0.5862767  0.8101109
## 3 -0.5304512 -0.5059226  0.8625789
## 4  1.8976216  0.9470663 -0.3210380
```

The reason is that it is more common to see expressions with function calls like this when the full data is also needed than when it is not. So by default f(g(.),h(.)) gets translated into f(.,g(.),h(.)). If you want to avoid this behavior, you can put curly brackets around the function call since {f(g(.),h(.))} is equivalent to f(g(.),h(.)). (I will explain the meaning of the curly brackets later). You can get both the behavior f(.,g(.),h(.)) and the behavior {f(g(.),h(.))} in function calls in a pipeline; the default is just the most common case.

Other Pipeline Operations

The %>% and |> operators are a very powerful mechanism for specifying data analysis pipelines, but there are some special cases where a slightly different behavior is needed, and the magrittr package provides some of these. To get them, you need to import the package with library(magrittr). If you use library(tidyverse) to load the tidyverse framework, you only get %>%; you need to explicitly load `magrittr` to get the other operators.

One case is when you need to refer to the parameters in a data frame you get from the left-hand side of the pipe expression directly. In many functions, you can get to the parameters of a data frame just by naming them, as we have seen earlier in `lm` and `plot`, but there are cases where that is not so simple.

You can do that by indexing "." like this:

```
d <- data.frame(x = rnorm(10), y = 4 + rnorm(10))
d %>% {data.frame(mean_x = mean(.$x), mean_y = mean(.$y))}

##         mean_x    mean_y
## 1 0.09496017 3.538881
```

but if you use the operator %$% instead of %>%, you can get to the variables just by naming them instead:

```
d %$% data.frame(mean_x = mean(x), mean_y = mean(y))

##         mean_x    mean_y
## 1 0.09496017 3.538881
```

Another common case is when you want to output or plot some intermediate result of a pipeline. You can, of course, write the first part of a pipeline, run data through it and store the result in a parameter, output or plot what you want, and then continue from the stored data. But you can also use the %T>% (tee) operator. It works like the %>% operator, but where %>% passes the result of the right-hand side of the expression on, %T>% passes on the result of the left-hand side. The right-hand side is computed but not passed on, which is perfect if you only want a step for its side effect, like printing some summary:

```
d <- data.frame(x = rnorm(10), y = rnorm(10))
d %T>% plot(y ~ x, data = .) %>% lm(y ~ x, data = .)
```

The final operator is %<>%, which does something I warned against earlier—it assigns the result of a pipeline back to a variable on the left. Sometimes, you do want this behavior—for instance, if you do some data cleaning right after loading the data and you never want to use anything between the raw and the cleaned data, you can use %<>%:

```
d <- read_my_data("/path/to/data")
d %<>% clean_data
```

I use it sparingly and would prefer just to pass this case through a pipeline:

```
d <- read_my_data("/path/to/data") %>% clean_data
```

Coding and Naming Conventions

People have been developing R code for a long time, and they haven't been all that consistent in how they do it. So as you use R packages, you will see many different conventions on how code is written and especially how variables and functions are named.

How you choose to write your code is entirely up to you as long as you are consistent with it. It helps somewhat if your code matches the packages you use, just to make everything easier to read, but it is up to you.

A few words on naming are worth going through, though. There are three ways people typically name their variables, data, or functions, and these are

```
underscore_notation(x, y)
camelBackNotation(x, y)
dot.notation(x, y)
```

You are probably familiar with the first two notations, but if you have used Python or Java or C/C++ before, the dot notation looks like method calls in object-oriented programming. It is not (although it is related to it). The dot in the name doesn't mean method call. R just allows you to use dots in variable and function names.

I will mostly use the underscore notation in this book, but you can do whatever you want. I would recommend that you stay away from the dot notation, though. There are good reasons for this. R put some interpretation into what dots mean in function names, as we will see when we visit object-oriented programming in the second part of the book, so you can get into some trouble. The built-in functions in R often use dots in function names, but it is a dangerous path so you should probably stay away from it unless you are absolutely sure that you are avoiding the pitfalls that are in it.

Exercises

Mean of Positive Values

You can simulate values from the normal distribution using the `rnorm()` function. Its first argument is the number of samples you want, and if you do not specify other values, it will sample from the $N(0, 1)$ distribution.

Write a pipeline that takes samples from this function as input, remove the negative values, and compute the mean of the rest. Hint: One way to remove values is to replace them with missing values (NA); if a vector has missing values, the `mean()` function can ignore them if you give it the option `na.rm = TRUE`.

Root Mean Square Error

If you have "true" values, $t = (t_1, \ldots, t_n)$, and "predicted" values, $y = (y_1, \ldots, y_n)$, then the root mean square error is defined as $\mathrm{RMSE}(t,y) = \sqrt{\dfrac{1}{n}\sum_{i=1}^{n}(t_i - y_i)^2}$.

Write a pipeline that computes this from a data frame containing the t and y values. Remember that you can do this by first computing the square difference in one expression, then computing the mean of that in the next step, and finally computing the square root of this. The R function for computing the square root is `sqrt()`.

CHAPTER 2

Reproducible Analysis

The typical data analysis workflow looks like this: you collect your data, and you put it in a file or spreadsheet or database. Then you run some analyses, written in various scripts, perhaps saving some intermediate results along the way or maybe always working on the raw data. You create some plots or tables of relevant summaries of the data, and then you go and write a report about the results in a text editor or word processor. This is the typical workflow in many organizations and in many research groups. Most people doing data analysis do variations thereof. But it is also a workflow that has many potential problems.

There is a separation between the analysis scripts and the data, and there is a separation between the analysis and the documentation of the analysis.

If all analyses are done on the raw data, then issue number one is not a major problem. But it is common to have scripts for different parts of the analysis, with one script saving intermediate results to files that are then read by the next script. The scripts describe a workflow of data analysis, and to reproduce an analysis, you have to run all the scripts in the right order. Often enough, this correct order is only described in a text file or even worse only in the head of the data scientist who wrote the workflow. And it gets worse; it won't stay there for long and is likely to be lost before it is needed again.

Ideally, you would always want to have your analysis scripts written in a way where you can rerun any part of your workflow, completely automatically, at any time.

For issue number two, the problem is that even if the workflow is automated and easy to run again, the documentation quickly drifts away from the actual analysis scripts. If you change the scripts, you won't necessarily remember to update the documentation. You probably don't forget to update figures and tables and such, but not necessarily the documentation of the exact analysis run—options to functions and filtering choices and such. If the documentation drifts far enough from the actual analysis, it becomes completely useless. You can trust automated scripts to represent the real data analysis at any time—that is the benefit of having automated analysis workflows in the first place— but the documentation can easily end up being pure fiction.

© Thomas Mailund 2022
T. Mailund, *Beginning Data Science in R 4*, https://doi.org/10.1007/978-1-4842-8155-0_2

What you want is a way to have dynamic documentation. Reports that describe the analysis workflow in a form that can be understood both by machines and humans. Machines use the report as an automated workflow that can redo the analysis at any time. We humans use it as documentation that always accurately describes the analysis workflow that we run.

Literate Programming and Integration of Workflow and Documentation

One way to achieve the goal of having automated workflows and documentation that is always up to date is something called "literate programming." Literate programming is an approach to software development, proposed by Stanford computer scientist Donald Knuth, which never became popular for programming, possibly because most programmers do not like to write documentation. But it has made a comeback in data science, where tools such as Jupyter Notebooks[1] and R Markdown (that we will explore later) are major components in many data scientists' daily work.

The idea in literate programming is that the documentation of a program—in the sense of the documentation of how the program works and how algorithms and data structures in the program work—is written together with the code implementing the program. Tools such as Javadoc[2] and Roxygen[3] do something similar. They have documentation of classes and methods written together with the code in the form of comments. Literate programming differs slightly from this. With Javadoc and Roxygen, the code is the primary document, and the documentation is comments added to it. With literate programming, the documentation is the primary text for humans to read, and the code is part of this documentation, included where it falls naturally to have it. The computer code is extracted automatically from this document when the program runs.

---

[1] https://jupyter.org
[2] https://en.wikipedia.org/wiki/Javadoc
[3] http://roxygen.org

Literate programming never became a huge success for writing programs, but for doing data science, it is having a comeback. The result of a data analysis project is typically a report describing models and analysis results, and it is natural to think of this document as the primary product. So the documentation is already the main focus. The only thing needed to use literate programming is a way of putting the analysis code inside the documentation report.

Many programming languages have support for this. Mathematica[4] has always had notebooks where you could write code together with documentation. Jupyter,[5] the descendant of iPython Notebook, lets you write notebooks with documentation and graphics interspersed with executable code. And in R there are several ways of writing documents that are used both as automated analysis scripts and for generating reports. The most popular of these approaches is R Markdown (for writing these documents) and `knitr` (for running the analysis and generating the reports), but R Notebooks, a variant of R Markdown, is also gaining popularity.

Creating an R Markdown/knitr Document in RStudio

To create a new R Markdown document, go to the File menu, pick New File and then R Markdown…. Now RStudio will bring up a dialog where you can decide which kind of document you want to make and add some information, such as title and author name. It doesn't matter so much what you do here, you can change it later, but try making an HTML document.

The result is a new file with some boilerplate text in it; see Figure 2-1. At the top of the file, between two lines containing just "---" is some meta-information for the document, and after the second "---" is the text proper. It consists of a mix of text, formatted in the Markdown language, and R code.

---

[4] `www.wolfram.com/mathematica`
[5] `http://jupyter.org`

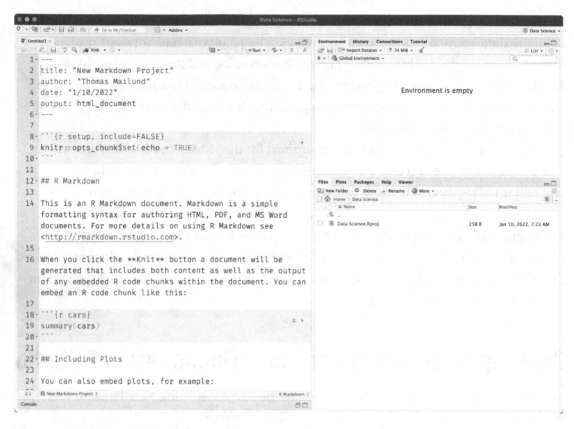

Figure 2-1. *New R Markdown file*

In the toolbar above the open file, there is a menu point saying Knit. If you click it, it will translate the R Markdown into an HTML document and open it; see Figure 2-2. You will have to save the file first, though. If you click the Knit HTML button before saving, you will be asked to save the file.

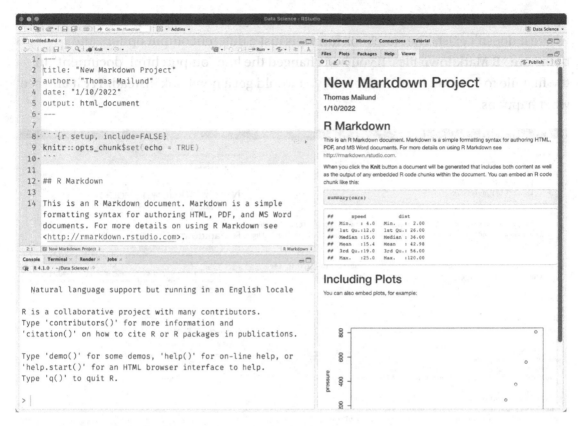

Figure 2-2. *Compiled Markdown file*

The newly created HTML file is also written to disk with a name taken from the name you gave the R Markdown file. The R Markdown file will have suffix .Rmd, and the HTML file will have the same prefix but suffix .html.

If you click the down-pointing arrow next to Knit, you get some additional options. You can ask to see the HTML document in the pane to the right in RStudio instead of in a new window. Having the document in a panel instead of a separate window can be convenient if you are on a laptop and do not have a lot of screen space. You can also generate a file or a Word file instead of an HTML file.

If you decide to produce a file in a different output format, RStudio will remember this. It will update the "output:" field in the metadata to reflect this. If you want, you can also change that line in your document and make the selection that way. Try it out.

If you had chosen File, New File, R Notebook instead of an R Markdown file, you would have gotten a very similar file; see Figure 2-3. The Knitr button is gone, and instead you have a Preview button. If you click it, you get the document shown on the right in Figure 2-3. The difference between the two types of files is tiny, to the point where it doesn't exist. The Notebook format is just a different output option, and both _are_ R Markdown files. If you had changed the line "output: html_document" in the first file to "output: html_notebook", you would get a notebook instead. Try it and see what happens.

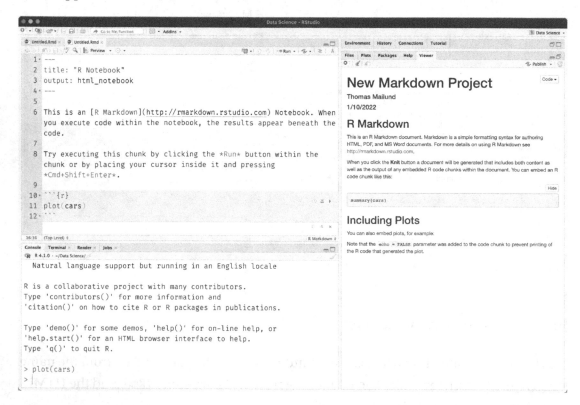

Figure 2-3. *R Notebook*

The R Markdown file is intended for creating reports, while the R Notebook format is more intended for interactive use. Knitting a large document can be slow, because all the analysis in your document will be run from scratch, but with the notebooks, you see the document as it is at any moment, without rerunning any analysis when you preview. Notebooks are thus faster, but they can be a little more dangerous to work with, if you evaluate code out of order (see the following). You can, however, always work with a Notebook while you do your analysis, and then change the "output: ..." format later, to generate a report from scratch. Because the formats are so similar, I will not distinguish between them in the following, and I will refer to the files as R Markdown files as that is the input format for both of them (as also apparent from their file suffix ".Rmd").

The actual steps in creating a document involve two tools and three languages, but it is all integrated so you typically will not notice. There is the R code embedded in the document. The R code is first processed by the `knitr` package that evaluates it and handles the results such as data and plots according to options you give it. The result is a Markdown document (notice no R). This Markdown document is then processed by the tool `pandoc` which is responsible for generating the output file. For this, it uses the metadata in the header, which is written in a language called YAML, and the actual formatting, written in the Markdown language.

You will usually never have to worry about `pandoc` working as the back end of document processing. If you just write R Markdown documents, then RStudio will let you compile them into different types of output documents. But because the pipeline goes from R Markdown via `knitr` to Markdown and then via `pandoc` to the various output formats, you do have access to a very powerful tool for creating documents. I have written this book in R Markdown where each chapter is a separate document that I can run through `knitr` independently. I then have `pandoc` with some options take the resulting Markdown documents, combining them, and produce both output and Epub output. With `pandoc`, it is possible to use different templates for setting up the formatting, and having it depend on the output document you create by using different templates for different formats. It is a very powerful, but also very complicated, tool, and it is far beyond what we can cover in this book. Just know that it is there if you want to take document writing in R Markdown further than what you can readily do in RStudio.

As I mentioned, there are actually three languages involved in an R Markdown document. We will handle them in order, first the header language, which is YAML, then the text formatting language, which is Markdown, and then finally how R is embedded in a document.

The YAML Language

YAML is a language for specifying key-value data. YAML stands for the (recursive) acronym YAML Ain't Markup Language. So yes, when I called this section the YAML language I shouldn't have included language since the L stands for language, but I did. I stand by that choice. The acronym used to stand for Yet Another Markup Language, but since "markup language" typically refers to commands used to mark up text for either specifying formatting or for putting structured information in a text, which YAML doesn't do, the acronym was changed. YAML is used for giving options in various forms to a computer program processing a document, not so much for marking up text, so it isn't really a markup language.

In your R Markdown document, the YAML is used in the header, which is everything that goes between the first and the second line with three dashes. In the document you create when you make a new R Markdown file, it can look like this:

```
---
title: "New Markdown Project"
author: "Thomas Mailund"
date: "1/10/2022"
output: html_document
---
```

You usually do not have to modify this header manually. If you use the GUI, it will adjust the header for you when you change options. You do need to alter it to include bibliographies, though, which we get to later. And you can always add anything you want to the header if you need to, and it can be easier than using the GUI. But you don't have to modify the header that often.

YAML gives you a way to specify key-value mappings. You write key: and then the value afterward. So in the preceding example, you have the key title referring to New Markdown Document, the key author to refer to "Thomas Mailund", and so on. You don't necessarily need to quote the values unless it has a colon in it, but you always can.

The YAML header is used both by RStudio and pandoc. RStudio uses the output key for determining which output format to translate your document into, and this choice is reflected in the Knit toolbar button—while pandoc uses the title, author, and date to put that information into the generated document.

You can have slightly more structure to the key-value specifications. If a key should refer to a list of values, you use "-", so if you have more than one author, you can use something like this:

```
---
...
author:
  - "Thomas Mailund"
  - "Tom Maygrove"
...
---
```

or you can have more key-value structure nested, so if you change the output theme (using Output Options… after clicking the tooth wheel in the toolbar next to the Knit button).

How the options are used depends on the toolchain used to format your document. The YAML header just provides specifications. Which options you have available and what they do are not part of the language.

For pandoc, it depends on the templates used to generate the final document (see later), so there isn't even a complete list that I can give you for pandoc. Anyone who writes a new template can decide on new options to use. The YAML header gives you a way to provide options to such templates, but there isn't a fixed set of keywords to use. It all depends on how tools later in the process interpret them.

The Markdown Language

The Markdown language is a markup language—the name is a pun. It was originally developed to make it easy to write web pages. HTML, the language we use to format web pages, is also a markup language but is not always easily human readable. Markdown intended to solve this by formatting text with very simple markup commands—familiar from emails back in the day before emails were also HTML documents—and then have tools for translating Markdown into HTML.

Markdown has gone far beyond just writing web pages, but it is still a very simple and intuitive language for writing human-readable text with markup commands that can then be translated into other document formats.

In Markdown, you write plain text as plain text. So the body of text is just written without any markup. You will need to write it in a text editor so the text is actually text, not a word processor where the file format usually already contains a lot of markup information that just isn't readily seen on screen. If you are writing code, you should already know about text editors. If not, just use RStudio to write R Markdown files, and you will be okay.

Markup commands are used when you want something else than just plain text. There aren't many commands to learn—the philosophy is that when writing you should focus on the text and not the formatting—so they are very quickly learned.

Formatting Text

First, there are section headers. You can have different levels of headers—think chapters, sections, subsections, etc.—and you specify them using # starting at the beginning of a new line:

```
# Header 1
## Header 2
### Header 3
```

For the first two, you can also use this format:

```
Header 1
========

Header 2
-------------
```

To have lists in your document, you write them as you have probably often seen them in raw text documents. A list with bullets (and not numbers) is written like this:

```
* this is a
* bullet
* list
```

and the result looks like this:

- this is a

- bullet

- list

You can have sublists just by indenting. You need to move the indented line in so there is a space between where the text starts at the outer level and where the bullet is at the next level. Otherwise, the line goes at the outer level. The output of this

```
* This is the first line
  * This is a sub-line
  * This is another sub-line
 * This actually goes to the outer level
* This is definitely at the outer level
```

is this list:

- This is the first line
    - This is a sub-line
    - This is another sub-line
    - This actually goes to the outer level
- Back to the outer level

If you prefer, you can use - instead of * for these lists, and you can mix the two:

```
- First line
* Second line
    - nested line
```

- First line
- Second line
    - nested line

To have numbered lists, just use numbers instead of * and -:

```
1. This is a
2. numbered
3. list
```

The result looks like this:

1. This is a
2. numbered
3. list

You don't actually need to get the numbers right, you just need to use numbers. So

```
1. This is a
3. numbered
2. list
```

would produce the same output. You will start counting at the first number, though, so

```
4. This is a
4. numbered
4. list
```

produces

4. This is a

5. numbered

6. list

To construct tables, you also use a typical text representation with vertical and horizontal lines. Vertical lines separate columns, and horizontal lines separate headers from the table body. This code

```
| First Header | Second Header | Third Header   |
| :----------- | :-----------: | -------------: |
| First row    | Centred text  | Right justified |
| Second row   | *Some data*   | *Some data*    |
| Third row    | *Some data*   | *Some data*    |
```

will result in this table:

First Header	Second Header	Third Header
First row	Centred text	Right justified
Second row	Some data	Some data
Third row	Some data	Some data

The : in the line separating the header from the body determines the justification of the column. Put it on the left to get left justification, on both sides to get the text centered, and on the right to get the text right justified.

Inside text, you use markup codes to make text italic or boldface. You use either *this* or _this_ to make "this" italic, while you use **this** or __this__ to make "this" boldface.

Since Markdown was developed to make HTML documents, it, of course, has an easy way to insert links. You use the notation [link text](link URL) to put "link text" into the document as a link to "link URL." This notation is also used to make cross-references inside a document—similar to how HTML documents have anchors and internal links—but more on that later.

To insert images into a document, you use a notation similar to the link notation; you just put a ! before the link, so `![Image description](URL to image)` will insert the image pointed to by "URL to image" with a caption saying "Image description." The URL here will typically be a local file, but it can be a remote file referred to via HTTP.

With long URLs, the marked-up text can be hard to read even with this simple notation, and it is possible to remove the URLs from the actual text and place it later in the document, for example, after the paragraph referring to the URL or at the very end of the document. For this, you use the notation `[link text][link tag]` and define the "link tag" as the URL you want later:

```
This is some text [with a link][1].
The link tag is defined below the paragraph.

[1]: interesting-url-of-some-sort-we-dont-want-inline
```

You can use a string here for the tag. Using numbers is easy, but for long documents, you won't be able to remember what each number refers to:

```
This is some text [with a link][interesting].
The link tag is defined below the paragraph.

[interesting]: interesting-url-of-some-sort-we-dont-want-inline
```

You can make block quotes in text using notation you will be familiar with from emails:

```
> This is a
> block quote
```

gives you this:

> This is a block quote

To put verbatim input as part of your text, you can either do it inline or as a block. In both cases, you use backticks `` ` ``. Inline in the text, you use single backticks `` `foo` ``. To create a block of text, you write

```
```

block of text
```
```

You can also just indent text with four spaces, which is how I managed to make a block of verbatim text that includes three backticks.

Markdown is used a lot by people who document programs, so there is a notation for getting code highlighted in verbatim blocks. The convention is to write the name of the programming language after the three backticks, then the program used for formatting the document will highlight the code when it can. For R code, you write r, so this block

```r
f <- function(x) ifelse(x %% 2 == 0, x**2, x**3)
f(2)
```

is formatted like this:

```
f <- function(x) ifelse(x %% 2 == 0, x**2, x**3)
f(2)
```

The only thing this markup of blocks does is highlighting the code. It doesn't try to evaluate the code. Evaluating code happens before the Markdown document is formatted, and we return to that shortly.

Cross-Referencing

Out of the box, there is not a lot of support for making cross-references in Markdown documents. You can make cross-references to sections but not figures or tables. There are ways of doing it with extensions to pandoc—I use it in this book—but out of the box from RStudio, you cannot yet.

However, with the work being done for making book-writing and lengthy reports in Bookdown,[6] that might change soon.[7]

The easiest way to reference a section is to put the name of the section in square brackets. If I write [Cross referencing] here, I get a link to this Cross referencing section. Of course, you don't always want the name of the section to be the text of the link, so you can also write [this section][Cross referencing] to get a link to the section "Cross referencing" but display the text "this section."

---

[6] https://bookdown.org/yihui/bookdown

[7] In any case, having cross-references to sections but not figures is still better than Word where the feature is there but buggy to the point of uselessness, in my experience...

This approach naturally only works if all section titles are unique. If they are not, then you cannot refer to them simply by their names. Instead, you can tag them to give them a unique identifier. You do this by writing the identifier after the title of the section. To put a name after a section header, you write

```
### Cross referencing {#section-cross-ref}
```

and then you can refer to the section using [this](#section-cross-ref). Here, you do need the # sign in the identifier—that markup is leftover from HTML where anchors use #.

Bibliographies

Often, you want to cite books or papers in a report. You can of course always handle citations manually, but a better approach is to have a file with the citation information and then refer to it using markup tags. To add a bibliography, you use a tag in the YAML header called bibliography:

```
---

...

bibliography: bibliography.bib

...

---
```

You can use several different formats here; see the R Markdown documentation[8] for a list. The suffix .bib is used for BibLaTeX. The format for the citation file is the same as BibTeX, and you get citation information in that format from nearly every site that will give you bibliography information.

To cite something from the bibliography, you use [@smith04] where smith04 is the identifier used in the bibliography file. You can cite more than one paper inside square brackets separated by a semicolon, [@smith04; doe99], and you can add text such as chapters or page numbers [@smith04, chapter 4]. To suppress the author name(s) in the citation, say when you mention the name already in the text, you put - before the @ so you write As Smith showed [-@smith04].... For in-text citations, similar to \citet{} in natbib, you just leave out the brackets, @smith04 showed that..., and you can combine that with additional citation information as @smith04 [chapter 4] showed that....

---

[8] http://rmarkdown.rstudio.com/authoring_bibliographies_and_citations.html

To specify the citation style to use, you use the `csl` tag in the YAML header:

```
---
...
bibliography: bibliography.bib
csl: biomed-central.csl
...
---
```

Check out the list of citation styles at `https://github.com/citation-style-language/styles` for a large number of different formats. There should be most if not all your heart desires.

Controlling the Output (Templates/Stylesheets)

The `pandoc` tool has a powerful mechanism for formatting the documents it generates. This is achieved using stylesheets in CSS for HTML and from using templates for how to format the output for all output formats. The template mechanism lets you write an HTML or LaTeX document, say, that determines where various part of the text goes and where variables from the YAML header are used. This mechanism is far beyond what we can cover in this chapter, but I just want to mention it if you want to start writing papers using R Markdown. You can do this; you just need to have a template for formatting the document in the style a journal wants. Often, they provide LaTeX templates, and you can modify these to work with Markdown.

There isn't much support for this in RStudio, but for HTML documents, you can use the Output Options… (click the tooth wheel) to choose different output formatting.

Running R Code in Markdown Documents

The formatting so far is all Markdown (and YAML). Where it combines with R and makes it R Markdown is through `knitr`. When you format a document, the first step evaluates R code to create a Markdown document—this translates an `.Rmd` document into an `.md` document, but this intermediate document is deleted afterward unless you explicitly tell RStudio not to do it. It does that by running all the R code you want to be executed and putting it into the Markdown document.

The simplest R code you can evaluate is part of a text. If you want an R expression evaluated, you use backticks but add r right after the first. So to evaluate 2 + 2 and put the result in your Markdown document, you write `r and then the expression 2 + 2 and get the result 4 inserted into the text. You can write any R expression there to get it evaluated. It is useful for inserting short summary statistics like means and standard deviations directly into the text and ensuring that the summaries are always up to date with the actual data you are analyzing.

For longer chunks of code, you use the block quotes, the three backticks. Instead of just writing

```
```r
2 + 2
```
```

which will only display the code (highlighted as R code), you put the r in curly brackets.

This will insert the code in your document but also show the result of evaluating it right after the code block. The boilerplate code you get when creating an R Markdown document in RStudio shows you examples of this (see Figure 2-4).

```
17
18 ▾ ```{r cars}
19    summary(cars)
20    ```
21
```

Figure 2-4. *Code chunk in RStudio*

You can name code chunks by putting a name right after r. You don't have to name all chunks—and if you have a lot of chunks, you probably won't bother naming all of them—but if you give them a name, they are easily located by clicking the structure button in the bar below the document (see Figure 2-5). You can also use the name to refer to chunks when caching results, which we will cover later.

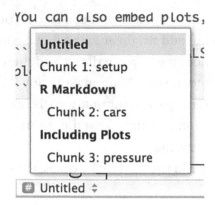

Figure 2-5. *Document structure with chunk names*

You should see a toolbar to the right on every code chunk (see Figure 2-6). The rightmost option, the "play" button, will let you evaluate the chunk. The results will be shown below the chunk unless you have disabled that option. The middle button evaluates all previous chunks down to and including the current one. This is useful when the current chunk depends on previous results. The tooth wheel lets you set options for the chunk.

```
17
18 ▾ ```{r cars}
19    summary(cars)
20    ```
21
```

Figure 2-6. *Code chunk toolbar*

The chunk options (see Figure 2-7) control the output you get when evaluating a code chunk. The Output drop-down selects what output the chunk should generate in the resulting document, while the Show warnings and Show messages selection buttons determine whether warnings and messages, respectively, should be included in the output. The "Use paged tables" changes how tables are displayed, splitting large tables into pages you can click through. The Use custom figure size is used to determine the size of figures you generate—but we return to these later.

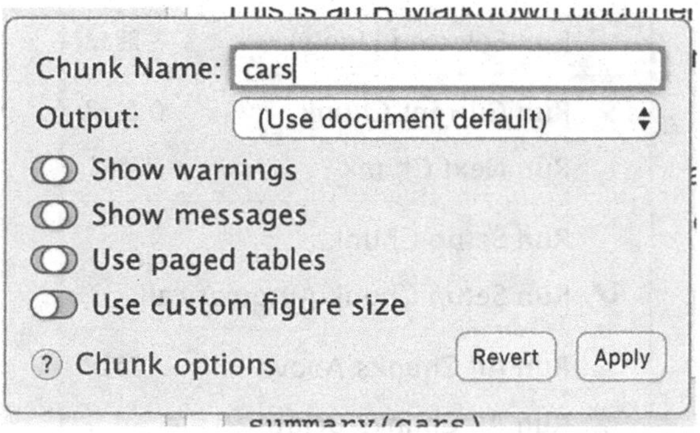

Figure 2-7. Code chunk options

If you modify these options, you will see that the options are included in the top line of the chunk. You can of course also manually control the options here, and there are more options than what you can control with the dialog in the GUI. You can read the knitr documentation[9] for all the details.

The dialog will handle most of your needs, though, except for displaying tables or when we want to cache results of chunks, both of which we return to later.

Using chunks when analyzing data (without compiling documents)

Before continuing, though, I want to stress that working with data analysis in an R Markdown document is useful for more than just creating documents. I personally do all my analysis in these documents because I can combine documentation and code, regardless of whether I want to generate a report at the end. The combination of explanatory text and analysis code is just convenient to have.

The way code chunks are evaluated as separate pieces of analysis is also part of this. You can evaluate chunks individually, or all chunks down to a point, and I find that very convenient when doing an analysis. There are keyboard shortcuts for evaluating all chunks, all previous chunks, or just the current chunk (see Figure 2-8), which makes it very easy to write a bit of code for an exploratory analysis and evaluating just that piece of code. If you are familiar with Jupyter, or similar notebooks, you will recognize the workflow.

---

[9]http://yihui.name/knitr/

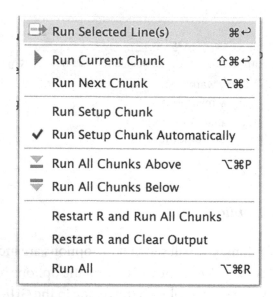

Figure 2-8. *Options for evaluating chunks*

Even without the option for generating final documents from a Markdown document, I would still be using them just for this feature.

Caching Results

Sometimes, part of an analysis is very time-consuming. Here, I mean in CPU time, not thinking time—it is also true for thinking time, but you don't need to think the same things over again and again. If you are not careful, however, you will need to run the same analysis on the computer again and again.

If you have such very time-consuming steps in your analysis, then compiling documents will be very slow. Each time you compile the document, all the analysis is done from scratch. This is the functionality you want since this makes sure that the analysis does not have results left over from code that isn't part of the document, but it limits the usability of the workflows if they take hours to compile.

To alleviate this, you can cache the results of a chunk. To cache the result of a chunk, you should add the option cache=TRUE to it. This means adding that in the header of the chunk similar to how output options are added. You will need to give the chunk a name to use this. Chunks without names are actually given a default name, but this name changes according to how many nameless chunks you have earlier in the document, and you can't have that if you use the name to remember results. So you need to name it.

A named chunk that is set to be cached will not only be when you compile a document if it has changed since the last time it was evaluated. If it hasn't been changed, the results of evaluating it will just be reused.

R cannot cache everything, so if you load libraries in a cached chunk, they won't be loaded unless the chunk is evaluating, so there are some limits to what you can do, but generally it is a very useful feature.

Since other chunks can depend on a cached chunk, there can also be problems if a cached chunk depends on another chunk, cached or not. The chunk will only be reevaluated if you have changed the code inside it, so if it depends on something you have changed, it will remember results based on outdated data. You have to be careful about that.

You can set up dependencies between chunks, though, to fix this problem. If a chunk is dependent on the results of another chunk, you can specify this using the chunk option dependson=other. Then, if the chunk other (and you need to name such chunks) is modified, the cache is considered invalid, and the depending chunk will be evaluated again.

Displaying Data

Since you are writing a report on data analysis, you naturally want to include some results. That means displaying data in some form or other.

You can simply include the results of evaluating R expressions in a code chunk, but often you want to display the data using tables or graphics, especially if the report is something you want to show to people not familiar with R. Luckily, both tables and graphics are easy to display.

To make a table, you can use the function kable() from the knitr package. Try adding a chunk like this to the boilerplate document you have:

```
library(knitr)
kable(head(cars))
```

The library(knitr) imports functions from the knitr package so you get access to the kable() function. You don't need to include it in every chunk you use kable() in, just in any chunk before you use the function—the setup chunk is a good place—but adding it in the chunk you write now will work.

The function `kable()` will create a table from a data frame in the Markdown format, so it will be formatted in the later step of the document compilation. Don't worry too much about the details about the code here; the `head()` function just picks out the first lines of the `cars` data so the table doesn't get too long.

Using `kable()` should generate a table in your output document. Depending on your setup, you might have to give the chunk the output option `result="asis"` to make it work, but it usually should give you a table even without this.

We will cover how to summarize data in later chapters. Usually, you don't want to make tables of full data sets, but for now, you can try just getting the first few lines of the `cars` data.

Adding graphics to the output is just as simple. You simply make a plot in a code chunk, and the result will be included in the document you generate. The boilerplate R Markdown document already gives you an example of this. We will cover plotting in much more detail later.

Exercises

Create an R Markdown Document

Go to the File... menu and create an R Markdown document. Read through the boilerplate text to see how it is structured. Evaluate the chunks. Compile the document.

Different Output

Create from the same R Markdown document an HTML document, a document, and a Word document.

Caching

Add a cached code chunk to your document. Make the code there sample random numbers, for example, using `rnorm()`. When you recompile the document, you should see that the random numbers do not change.

Make another cached chunk that uses the results of the first cached chunk. Say, compute the mean of the random numbers. Set up dependencies and see that if you modify the first chunk the second chunk gets evaluated.

CHAPTER 3

Data Manipulation

Data science is as much about manipulating data as it is about fitting models to data. Data rarely arrives in a form that we can directly feed into the statistical models or machine learning algorithms we want to analyze them with. The first stages of data analysis are almost always figuring out how to load the data into R and then figuring out how to transform it into a shape you can readily analyze.

Data Already in R

There are some data sets already built into R or available in R packages. Those are useful for learning how to use new methods—if you already know a data set and what it can tell you, it is easier to evaluate how a new method performs—or for benchmarking methods you implement. They are of course less helpful when it comes to analyzing new data.

Distributed together with R is the package dataset. We can load the package into R using the library() function and get a list of the data sets within it, together with a short description of each, like this:

```
library(datasets)
library(help = "datasets")
```

To load an actual data set into R's memory, use the data() function. The data sets are all relatively small, so they are ideal for quickly testing code you are working with. For example, to experiment with plotting x-y plots (Figure 3-1), you could use the cars data set that consists of only two columns, a speed and a breaking distance:

```
data(cars)
head(cars)
```

© Thomas Mailund 2022
T. Mailund, *Beginning Data Science in R 4*, https://doi.org/10.1007/978-1-4842-8155-0_3

```
##  speed  dist
## 1     4     2
## 2     4    10
## 3     7     4
## 4     7    22
## 5     8    16
## 6     9    10
```

```
cars %>% plot(dist ~ speed, data = .)
```

I used the %>% pipe operator here, because we need to pass the left-hand side to the data argument in plot. With |> we can only pass the left-hand side to the first argument in the right-hand side function call, but with %>% we can use "." to move the input to another parameter. Generally, I will use the two pipe operators interchangeably in this chapter, except for cases where one is more convenient than another, like before, and then point out why that is.

Don't worry about the plotting function for now; we will return to plotting in the next chapter.

If you are developing new analysis or plotting code, usually one of these data sets is useful for testing it.

Another package with several useful data sets is mlbench. It contains data sets for machine learning benchmarks, so these data sets are aimed at testing how new methods perform on known data sets. This package is not distributed together with R, but you can install it, load it, and get a list of the data sets within it like this:

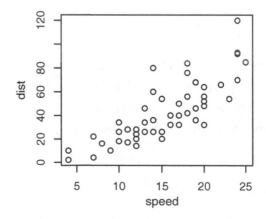

Figure 3-1. *Plot of the cars data set*

```
install.packages("mlbench")
library(mlbench)
library(help = "mlbench")
```

In this book, I will use data from one of those two packages when giving examples of data analyses.

The packages are convenient for me for giving examples, and if you are developing new functionality for R, they are suitable for testing, but if you are interested in data analysis, presumably you are interested in your own data, and there they are of course useless. You need to know how to get your own data into R. We get to that shortly, but first I want to say a few words about how you can examine a data set and get a quick overview.

Quickly Reviewing Data

Earlier, I have already used the function head(). This function shows the first *n* lines of a data frame where *n* is an option with default 6. You can use another *n* to get more or less:

```
cars |> head(3)
```

```
##   speed dist
## 1     4    2
## 2     4   10
## 3     7    4
```

The similar function tail() gives you the last *n* lines:

```
cars %>% tail(3)
```

```
##    speed dist
## 48    24   93
## 49    24  120
## 50    25   85
```

To get summary statistics for all the columns in a data frame, you can use the summary() function:

```
cars %>% summary

##     speed          dist
## Min.   : 4.0  Min.   :  2.00
## 1st Qu.:12.0  1st Qu.: 26.00
## Median :15.0  Median : 36.00
## Mean   :15.4  Mean   : 42.98
## 3rd Qu.:19.0  3rd Qu.: 56.00
## Max.   :25.0  Max.   :120.00
```

It isn't that exciting for the cars data set, so let us see it on another built-in data set:

```
data(iris)
iris |> summary()

##   Sepal.Length    Sepal.Width     Petal.Length
## Min.   :4.300   Min.   :2.000   Min.   :1.000
## 1st Qu.:5.100   1st Qu.:2.800   1st Qu.:1.600
## Median :5.800   Median :3.000   Median :4.350
## Mean   :5.843   Mean   :3.057   Mean   :3.758
## 3rd Qu.:6.400   3rd Qu.:3.300   3rd Qu.:5.100
## Max.   :7.900   Max.   :4.400   Max.   :6.900
##   Petal.Width          Species
## Min.   :0.100   setosa    :50
## 1st Qu.:0.300   versicolor:50
## Median :1.300   virginica :50
## Mean   :1.199
## 3rd Qu.:1.800
## Max.   :2.500
```

The summary you get depends on the types the columns have. Numerical data is summarized by their quantiles, while categorical and boolean data are summarized by counts of each category or TRUE/FALSE values. In the iris data set, there is one column, Species, that is categorical, and its summary is the count of each level.

To see the type of each column, you can use the str() function. This gives you the structure of a data type and is much more general than we need here, but it does give you an overview of the types of columns in a data frame and is very useful for that:

```
data(iris)
iris |> str()

## 'data.frame':    150 obs. of  5 variables:
## $ Sepal.Length: num  5.1 4.9 4.7 4.6 5 5.4 4.6 ..
## $ Sepal.Width : num  3.5 3 3.2 3.1 3.6 3.9 3.4 ..
## $ Petal.Length: num  1.4 1.4 1.3 1.5 1.4 1.7 1...
## $ Petal.Width : num  0.2 0.2 0.2 0.2 0.2 0.4 0...
## $ Species     : Factor w/ 3 levels "setosa","v"..
```

Reading Data

There are several packages for reading data in different file formats, from Excel to JSON to XML and so on. If you have data in a particular format, try to Google for how to read it into R. If it is a standard data format, the chances are that there is a package that can help you.

Quite often, though, data is available in a text table of some kind. Most tools can import and export those. R has plenty of built-in functions for reading such data. Use

```
?read.table
```

to get a list of them. These functions are all variations of the read.table() function, just using different default options. For instance, while read.table() assumes that the data is given in whitespace-separated columns, the read.csv() function assumes that the data is represented as comma-separated values, so the difference between the two functions is in what they consider being separating data columns.

The read.table() function takes a lot of arguments. These are used to adjust it to the specific details of the text file you are reading. (The other functions take the same arguments, they just have different defaults.) The options I find I use the most are these:

- header: This is a boolean value telling the function whether it should consider the first line in the input file a header line. If set to true, it uses the first line to set the column names of the data frame it constructs; if it is set to false, the first line is interpreted as the first row in the data frame.

77

- `col.names`: If the first line is not used to specify the header, you can use this option to name the columns. You need to give it a vector of strings with a string for each column in the input.

- `dec`: This is the decimal point used in numbers. I get spreadsheets that use both "." and "," for decimal points, so this is an important parameter for me. How important it will be for you probably depends on how many nationalities you collaborate with.

- `comment.char`: By default, the function assumes that "#" is the start of a comment and ignores the rest of a line when it sees it. If "#" is actually used in your data, you need to change this. The same goes if comments are indicated with a different symbol.

- `colClasses`: This lets you specify which type each column should have, so here you can specify that some columns should be factors, and others should be strings. You have to specify all columns, though, which is cumbersome and somewhat annoying since R, in general, is pretty good at determining the right types for a column. The option will only take you so far in any case. You can tell it that a column should be an ordered factor but not what the levels should be and such. I mainly use it for specifying which columns should be factors and which should be strings, but using it will also speed up the function for large data sets since R then doesn't have to figure out the column types itself.

For reading in tables of data, `read.table()` and friends will usually get you there with the right options. If you are having problems reading data, check the documentation carefully to see if you cannot tweak the functions to get the data loaded. It isn't always possible, but it usually is. When it really isn't, I usually give up and write a script in another language to format the data into a form I can load into R. For raw text processing, R isn't really the right tool, and rather than forcing all steps in an analysis into R, I will be pragmatic and choose the best tools for the task, and R isn't always it. But before taking drastic measures, and go programming in another language, you should carefully check if you cannot tweak one of the `read.table()` functions first.

Examples of Reading and Formatting Data Sets

Rather than discussing the import of data in the abstract, let us now see a couple of examples of how data can be read in and formatted.

Breast Cancer Data set

As a first example of reading data from a text file, we consider the BreastCancer data set from mlbench. Then we have something to compare our results with. The first couple of lines from this data set are

```
library(mlbench)
data(BreastCancer)
BreastCancer %>% head(3)
```

```
##         Id Cl.thickness Cell.size Cell.shape
## 1 1000025            5         1          1
## 2 1002945            5         4          4
## 3 1015425            3         1          1
##   Marg.adhesion Epith.c.size Bare.nuclei
## 1             1            2           1
## 2             5            7          10
## 3             1            2           2
##   Bl.cromatin Normal.nucleoli Mitoses  Class
## 1           3               1       1 benign
## 2           3               2       1 benign
## 3           3               1       1 benign
```

The data can be found at https://archive.ics.uci.edu/ml/datasets/ Breast+C ancer+Wisconsin+(Original) where there is also a description of the data. The URL to the actual data is https://archive.ics.uci.edu/ml/machine-learning-databases/ breast-cancer-wisconsin/breast-cancer-wisconsin.data, but since this URL is too long to fit on the pages of this book, I have saved it in a variable, data_url, that I will use in the following code. To run the code yourself, you simply need to set the variable to the URL:

```
data_url <- "https://..."
```

To get the data downloaded, we could go to the URL and save the file. Explicitly downloading data outside of our R code has pros and cons. It is pretty simple, and we can have a look at the data before we start parsing it, but on the other hand, it gives us a step in the analysis workflow that is not automatically reproducible. Even if the URL is described in our documentation and at a link that doesn't change over time, it is a manual step in the workflow—and a step that people could make mistakes in.

Instead, I am going to read the data directly from the URL. Of course, this is also a risky step in a workflow because I am not in control of the server the data is on, and I cannot guarantee that the data will always be there and that it won't change over time. It is a bit of a risk either way. I will usually add the code to my workflow for downloading the data, but I will also store the data in a file. If I leave the code for downloading the data and saving it to my local disk in a cached Markdown chunk, it will only be run the one time I need it.

I can read the data and get it as a vector of lines using the readLines() function. I can always use that to scan the first one or two lines to see what the file looks like:

```
lines <- readLines(data_url) lines[1:5]
```

```
## [1] "1000025,5,1,1,1,2,1,3,1,1,2"
## [2] "1002945,5,4,4,5,7,10,3,2,1,2"
## [3] "1015425,3,1,1,1,2,2,3,1,1,2"
## [4] "1016277,6,8,8,1,3,4,3,7,1,2"
## [5] "1017023,4,1,1,3,2,1,3,1,1,2"
```

For this data, it seems to be a comma-separated values file without a header line. So I save the data with the ".csv" suffix. None of the functions for writing or reading data in R cares about the suffixes, but it is easier for myself to remember what the file contains that way:

```
writeLines(lines, con = "data/raw-breast-cancer.csv")
```

For that function to succeed, I first need to make a data/ directory. I suggest you have a data/ directory for all your projects, always, since you want your directories and files structured when you are working on a project.

The file I just wrote to disk can then read in using the read.csv() function:

```
raw_breast_cancer <- read.csv("data/raw-breast-cancer.csv") raw_breast_
cancer |> head(3)
```

```
##     X1000025 X5 X1 X1.1 X1.2 X2 X1.3 X3 X1.4 X1.5
## 1   1002945  5  4   4    5   7  10  3   2    1
## 2   1015425  3  1   1    1   2   2  3   1    1
## 3   1016277  6  8   8    1   3   4  3   7    1
##     X2.1
## 1    2
## 2    2
## 3    2
```

Of course, I wouldn't write exactly these steps into a workflow. Once I have discovered that the data at the end of the URL is a ".csv" file, I would just read it directly from the URL:

```
raw_breast_cancer <- read.csv(data_url)
raw_breast_cancer |> head(3)
```

```
##     X1000025 X5 X1 X1.1 X1.2 X2 X1.3 X3 X1.4 X1.5
## 1   1002945  5  4   4    5   7  10  3   2    1
## 2   1015425  3  1   1    1   2   2  3   1    1
## 3   1016277  6  8   8    1   3   4  3   7    1
##     X2.1
## 1    2
## 2    2
## 3    2
```

The good news is that this data looks similar to the BreastCancer data. The bad news is that it appears that the first line in BreastCancer seems to have been turned into column names in raw_breast_cancer. The read.csv() function interpreted the first line as a header. We can fix this using the header parameter:

```
raw_breast_cancer <- read.csv(data_url, header = FALSE)
raw_breast_cancer |> head(3)
```

```
##          V1 V2 V3 V4 V5 V6 V7 V8 V9 V10 V11
## 1 1000025  5  1  1  1  2  1  3  1   1   2
## 2 1002945  5  4  4  5  7 10  3  2   1   2
## 3 1015425  3  1  1  1  2  2  3  1   1   2
```

Now the first line is no longer interpreted as header names. That is good, but the names we actually get are not that informative about what the columns contain.

If you read the description of the data from the website, you can see what each column is and choose names that are appropriate. I am going to cheat here and just take the names from the BreastCancer data set.

I can set the names explicitly like this:

```
names(raw_breast_cancer) <- names(BreastCancer)
raw_breast_cancer |> head(3)
```

```
##          Id Cl.thickness Cell.size Cell.shape
## 1 1000025            5         1          1
## 2 1002945            5         4          4
## 3 1015425            3         1          1
##  Marg.adhesion Epith.c.size Bare.nuclei
## 1             1            2           1
## 2             5            7          10
## 3             1            2           2
##  Bl.cromatin Normal.nucleoli Mitoses Class
## 1           3               1       1     2
## 2           3               2       1     2
## 3           3               1       1     2
```

or I could set them where I load the data:

```
raw_breast_cancer <- read.csv(data_url, header = FALSE,
                           col.names = names(BreastCancer))
raw_breast_cancer |> head(3)
```

```
##          Id Cl.thickness Cell.size Cell.shape
## 1 1000025            5         1          1
## 2 1002945            5         4          4
## 3 1015425            3         1          1
##    Marg.adhesion Epith.c.size Bare.nuclei
## 1              1            2           1
## 2              5            7          10
## 3              1            2           2
```

```
##    Bl.cromatin Normal.nucleoli Mitoses Class
## 1            3               1       1     2
## 2            3               2       1     2
## 3            3               1       1     2
```

Okay, we are getting somewhere. The `Class` column is not right. It encodes the classes as numbers (the web page documentation specifies 2 for benign and 4 for malignant), but in R it would be more appropriate with a factor.

We can translate the numbers into a factor by first translating the numbers into strings and then the strings into factors. I don't like modifying the original data—even if I have it in a file—so I am going to copy it first and then do the modifications:

```
formatted_breast_cancer <- raw_breast_cancer
```

It is easy enough to map the numbers to strings using `ifelse()`:

```
map_class <- function(x) {
    ifelse(x == 2, "bening",
    ifelse(x == 4, "malignant",
        NA))
}
mapped <- formatted_breast_cancer$Class %>% map_class
mapped |> table()
```

```
## mapped
##    bening malignant
##       458       241
```

I could have made it simpler with

```
map_class <- function(x) {
ifelse(x == 2, "bening", "malignant")
}
mapped <- formatted_breast_cancer$Class %>% map_class
mapped |> table()
```

```
## mapped
##    bening malignant
##       458       241
```

since 2 and 4 are the only numbers in the data

```
formatted_breast_cancer$Class |> unique()
```

```
## [1] 2 4
```

but it is always a little risky to assume that there are no unexpected values, so I always prefer to have "weird values" as something I handle explicitly by setting it to NA.

Nested ifelse() are easy enough to program, but if there are many different possible values, it also becomes somewhat cumbersome. Another option is to use a table to map between values. To avoid confusion between a table as the one we are going to implement and the function table(), which counts how many times a given value appears in a vector, I am going to call the table we create a dictionary. A dictionary is a table where you can look up words, and that is what we are implementing.

For this, we can use named values in a vector. Remember that we can index in a vector both using numbers and using names.

You can create a vector where you use names as the indices. Use the keys you want to map from as the indices and the names you want as results as the values. We want to map from numbers to strings which pose a small problem. If we index into a vector with numbers, R will think we want to get positions in the vector. If we make the vector v <- c(2 = "benign", 4 = "malignant")—which we can't, it is a syntax error and for good reasons—then how should v[2] be interpreted? Do we want the value at index 2, "malignant", or the value that has key 2, "benign"? When we use a vector as a table, we need to have strings as keys. That also means that the numbers in the vector we want to map from should be converted to strings before we look up in the dictionary. The code looks like this:

```
dict <- c("2" = "benign", "4" = "malignant")
map_class <- function(x) dict[as.character(x)]

mapped <- formatted_breast_cancer$Class |> map_class()
mapped |> table()
```

```
## mapped
##    benign malignant
##       458       241
```

That worked fine, but if we look at the actual vector instead of summarizing it, we will see that it looks a little strange:

```
mapped[1:5]
```

```
##         2         2         2         2         2
## "benign" "benign" "benign" "benign" "benign"
```

This is because when we create a vector by mapping in this way, we preserve the names of the values. Remember that the dictionary we made to map our keys to values has the keys as names; these names are passed on to the resulting vector. We can get rid of them using the unname() function:

```
library(magrittr)
mapped %<>% unname
mapped[1:5]
```

```
## [1] "benign" "benign" "benign" "benign" "benign"
```

Here, I used the magrittr %<>% operator to both pipe and rename mapped; alternatively, we could have used mapped <- mapped %>% unname or mapped <- mapped |> unname().

You don't need to remove these names, they are not doing any harm in themselves, but some data manipulation can be slower when your data is dragging names along.

Now we just need to translate this vector of strings into a factor, and we will have our Class column.

The entire reading of data and formatting can be done like this:

```
# Download data and put it in a variable
raw_breast_cancer <- read.csv(
  data_url, header = FALSE,
  col.names = names(BreastCancer))

# Get a copy of the raw data that we can transform
formatted_breast_cancer <- raw_breast_cancer

# Reformat the Class variable
formatted_breast_cancer$Class <-
  formatted_breast_cancer$Class %>% {
    c("2" = "benign", "4" = "malignant")[as.character(.)]
  } |> factor(levels = c("benign", "malignant"))
```

In the last statement, we use the %>% operator so we can put an expression in curly braces. In there, the incoming class is the "." that we translate to a character and then use to look up the name that we want. Then we pipe the names through factor() to get the factor that we went for. We can use either %>% or |> here. We don't have to remove the names with unname() when we put the result back into formatted_breast_cancer, so we don't bother.

It is not strictly necessary to specify the levels in the factor() call, but I prefer always to do so explicitly. If there is an unexpected string in the input to factor(), it would end up being one of the levels, and I wouldn't know about it until much later. Specifying the levels explicitly alleviates that problem.

Now, you don't want to spend time parsing input data files all the time, so I would recommend putting all the code you write to read in data and transforming it into the form you want in a cached code chunk in an R Markup document. This way, you will only evaluate the code when you change it.

You can also explicitly save data using the save() function:

```
formatted_breast_cancer %>%
    save(file = "data/formatted-breast-cancer.rda")
```

Here, I use the suffix ".rda" for the data. It stands for R data, and your computer will probably recognize it. If you click a file with that suffix, it will be opened in RStudio (or whatever tool you use to work on R). The actual R functions for saving and loading data do not care what suffix you use, but it is easier to recognize the files for what they are if you stick to a fixed suffix.

The data is saved together with the name of the data frame, so when you load it again, using the load() function, you don't have to assign the loaded data to a variable. It will be loaded into the name you used when you saved the data:

```
load("data/formatted-breast-cancer.rda")
```

This is both good and bad. I would probably have preferred to control which name the data is assigned to so I have explicit control over the variables in my code, but save() and load() are designed to save more than one variable, so this is how they work.

I personally do not use these functions that much. I prefer to write my analysis pipelines in Markdown documents, and there it is easier just to cache the import code.

Boston Housing Data Set

For the second example of loading data, we take another data set from the mlbench package, the BostonHousing data, which contains information about crime rates and some explanatory variables we can use to predict crime rates:

```
library(mlbench)
data(BostonHousing)
str(BostonHousing)
```

```
## 'data.frame':    506 obs. of  14 variables:
## $ crim    : num  0.00632 0.02731 0.02729 0.03237..
## $ zn      : num  18 0 0 0 0 0 12.5 12.5 ...
## $ indus   : num  2.31 7.07 7.07 2.18 2.18 2.18 7..
## $ chas    : Factor w/ 2 levels "0","1": 1 1 1 1 ..
## $ nox     : num  0.538 0.469 0.469 0.458 0.458 0..
## $ rm      : num  6.58 6.42 7.18 7 ...
## $ age     : num  65.2 78.9 61.1 45.8 54.2 58.7 6..
## $ dis     : num  4.09 4.97 4.97 6.06 ...
## $ rad     : num  1 2 2 3 3 3 5 5 ...
## $ tax     : num  296 242 242 222 222 222 311 311..
## $ ptratio : num  15.3 17.8 17.8 18.7 18.7 18.7 1..
## $ b       : num  397 397 393 395 ...
## $ lstat   : num  4.98 9.14 4.03 2.94 ...
## $ medv    : num  24 21.6 34.7 33.4 36.2 28.7 22...
```

As before, the link to the actual data is pretty long, so I will give you a tinyURL to it, http://tinyurl.com/zq2u8vx, and I have saved the original URL in the variable data_url.

I have already looked at the file at the end of the URL and seen that it consists of whitespace-separated columns of data, so the function we need to load it is read. table():

```
boston_housing <- read.table(data_url)
str(boston_housing)
```

```
## 'data.frame':    506 obs. of  14 variables:
##   $ V1 : num  0.00632 0.02731 0.02729 0.03237 ...
##   $ V2 : num  18 0 0 0 0 0 12.5 12.5 ...
##   $ V3 : num  2.31 7.07 7.07 2.18 2.18 2.18 7.87 ..
##   $ V4 : int  0 0 0 0 0 0 0 0 ...
##   $ V5 : num  0.538 0.469 0.469 0.458 0.458 0.458..
##   $ V6 : num  6.58 6.42 7.18 7 ...
##   $ V7 : num  65.2 78.9 61.1 45.8 54.2 58.7 66.6 ..
##   $ V8 : num  4.09 4.97 4.97 6.06 ...
##   $ V9 : int  1 2 2 3 3 3 5 5 ...
##   $ V10: num  296 242 242 222 222 222 311 311 ...
##   $ V11: num  15.3 17.8 17.8 18.7 18.7 18.7 15.2 ..
##   $ V12: num  397 397 393 395 ...
##   $ V13: num  4.98 9.14 4.03 2.94 ...
##   $ V14: num  24 21.6 34.7 33.4 36.2 28.7 22.9 27..
```

If we compare the data that we have loaded with the data from mlbench

```
str(BostonHousing)
```

```
## 'data.frame':    506 obs. of  14 variables:
##   $ crim  : num  0.00632 0.02731 0.02729 0.03237..
##   $ zn    : num  18 0 0 0 0 0 12.5 12.5 ...
##   $ indus : num  2.31 7.07 7.07 2.18 2.18 2.18 7..
##   $ chas  : Factor w/ 2 levels "0","1": 1 1 1 1 ..
##   $ nox   : num  0.538 0.469 0.469 0.458 0.458 0..
##   $ rm    : num  6.58 6.42 7.18 7 ...
##   $ age   : num  65.2 78.9 61.1 45.8 54.2 58.7 6..
##   $ dis   : num  4.09 4.97 4.97 6.06 ...
##   $ rad   : num  1 2 2 3 3 3 5 5 ...
##   $ tax   : num  296 242 242 222 222 222 311 311..
```

```
##   $ ptratio: num  15.3 17.8 17.8 18.7 18.7 18.7 1..
##   $ b       : num  397 397 393 395 ...
##   $ lstat   : num  4.98 9.14 4.03 2.94 ...
##   $ medv    : num  24 21.6 34.7 33.4 36.2 28.7 22...
```

we see that we have integers and numeric data in our imported data but that it should be a factor for the chas variable and numeric for all the rest. We can use the colClasses parameter for read.table() to fix this. We just need to make a vector of strings for the classes, a vector that is "numeric" for all columns except for the "chas" column, which should be "factor":

```
col_classes <- rep("numeric", length(BostonHousing))
col_classes[which("chas" == names(BostonHousing))] <- "factor"
```

We should also name the columns, but again we can cheat and get the names from BostonHousing:

```
boston_housing <- read.table(data_url,
                             col.names = names(BostonHousing),
                             colClasses = col_classes)
str(boston_housing)
```

```
## 'data.frame':    506 obs. of  14 variables:
##   $ crim    : num  0.00632 0.02731 0.02729 0.03237..
##   $ zn      : num  18 0 0 0 0 12.5 12.5 ...
##   $ indus   : num  2.31 7.07 7.07 2.18 2.18 2.18 7..
##   $ chas    : Factor w/ 2 levels "0","1": 1 1 1 1 ..
##   $ nox     : num  0.538 0.469 0.469 0.458 0.458 0..
##   $ rm      : num  6.58 6.42 7.18 7 ...
##   $ age     : num  65.2 78.9 61.1 45.8 54.2 58.7 6..
##   $ dis     : num  4.09 4.97 4.97 6.06 ...
##   $ rad     : num  1 2 2 3 3 3 5 5 ...
##   $ tax     : num  296 242 242 222 222 222 311 311..
##   $ ptratio: num  15.3 17.8 17.8 18.7 18.7 18.7 1..
##   $ b       : num  397 397 393 395 ...
##   $ lstat   : num  4.98 9.14 4.03 2.94 ...
##   $ medv    : num  24 21.6 34.7 33.4 36.2 28.7 22...
```

The levels in the `"chas"` factor are "0" and "1." It is not really good levels as they are very easily confused with numbers—they will print like numbers—but they are not. The numerical values in the factor are actually 1 for "0" and 2 for "1", so that can be confusing. But it is the same levels as the `mlbench` data frame, so I will just leave it the way it is as well.

The `readr` Package

The `read.table()` class of functions will usually get you to where you want to go with importing data. I use these in almost all my work. But there is a package aimed at importing data that tries to both speed up the importing and being more consistent in how data is imported, so I think I should mention it.

That package is the `readr` package:

```
library(readr)
```

It implements the same class of import functions as the built-in functions. It just uses underscores except for dots in the function names. So where you would use `read.table()`, the `readr` package gives you `read_table()`. Similarly, it gives you `read_csv()` as a substitute for `read.csv()`.

The `readr` package has different defaults for how to read data. Other than that, its main call to fame is being faster than the built-in R functions. This shouldn't concern you much if you put your data import code in a cached code chunk, and in any case if loading data is an issue, you need to read Chapter 5. The functions from `readr` do not return data frames but the tibble data structure we briefly discussed before. For most purposes, this makes no difference, so we can still treat the loaded data the same way.

Let us look at how to import data using the functions in the package. We return to the breast cancer data we imported earlier. We downloaded the breast cancer data and put it in a file called `"data/raw-breast-cancer.csv"`, so we can try to read it from that file. Obviously, since it is a CSV file, we will use the `read_csv()` function:

```
raw_breast_cancer <- read_csv("data/raw-breast-cancer.csv",
                              show_col_types = FALSE)
```

```
## New names:
## • `1` -> `1...3`
## • `1` -> `1...4`
## • `1` -> `1...5`
## • `2` -> `2...6`
## • `1` -> `1...7`
## • `1` -> `1...9`
## • `1` -> `1...10`
## • `2` -> `2...11`
```

```
raw_breast_cancer %>% head(3)
```

```
## # A tibble: 3 × 11
##   `1000025`   `5`  `1...3`  `1...4`  `1...5`  `2...6`
##       <dbl> <dbl>   <dbl>    <dbl>    <dbl>    <dbl>
## 1   1002945     5       4        4        5        7
## 2   1015425     3       1        1        1        2
## 3   1016277     6       8        8        1        3
## # ... with 5 more variables: `1...7` <chr>,
## #   `3` <dbl>, `1...9` <dbl>, `1...10` <dbl>,
## #   `2...11` <dbl>
```

(The show_col_types = FALSE option just says that we don't want to see inferred column types; it removes a bunch of output lines that we aren't interested in here. If you want to see what the function prints, you can try it out for yourself.)

The function works similar to the read.csv() function and interprets the first line as the column names. The warning we get, and the weird column names, is because of that. We don't want this, but this function doesn't have the option to tell it that the first line is not the names of the columns. Instead, we can inform it what the names of the columns are, and then it will read the first line as actual data:

```
raw_breast_cancer <- read_csv("data/raw-breast-cancer.csv",
                    col_names = names(BreastCancer),
                    show_col_types = FALSE)
raw_breast_cancer %>% head(3)
```

```
## # A tibble: 3 × 11
##         Id Cl.thickness Cell.size Cell.shape
##      <dbl>        <dbl>     <dbl>      <dbl>
## 1 1000025            5         1          1
## 2 1002945            5         4          4
## 3 1015425            3         1          1
## # ... with 7 more variables: Marg.adhesion <dbl>,
## #  Epith.c.size <dbl>, Bare.nuclei <chr>,
## #  Bl.cromatin <dbl>, Normal.nucleoli <dbl>,
## #  Mitoses <dbl>, Class <dbl>
```

Which functions you use to import data doesn't much matter. You can use the built-in functions or the readr functions. It is all up to you.

Manipulating Data with dplyr

Data frames are ideal for representing data where each row is an observation of different parameters you want to fit in models. Nearly all packages that implement statistical models or machine learning algorithms in R work on data frames. But to actually manipulate a data frame, you often have to write a lot of code to filter data, rearrange data, summarize it in various ways, and such. A few years ago, manipulating data frames required a lot more programming than actually analyzing data. That has improved dramatically with the dplyr package (pronounced "d plier" where "plier" is pronounced as "pliers").

This package provides a number of convenient functions that let you modify data frames in various ways and string them together in pipes using the %>% or |> operator. If you import dplyr, you get a large selection of functions that let you build pipelines for data frame manipulation using pipelines.

Earlier, we have formatted data by manipulating data columns by directly assigning to them, like we did with formatted_breast_cancer$Class <- formatted_breast_cancer$Class %>% ... statements. There is nothing inherently wrong with this approach, but it breaks the idea of using pipelines to send data through a series of transformations, if we have to assign the results to old or new columns from time to time. With dplyr, you get functions that manipulate data frames as a whole, but where you can still modify, add, or remove columns as you please.

The transformations we did on the breast cancer set you can also do with dplyr, going directly from the raw data set to the formatted data, without first creating the data frame for the formatted data and then updating it. The way you would do it could look like this:

```
library(dplyr)

# Download data and put it in a variable
raw_breast_cancer <- read.csv("data/raw-breast-cancer.csv",
                              header = FALSE,
                              col.names = names(BreastCancer))
# Reformat the Class column as a benign/malignant factor
formatted_breast_cancer <- raw_breast_cancer |>
  mutate(
    Class =
        case_when(Class == 2 ~ "benign", Class == 4 ~ "malignant") |>
        factor(levels = c("benign", "malignant"))
  )
```

We still get the raw data by downloading it, we don't have any other choice, but then we take this raw data and pipe it through a function called mutate: raw_breast_cancer |> mutate(...) to change the data frame. The mutate() function (see later) can modify the columns in the data frame, and in the preceding expression, that is what we do. With raw_breast_cancer |> mutate(Class = ...), we say that we want to change the Class column to what is to the right of the = inside mutate. You can modify more than one column with the same mutate, but we only change one. The right-hand side of = is a bit complex because we are doing all the work we did earlier, translating 2 and 4 into the strings "benign" and "malignant" and then making a factor out of that. It is not that bad if we split the operation into two pieces, though. First, we use a dplyr function called case_when. It works a bit like ifelse but is more general. Its input is a sequence of so-called formulas—expressions on the form lhs ~ rhs—where the left-hand side (lhs) should be a boolean expression and the right-hand side (rhs) the value you want if the left-hand side is TRUE. Here, the expressions are simple: we test if Class is two or four and return "benign" or "malignant" accordingly. This maps the 2/4 encoding to strings, and we then pipe that through factor to get a factor out of that. The result is written to the Class column. Notice that we can refer directly to the values in a column, here

Class, using just the column name. We couldn't do that in the preceding section where we worked with the data frame's Class column, where we had to write raw_breast_cancer$Class to get at it. The functions in dplyr do some magic that lets you treat data frame columns as if they were variables, and this makes the expressions you have to write a bit more manageable.

Some Useful dplyr Functions

I will not be able to go through all of the dplyr functionality in this chapter. In any case, it is updated frequently enough that, by the time you read this, there is probably more functionality than at the time I write. So you will need to check the documentation for the package.

The following are just the functions I use on a regular basis. They all take a data frame or equivalent as the first argument, so they work perfectly with pipelines. When I say "data frame equivalent," I mean that they take as an argument anything that works like a data frame. Quite often, there are better representations of data frames than the built-in data structure. For large data sets, it is often better to use a different representation than the built-in data frame, something we will return to in Chapter 5. Some alternative data structures are better because they can work with data on disk—R's data frames have to be loaded into memory, and others are just faster to do some operations on. Or maybe they just print better. If you write the name of a data frame into the R terminal, it will print the entire data. Other representations will automatically give you just the head of the data.

The dplyr package has several representations. The tibbles, mentioned a few times in the previous chapters, are a favorite of mine that I use just because I prefer the output when I print such tables. You can translate a base data frame into a tibble using the as_tibble function:

```
iris %>% as_tibble()

## # A tibble: 150 × 5
##     Sepal.Length Sepal.Width Petal.Length
##            <dbl>       <dbl>       <dbl>
## 1           5.1         3.5         1.4
## 2           4.9         3           1.4
## 3           4.7         3.2         1.3
```

```
## 4           4.6        3.1        1.5
## 5           5          3.6        1.4
## 6           5.4        3.9        1.7
## 7           4.6        3.4        1.4
## 8           5          3.4        1.5
## 9           4.4        2.9        1.4
## 10          4.9        3.1        1.5
## # ... with 140 more rows, and 2 more variables:
## #   Petal.Width <dbl>, Species <fct>
```

It only prints the first ten rows, and it doesn't print all columns. The output is a little easier to read than if you get the entire data frame.

Anyway, let us get to the dplyr functions.

select—Pick selected columns and get rid of the rest.

The select() function selects columns of the data frame. It is equivalent to indexing columns in the data.

You can use it to pick out a single column:

```
iris %>% as_tibble() %>% select(Petal.Width) %>% head(3)
```

```
## # A tibble: 3 × 1
##    Petal.Width
##          <dbl>
## 1          0.2
## 2          0.2
## 3          0.2
```

Or several columns:

```
iris %>% as_tibble() %>%
  select(Sepal.Width, Petal.Length) %>% head(3)
```

```
## # A tibble: 3 × 2
##    Sepal.Width Petal.Length
##          <dbl>        <dbl>
## 1          3.5          1.4
## 2          3            1.4
## 3          3.2          1.3
```

You can even give it ranges of columns:

```
iris %>% as_tibble() %>%
  select(Sepal.Length:Petal.Length) %>% head(3)
```

```
## # A tibble: 3 × 3
##    Sepal.Length Sepal.Width Petal.Length
##           <dbl>       <dbl>        <dbl>
## 1           5.1         3.5          1.4
## 2           4.9         3            1.4
## 3           4.7         3.2          1.3
```

but how that works depends on the order the columns are in for the data frame, and it is not something I find all that useful.

I pipe `iris` through `as_tibble()` in these pipelines, because I like the formatting more. You don't have to, to use the `dplyr` functions, but if you don't, you are working on a base data frame instead of a tibble, and R will print your data slightly differently.

The real usefulness comes with pattern matching on column names. There are different ways to pick columns based on the column names:

```
iris |> as_tibble() |> select(starts_with("Petal")) |> head(3)
```

```
## # A tibble: 3 × 2
##    Petal.Length Petal.Width
##           <dbl>       <dbl>
## 1           1.4         0.2
## 2           1.4         0.2
## 3           1.3         0.2
```

```
iris |> as_tibble() |> select(ends_with("Width")) |> head(3)
```

```
## # A tibble: 3 × 2
##    Sepal.Width Petal.Width
##          <dbl>       <dbl>
## 1          3.5         0.2
## 2          3           0.2
## 3          3.2         0.2
```

```
iris |> as_tibble() |> select(contains("etal")) |> head(3)
```

```
## # A tibble: 3 × 2
##   Petal.Length Petal.Width
##          <dbl>       <dbl>
## 1          1.4         0.2
## 2          1.4         0.2
## 3          1.3         0.2
```

```
iris |> as_tibble() |> select(matches(".t.")) |> head(3)
```

```
## # A tibble: 3 × 4
##   Sepal.Length Sepal.Width Petal.Length
##          <dbl>       <dbl>        <dbl>
## 1          5.1         3.5          1.4
## 2          4.9         3            1.4
## 3          4.7         3.2          1.3
## # ... with 1 more variable: Petal.Width <dbl>
```

The matches function searches for a regular expression and in this example will select any name that contains a t except if it is the first or last letter.

Check out the documentation for dplyr to see which options you have for selecting columns.

You can also use select() to remove columns. The preceding examples select the columns you want to include, but if you use "-" before the selection criteria, you will remove, instead of include, the columns you specify:

```
iris %>% as_tibble() %>%
  select(-starts_with("Petal")) %>% head(3)
```

```
## # A tibble: 3 × 3
##   Sepal.Length Sepal.Width Species
##          <dbl>       <dbl> <fct>
## 1          5.1         3.5 setosa
## 2          4.9         3   setosa
## 3          4.7         3.2 setosa
```

mutate—Add computed values to your data frame.

The `mutate()` function lets you add a column to your data frame by specifying an expression for how to compute it:

```
iris %>% as_tibble() %>%
  mutate(Petal.Width.plus.Length = Petal.Width + Petal.Length) %>%
  select(Species, Petal.Width.plus.Length) %>%
  head(3)
```

```
## # A tibble: 3 × 2
##    Species Petal.Width.plus.Length
##    <fct>                     <dbl>
## 1 setosa                      1.6
## 2 setosa                      1.6
## 3 setosa                      1.5
```

You can add more columns than one by specifying them in the `mutate()` function:

```
iris %>% as_tibble() %>%
  mutate(Petal.Width.plus.Length = Petal.Width + Petal.Length,
         Sepal.Width.plus.Length = Sepal.Width + Sepal.Length) %>%
  select(Petal.Width.plus.Length, Sepal.Width.plus.Length) %>%
  head(3)
```

```
## # A tibble: 3 × 2
##    Petal.Width.plus.Length Sepal.Width.plus.Length
##                      <dbl>                   <dbl>
## 1                      1.6                     8.6
## 2                      1.6                     7.9
## 3                      1.5                     7.9
```

but you could of course also just call `mutate()` several times in your pipeline.

transmute—Add computed values to your data frame and get rid of all other columns.

The `transmute()` function works just like the `mutate()` function, except it combines it with a `select()` so the result is a data frame that only contains the new columns you make:

```
iris %>% as_tibble() %>%
  transmute(Petal.Width.plus.Length = Petal.Width + Petal.Length) %>%
  head(3)
```

```
## # A tibble: 3 × 1
##   Petal.Width.plus.Length
##                     <dbl>
## 1                     1.6
## 2                     1.6
## 3                     1.5
```

arrange—Reorder your data frame by sorting columns.

The `arrange()` function just reorders the data frame by sorting columns according to what you specify:

```
iris %>% as_tibble() %>%
  arrange(Sepal.Length) %>%
  head(3)
```

```
## # A tibble: 3 × 5
##   Sepal.Length Sepal.Width Petal.Length
##          <dbl>       <dbl>        <dbl>
## 1          4.3         3            1.1
## 2          4.4         2.9          1.4
## 3          4.4         3            1.3
## # ... with 2 more variables: Petal.Width <dbl>,
## #   Species <fct>
```

By default, it orders numerical values in increasing order, but you can ask for decreasing order using the `desc()` function:

```
iris %>% as_tibble() %>%
  arrange(desc(Sepal.Length)) %>%
  head(3)
```

```
## # A tibble: 3 × 5
##   Sepal.Length Sepal.Width Petal.Length
##          <dbl>       <dbl>        <dbl>
## 1          7.9         3.8          6.4
## 2          7.7         3.8          6.7
## 3          7.7         2.6          6.9
## # ... with 2 more variables: Petal.Width <dbl>,
## #   Species <fct>
```

filter—Pick selected rows and get rid of the rest.

The `filter()` function lets you pick out rows based on logical expressions. You give the function a predicate specifying what a row should satisfy to be included:

```
iris %>% as_tibble() %>%
  filter(Sepal.Length > 5) %>%
  head(3)
```

```
## # A tibble: 3 × 5
##    Sepal.Length Sepal.Width Petal.Length
##           <dbl>       <dbl>        <dbl>
## 1           5.1         3.5          1.4
## 2           5.4         3.9          1.7
## 3           5.4         3.7          1.5
## # ... with 2 more variables: Petal.Width <dbl>,
## #   Species <fct>
```

You can get as inventive as you want here with the logical expressions:

```
iris %>% as_tibble() %>%
  filter(Sepal.Length > 5 & Species == "virginica") %>%
  select(Species, Sepal.Length) %>%
  head(3)
```

```
## # A tibble: 3 × 2
##    Species   Sepal.Length
##    <fct>            <dbl>
##       1  virginica   6.3
##       2  virginica   5.8
##       3  virginica   7.1
```

group_by—Split your data into subtables based on values of some of the columns.

The `group_by()` function tells `dplyr` that you want to work on data separated into different subsets.

By itself, it isn't that useful. It just tells `dplyr` that, in future computations, it should consider different subsets of the data as separate data sets. It is used with the `summarise()` function where you want to compute summary statistics.

You can group by one or more variables; you just specify the columns you want to group by as separate arguments to the function. It works best when grouping by factors or discrete numbers; there isn't much fun in grouping by real numbers:

```
iris %>% as_tibble() %>% group_by(Species) %>% head(3)
```

```
## # A tibble: 3 × 5
## # Groups:    Species [1]
##    Sepal.Length Sepal.Width Petal.Length
##           <dbl>       <dbl>        <dbl>
## 1           5.1         3.5          1.4
## 2           4.9         3            1.4
## 3           4.7         3.2          1.3
## # ... with 2 more variables: Petal.Width <dbl>,
## #   Species <fct>
```

Not much is happening here. You have restructured the data frame such that there are groupings, but until you do something with the new data, there isn't much to see. The power of `group_by()` is the combination with the `summarise()` function.

summarise/summarize—Calculate summary statistics.

The spelling of this function depends on which side of the pond you are on. It is the same function regardless of how you spell it.

The `summarise()` function is used to compute summary statistics from your data frame. It lets you compute different statistics by expressing what you want to summarize. For example, you can ask for the mean of values:

```
iris %>%
  summarise(Mean.Petal.Length = mean(Petal.Length),
            Mean.Sepal.Length = mean(Sepal.Length))
```

```
##    Mean.Petal.Length Mean.Sepal.Length
## 1              3.758          5.843333
```

Where it is really powerful is in the combination with `group_by()`. There, you can split the data into different groups and compute the summaries for each group:

```
iris %>%
  group_by(Species) %>%
  summarise(Mean.Petal.Length = mean(Petal.Length))
```

```
## # A tibble: 3 × 2
##   Species     Mean.Petal.Length
##   <fct>                   <dbl>
## 1 setosa                   1.46
## 2 versicolor               4.26
## 3 virginica                5.55
```

Depending on your version of dplyr, you might get a warning here that summarise is "ungrouping" the output. This has to do with which groupings a data frame has when we give it to summarise and which groupings the output has. Until dplyr 1.0.0, there was only one option, which of course you couldn't change, but now there are four and a default behavior, which may or may not be what you want (but is probably there to at least make the function backward compatible with older code). If you get a warning, it is because you have a version of dplyr that has the new behavior, but that is still old enough to warn you that you are using the default behavior. It will then tell you that you can change this using the .groups argument.

The four behaviors we can get are

- .groups = "drop_last" removes the last grouping we introduced. If we are summarizing over some grouping, we end up with a table with one row per group, and there is no need to keep the group information for that any longer. This is the old default behavior.

- .groups = "drop" removes all groupings, so we have called group_ by multiple times we lose all the groups; with drop_last, we would only lose the last grouping.

- .groups = "keep" keeps all the groups exactly as they are.

- .groups = "rowwise" makes each row in the output its own group.

You can check which columns in the data frame are part of a grouping using the group_vars, so we can see which groups the output of a summarise has for each of the four options. First, we make a grouping for the iris data and check the group variables:

```
grouped_iris <- iris %>% as_tibble() %>%
  group_by(Species, Petal.Length)
grouped_iris %>% group_vars()
```

```
## [1] "Species"      "Petal.Length"
```

With drop_last, the second grouping variable, Petal.Length, is removed, but the first is not; we are dropping the last variable and only the last:

```
grouped_iris %>%
  summarise(Mean.Petal.Length = mean(Petal.Length),
            .groups = "drop_last") %>%
  group_vars()
```

```
## [1] "Species"
```

If we use drop instead, all the grouping variables are removed, and group_vars gives us character(0) which is the empty string vector:

```
grouped_iris %>%
  summarise(Mean.Petal.Length = mean(Petal.Length),
            .groups = "drop") %>%
  group_vars()
```

```
## character(0)
```

If we use keep, none of the grouping variables are removed:

```
grouped_iris %>%
  summarise(Mean.Petal.Length = mean(Petal.Length),
            .groups = "keep") %>%
  group_vars()
```

```
## [1] "Species"      "Petal.Length"
```

At first glance, the rowwise option looks a lot like keep:

```
grouped_iris %>%
  summarise(Mean.Petal.Length = mean(Petal.Length),
            .groups = "rowwise") %>%
  group_vars()
```

```
## [1] "Species"      "Petal.Length"
```

It isn't the same thing, though. If we summarize with `rowwise`, we get a different kind of data structure, a so-called `rowwise_df` (a row-wise data frame), and we can see this if we ask for the "class" of the result:

```
grouped_iris %>%
  summarise(Mean.Petal.Length = mean(Petal.Length),
            .groups = "rowwise") %>%
  class()
```

```
## [1] "rowwise_df" "tbl_df"     "tbl"
## [4] "data.frame"
```

The class is a list of types, and we will see how this works in the second half of the book. The purpose of this rowwise data type is a different kind of manipulation, where we manipulate rows of data frames. They are beyond the scope of this book, and I will leave it at that.

For the `summarise()` function, the `.groups` option is only important if you do more than one summary. In most usage, you don't. You manipulate your data, then you compute some summary statistics, and then that is your result. In cases such as those, it doesn't matter if you keep or drop any grouping variables. If you don't care about the grouping in the result of a summary, you can turn the warning off using the option `dplyr.summarise.inform = FALSE`:

```
options(dplyr.summarise.inform = FALSE)
summary <- grouped_iris %>%
  summarise(Mean.Petal.Length = mean(Petal.Length))
```

Here, even though you didn't use `.groups`, you shouldn't get a warning.

If you turn the option on again, you should get the warning:

```
options(dplyr.summarise.inform = TRUE)
```

Although you usually don't care, it is safer to specify `.groups`. At least you have to think about whether you want the result to have any groups, and if you do, you have to consider if you get the right ones.

A summary function worth mentioning here is n() which just counts how many observations you have in a subset of your data:

```
iris %>%
  summarise(Observations = n())

## Observations
## 1          150
```

Again, this is more interesting when combined with group_by():

```
iris %>%
  group_by(Species) %>%
  summarise(Number.Of.Species = n(), .groups = "drop")

## # A tibble: 3 × 2
##   Species    Number.Of.Species
##   <fct>                  <int>
## 1 setosa                    50
## 2 versicolor                50
## 3 virginica                 50
```

You can combine summary statistics simply by specifying more than one in the summary() function:

```
iris %>%
  group_by(Species) %>%
  summarise(Number.Of.Samples = n(),
            Mean.Petal.Length = mean(Petal.Length),
            .groups = "drop")

## # A tibble: 3 × 3
##   Species    Number.Of.Samples Mean.Petal.Length
##   <fct>                  <int>             <dbl>
## 1 setosa                    50              1.46
## 2 versicolor                50              4.26
## 3 virginica                 50              5.55
```

Breast Cancer Data Manipulation

To get a little more feeling for how the `dplyr` package can help us explore data, let us see it in action.

Let us return to the breast cancer data. We start with the modifications we used to transform the raw data we imported from the CSV file (stored in the variable `raw_breast_cancer`). Using `dplyr` functions, we could create the `formatted_breast_cancer` data, as we saw earlier, like this:

```
formatted_breast_cancer <- raw_breast_cancer |>
  as_tibble() |>
  mutate(
    Class =
      case_when(Class == 2 ~ "benign", Class == 4 ~ "malignant") |>
      factor(levels = c("benign", "malignant"))
  )
```

I piped the raw data through `as_tibble` first, this time, because I prefer to work with tibbles. Otherwise, it is the same code as earlier.

We can check if things look the way they should using a `select` and a `head`:

```
formatted_breast_cancer |> select(Normal.nucleoli:Class) |> head(5)
```

```
## # A tibble: 5 × 3
##    Normal.nucleoli Mitoses Class
##              <int>   <int> <fct>
## 1                1       1 benign
## 2                2       1 benign
## 3                1       1 benign
## 4                7       1 benign
## 5                1       1 benign
```

Now let us look a little at the actual data. This is a very crude analysis of the data we can do for exploratory purposes. It is not a proper analysis, but we will return to that in Chapter 6 later.

We could be interested in how the different parameters affect the response variable, the `Class` variable. For instance, is cell thickness different for benign and malignant tumors? To check that, we can group the data by the `Cell` parameter and look at the mean cell thickness:

```
formatted_breast_cancer %>%
  group_by(Class) %>%
  summarise(mean.thickness = mean(Cl.thickness), .groups = "drop")
```

```
## # A tibble: 2 × 2
##   Class      mean.thickness
##   <fct>               <dbl>
## 1 benign               2.96
## 2 malignant            7.20
```

It looks like there is a difference. Now whether this difference is significant requires a proper test—after all, we are just comparing means here, and the variance could be huge. But just exploring the data, it gives us a hint that there might be something to work with here.

We could ask the same question for other variables, like cell size:

```
formatted_breast_cancer %>%
  group_by(Class) %>%
  summarise(mean.size = mean(Cell.size), .groups = "drop")
```

```
## # A tibble: 2 × 2
##   Class      mean.size
##   <fct>          <dbl>
## 1 benign          1.33
## 2 malignant       6.57
```

Another way of looking at this could be to count, for each cell size, how many benign tumors and how many malignant tumors we see. Here, we would need to group by both cell size and class and then count, and we would probably want to arrange the data so we get the information in order of increasing or decreasing cell size:

```
formatted_breast_cancer %>%
  arrange(Cell.size) %>%
  group_by(Cell.size, Class) %>%
  summarise(ClassCount = n(), .groups = "drop")
```

```
## # A tibble: 18 × 3
##    Cell.size Class    ClassCount
##        <int> <fct>         <int>
## 1         1 benign          380
## 2         1 malignant         4
## 3         2 benign           37
## 4         2 malignant         8
## 5         3 benign           27
## 6         3 malignant        25
## 7         4 benign            9
## 8         4 malignant        31
## 9         5 malignant        30
## 10        6 benign            2
## 11        6 malignant        25
## 12        7 benign            1
## 13        7 malignant        18
## 14        8 benign            1
## 15        8 malignant        28
## 16        9 benign            1
## 17        9 malignant         5
## 18       10 malignant        67
```

Here again, we get some useful information. It looks like there are more benign tumors compared to malignant tumors when the cell size is small and more malignant tumors when the cell size is large. Again something we can start to work from when we later want to build statistical models.

This kind of grouping only works because the cell size is measured as discrete numbers. It wouldn't be helpful to group by a floating-point number. There, plotting is more useful. But for this data, we have the cell size as integers, so we can explore the data just by building tables in this way.

We can also try to look at combined parameters. We have already seen that both cell size and cell thickness seem to be associated with how benign or malignant a tumor is, so let us try to see how the cell thickness behaves as a function of both class and cell size:

```
formatted_breast_cancer %>%
  group_by(Class, as.factor(Cell.size)) %>%
  summarise(mean.thickness = mean(Cl.thickness),
            .groups = "drop")
```

```
## # A tibble: 18 × 3
##    Class     `as.factor(Cell.size)` mean.thickness
##    <fct>     <fct>                          <dbl>
##  1 benign    1                               2.76
##  2 benign    2                               3.49
##  3 benign    3                               3.81
##  4 benign    4                               5.11
##  5 benign    6                               5
##  6 benign    7                               5
##  7 benign    8                               6
##  8 benign    9                               6
##  9 malignant 1                               7.25
## 10 malignant 2                               6.75
## 11 malignant 3                               6.44
## 12 malignant 4                               7.71
## 13 malignant 5                               6.87
## 14 malignant 6                               6.88
## 15 malignant 7                               6.89
## 16 malignant 8                               7.18
## 17 malignant 9                               8.8
## 18 malignant 10                              7.52
```

I am not sure how much I learn from this. It seems that for the benign tumors, the thickness increases with the cell size, but for the malignant, there isn't that pattern.

Maybe we can learn more by ordering the data in a different way. What if we look at the numbers of benign and malignant tumors for each cell size and see what the thickness is?

```
formatted_breast_cancer %>%
  group_by(as.factor(Cell.size), Class) %>%
  summarise(mean.thickness = mean(Cl.thickness),
            .groups = "drop")
```

```
## # A tibble: 18 × 3
##    `as.factor(Cell.size)` Class    mean.thickness
##    <fct>                  <fct>             <dbl>
##  1 1                      benign             2.76
##  2 1                      malignant          7.25
##  3 2                      benign             3.49
##  4 2                      malignant          6.75
##  5 3                      benign             3.81
##  6 3                      malignant          6.44
##  7 4                      benign             5.11
##  8 4                      malignant          7.71
##  9 5                      malignant          6.87
## 10 6                      benign             5
## 11 6                      malignant          6.88
## 12 7                      benign             5
## 13 7                      malignant          6.89
## 14 8                      benign             6
## 15 8                      malignant          7.18
## 16 9                      benign             6
## 17 9                      malignant          8.8
## 18 10                     malignant          7.52
```

I am not sure how much we learned from that either, but at least it looks like for each cell size where we have both benign and malignant tumors the thickness is higher for the malignant than the benign. That is something at least. A place to start the analysis. But we can learn more when we start plotting data and when we do a proper statistical analysis of them. We will return to that in later chapters. For now, we leave it at that.

Tidying Data with `tidyr`

I am not really sure where the concept of "tidy data" comes from. Hadley Wickham, the author of many of the essential packages you will use in your R data analysis, describes tidy data as such:

> Tidy data is a standard way of mapping the meaning of a data set
> to its structure. A data set is messy or tidy depending on how rows,
> columns and tables are matched up with observations, variables
> and types.

In my experience, tidy data means that I can plot or summarize the data efficiently. It mostly comes down to what data is represented as columns in a data frame and what is not.

In practice, this means that I have columns in my data frame that I can work with for the analysis I want to do. For example, if I want to look at the iris data set and see how the Petal.Length varies among species, then I can look at the Species column against the Petal.Length column:

```
iris |>
  as_tibble() |>
  select(Species, Petal.Length) |>
  head(3)
```

```
## # A tibble: 3 × 2
##     Species Petal.Length
##     <fct>          <dbl>
## 1   setosa           1.4
## 2   setosa           1.4
## 3   setosa           1.3
```

I have a column specifying the Species and another specifying the Petal.Length, and it is easy enough to look at their correlation. I can plot one against the other (we will cover visualization in the next chapter). I can let the x-axis be species and the y-axis be Petal.Length (see Figure 3-2).

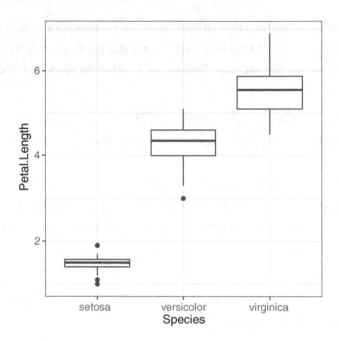

Figure 3-2. *Plot species vs. petal length*

```
library(ggplot2)

iris %>%
  select(Species, Petal.Length) %>%
  qplot(Species, Petal.Length, geom = "boxplot", data = .)
```

(In this pipeline, we need to use the %>% operator rather than |> because we need the input of qplot to go into the data argument, where the "." is, and |> cannot do this.)

This works because I have a column for the x-axis and another for the y-axis. But what happens if I want to plot the different measurements of the irises to see how those are? Each measurement is a separate column. They are Petal.Length, Petal.Width, and so on.

Now I have a bit of a problem because the different measurements are in different columns in my data frame. I cannot easily map them to an x-axis and a y-axis.

The tidyr package lets me fix that:

```
library(tidyr)
```

It has a function, pivot_longer(), that modifies the data frame, so columns become names in a factor and other columns become values.

What it does is essentially transforming the data frame such that you get one column containing the name of your original columns and another column containing the values in those columns.

In the `iris` data set, we have observations for sepal length and sepal width. If we want to examine `Species` vs. `Sepal.Length` or `Sepal.Width`, we can readily do this. We have more of a problem if we want to examine for each species both measurements at the same time. The data frame just doesn't have the structure we need for that.

If we want to see `Sepal.Length` and `Sepal.Width` as two measurements, we can plot against their values, and we would need to make a column in our data frame that tells us if a measurement is a length or a width and another column that shows us what the measurement actually is. The `pivot_longer()` function from `tidyr` lets us do that:

```
iris |>
  pivot_longer(
    c(Sepal.Length, Sepal.Width),
    names_to = "Attribute",
    values_to = "Measurement"
  ) |>
  head()

## # A tibble: 6 × 5
##   Petal.Length Petal.Width Species Attribute
##          <dbl>       <dbl> <fct>   <chr>
## 1          1.4         0.2 setosa  Sepal.Length
## 2          1.4         0.2 setosa  Sepal.Width
## 3          1.4         0.2 setosa  Sepal.Length
## 4          1.4         0.2 setosa  Sepal.Width
## 5          1.3         0.2 setosa  Sepal.Length
## 6          1.3         0.2 setosa  Sepal.Width
## # ... with 1 more variable: Measurement <dbl>
```

The preceding code tells `pivot_longer()` to take the columns `Sepal.Length` and `Sepal.Width` and make them names and values in two new columns. Here, you should read "names" as the name of the input columns a value comes from, and you should read "values" as the actual value in that column. We are saying that we want two new columns that hold all the values from the two columns `Sepal.Length` and `Sepal.Width`. The first of these columns, we get to name it with the `names_to` parameter and we give it the name `Attribute`, will contain the original column name we got the value from, and the second column, `values_to` that we name `Measurement` gets the values from the original columns. The original columns that we didn't specify are still there, but values are duplicated to match that the `Sepal.Length` and `Sepal.Width` values now are merged into a single column.

We don't necessarily want to keep all columns after a transformation like this. If we just want to plot `Sepal.Length` against `Sepal.Width`, maybe colored by `Species`, we can use `select` to pick the columns we want to keep. That would be `Species` for the colors, `Attributes` so we can tell which values are `Sepal.Length` and which are `Sepal.Width`, and then `Measurement` for the actual values:

```
iris |>
  pivot_longer(
    c(Sepal.Length, Sepal.Width),
    names_to = "Attribute",
    values_to = "Measurement"
  ) |>
  select(Species, Attribute, Measurement) |>
  head(3)
```

```
## # A tibble: 3 × 3
##    Species Attribute   Measurement
##    <fct>   <chr>             <dbl>
## 1 setosa  Sepal.Length        5.1
## 2 setosa  Sepal.Width         3.5
## 3 setosa  Sepal.Length        4.9
```

This transforms the data into a form where we can plot the attributes against measurements (see Figure 3-3 for the result):

```
iris |>
  pivot_longer(
    c(Sepal.Length, Sepal.Width),
    names_to = "Attribute",
    values_to = "Measurement"
  ) |>
```

```
select(Species, Attribute, Measurement) %>%
qplot(Attribute, Measurement,
        geom = "boxplot",
        facets = . ~ Species, data = .)
```

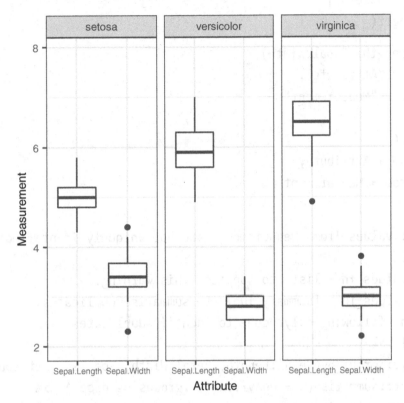

Figure 3-3. *Plot measurements vs. values*

The tidyr package also has a function, pivot_wider, for transforming data frames in the other direction. It is not a reverse of pivot_longer, because pivot_longer removes information about correlations in the data; we cannot, after pivoting, see which lengths originally sat in the same rows as which width, unless the remaining columns uniquely identified this, which they are not guaranteed to do (and do not for the iris data set). With pivot_wider, you get columns back that match the names you used with pivot_wider, and you get the values you specified there as well, but if the other variables do not uniquely identify how they should match up, you can get multiple values in the same column:

```
iris |> as_tibble() |>
  pivot_longer(

  c(Sepal.Length, Sepal.Width),
  names_to = "Attribute",
  values_to = "Measurement"
) |>
pivot_wider(
  names_from = Attribute,
  values_from = Measurement
)

## Warning: Values from `Measurement` are not uniquely identified;
output will
## * Use `values_fn = list` to suppress this warning.
## * Use `values_fn = {summary_fun}` to summarise duplicates.
## * Use the following dplyr code to identify duplicates.
##    {data} %>%
##      dplyr::group_by(Petal.Length, Petal.Width, Species, Attribute) %>%
##      dplyr::summarise(n = dplyr::n(), .groups = "drop") %>%
##      dplyr::filter(n > 1L)

## # A tibble: 103 × 5
##    Petal.Length Petal.Width Species Sepal.Length
##           <dbl>       <dbl> <fct>   <list>
## 1          1.4         0.2 setosa  <dbl [8]>
## 2          1.3         0.2 setosa  <dbl [4]>
## 3          1.5         0.2 setosa  <dbl [7]>
## 4          1.7         0.4 setosa  <dbl [1]>
## 5          1.4         0.3 setosa  <dbl [3]>
## 6          1.5         0.1 setosa  <dbl [2]>
## 7          1.6         0.2 setosa  <dbl [5]>
```

```
## 8            1.4        0.1 setosa  <dbl [2]>
## 9            1.1        0.1 setosa  <dbl [1]>
## 10           1.2        0.2 setosa  <dbl [2]>
## # ... with 93 more rows, and 1 more variable:
## #   Sepal.Width <list>
```

You should get some warnings here; after the first transformation, we do not have enough information to transform back. Instead, R has to map some keys to multiple values. A column with type <list> is where you have multiple values, and the values are printed as their type and the number, for example, <dbl [8]>. Working with this kind of data is beyond the scope of this book, since it is atypical data for most data science applications. We are only ending up in this situation now, because we are trying to reverse an operation we did, when that operation isn't reversible. You would never try to reverse a pivot_longer() with a pivot_wider() in this way, and I have never had to attempt it myself.

Still, there are cases where your data is in a tidy format and you want to transform it back to a variable per column, and then pivot_wider() is your function of choice. You just have to deal with cases that might arise, where remaining columns do not provide enough information to do it. One option is to summarize all the values that would have to go into the same row, that is, those that we got as lists in the iris example earlier. You can provide a function for that using the values_fn parameter. I don't know if it makes sense to summarize the values here by their mean, but we can if we want to:

```
iris |> as_tibble() |>
  pivot_longer(
    c(Sepal.Length, Sepal.Width),
    names_to = "Attribute",
    values_to = "Measurement"
  ) |>
  pivot_wider(
    names_from = Attribute,
    values_from = Measurement,
    values_fn = mean
  ) |>
  # Let's just look at the columns we summarised...
  select(Sepal.Length, Sepal.Width)
```

```
## # A tibble: 103 × 2
##    Sepal.Length Sepal.Width
##           <dbl>       <dbl>
## 1          4.96        3.39
## 2          4.75        3.22
## 3          5.07        3.41
## 4          5.4         3.9
## 5          4.83        3.3
## 6          5.05        3.6
## 7          4.88        3.3
## 8          4.85        3.3
## 9          4.3         3
## 10         5.4         3.6
## # ... with 93 more rows
```

Exercises

It is time to put what we have learned into practice. There are only a few exercises, but I hope you will do them. You can't learn without doing exercises after all.

Importing Data

To get a feeling of the steps in importing and transforming data, you need to try it yourself. So try finding a data set you want to import. You can do that from one of the repositories I listed in the first chapter:

- RDataMining.com

- UCI Machine Learning Repository

- KDNuggets

- Reddit r/data sets

- GitHub Awesome Public Data sets

Or maybe you already have a data set you would like to analyze.

Have a look at it and figure out which import function you need. You might have to set a few parameters in the function to get the data loaded correctly, but with a bit of effort, you should be able to. For column names, you should choose some appropriate ones from reading the data description, or if you are loading something in that is already in `mlbench`, you can cheat as I did in the preceding examples.

Using `dplyr`

Now take the data you just imported and examine various summaries. It is not so important what you look at in the data as it is that you try summarizing different aspects of it. We will look at proper analyses later. For now, just use `dplyr` to explore your data.

Using `tidyr`

Look at the preceding `dplyr` example. There, I plotted `Sepal.Length` and `Sepal.Width` for each species. Do the same thing for `Petal.Length` and `Petal.Width`.

If there is something similar to do with the data set you imported in the first exercise, try doing it with that.

CHAPTER 4

Visualizing Data

Nothing tells a story about your data as powerfully as good plots. Graphics captures your data much better than summary statistics and often shows you features that you would not be able to glean from summaries alone.

R has excellent tools for visualizing data. Unfortunately, it also has more tools than you really know what to do with. There are several different frameworks for visualizing data, and they are usually not compatible, so you cannot easily combine the various approaches.

In this chapter, we look at graphics in R, and we cannot possibly cover all the plotting functionality, so I will focus on two frameworks. The first is the basic graphics framework. It is not something I frequently use or recommend that you use, but it is the default for many packages, so you need to know about it. The second is the `ggplot2` framework, which is my preferred approach to visualizing data. It defines a small domain-specific language for constructing data and is perfect for exploring data as long as you have it in a data frame (and with a little bit more work, for creating publication-ready plots).

Basic Graphics

The basic plotting system is implemented in the `graphics` package. You usually do not have to include the package

```
library(graphics)
```

since it is already loaded when you start up R, but you can use

```
library(help = "graphics")
```

121

© Thomas Mailund 2022
T. Mailund, *Beginning Data Science in R 4*, https://doi.org/10.1007/978-1-4842-8155-0_4

to get a list of the functions implemented in the package. This list isn't exhaustive, though, since the primary plotting function, plot(), is generic and many packages write extensions to it to specialize plots.

In any case, you create basic plots using the plot() function. This function is a so-called generic function, which means that what it does depends on the input it gets. So you can give it different first arguments to get plots of various objects.

The simplest plot you can make is a scatter plot, plotting points for *x* and *y* values (see Figure 4-1):

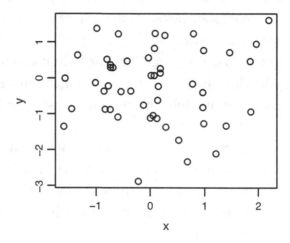

Figure 4-1. *Scatter plot*

```
x <- rnorm(50)
y <- rnorm(50)
plot(x, y)
```

The plot() function takes a data argument you can use to plot data from a data frame, but you cannot write code like this to plot the cars data from the datasets package:

```
data(cars)
cars %>% plot(speed, dist, data = .)
```

Despite giving plot() the data frame, it will not recognize the variables for the x and y parameters, and so adding plots to pipelines requires that you use the %$% operator to give plot() access to the variables in a data frame. So, for instance, we can plot the cars data like this (see Figure 4-2):

```
cars %$% plot(speed, dist, main="Cars data",
              xlab="Speed", ylab="Stopping distance")
```

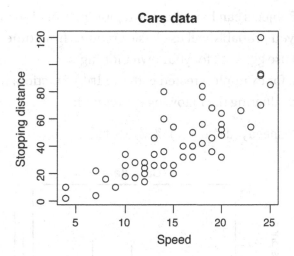

Figure 4-2. *Scatter plot of speed and distance for cars*

Here, we use `main` to give the figure a title and `xlab` and `ylab` to specify the axis labels.

The data argument of `plot()` is used when the variables to the plot are specified as a formula. The `plot()` function then interprets the formula as specifying how the data should be plotted. If the x and y values are specified in a formula, you can give the function a data frame that holds the variables and plot from that:

```
cars %>% plot(dist ~ speed, data = .)
```

Here, you must use the `%>%` operator and not `|>` since you need the left-hand side to go into the `data` argument and not the first argument.

By default, the plot shows the data as points, but you can specify a `type` parameter to display the data in other ways such as lines or histograms (see Figure 4-3):

```
cars %$% plot(speed, dist, main="Cars data", type="h",
              xlab="Speed", ylab="Stopping distance")
```

To get a histogram for a single variable, you should use the function `hist()` instead of `plot()` (see Figure 4-4):

```
cars %$% hist(speed)
```

What is meant by `plot()` being a generic function (something we will cover in much greater detail in Chapter 12) is that it will have different functionality depending on what parameters you give it.

123

Different kinds of objects can have their own plotting functionality, though, and often do. This is why you probably will use basic graphics from time to time even if you follow my advice and use ggplot2 for your own plotting.

Linear regression, for example, created with the lm() function, has its specialized plotting routine. Try evaluating the following expression:

```
cars %>% lm(dist ~ speed, data = .) %>% plot()
```

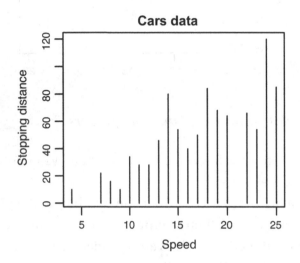

Figure 4-3. *Histogram plot of speed and distance for cars*

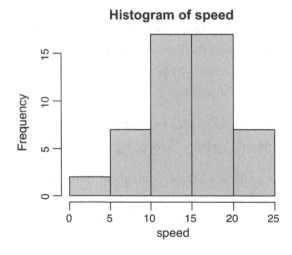

Figure 4-4. *Histogram for cars speed*

It will give you several summary plots for visualizing the quality of the linear fit.

Many model fitting algorithms return a fitted object that has specialized plotting functionality like this, so when you have fitted a model, you can always try to call plot() on it and see if you get something useful out of that.

Functions like plot() and hist() and a few more create new plots, but there is also a large number of functions for annotating a plot. Functions such as lines() and points() add lines and points, respectively, to the current plot rather than making a new plot.

We can see them in action if we want to plot the longley data set and want to see both the unemployment rate and people in the armed forces over the years:

```
data(longley)
```

Check the documentation for longley (?longley) for a description of the data. The data has various statistics for each year from 1947 to 1962 including the number of people unemployed (variable Unemployed) and the number of people in the armed forces (variable Armed.Forces). To plot both of these on the same plot, we can first plot Unemployed against years (variable Year) and then add lines for Armed.Forces. See Figure 4-5.

```
longley %>% plot(Unemployed ~ Year, data = ., type = 'l')
longley %>% lines(Armed.Forces ~ Year, data = ., col = "blue")
```

This almost gets us what we want, but the y-axis is chosen by the plot() function to match the range of y values in the call to plot(), and the Armed.Forces doesn't quite fit into this range. To fit both, we have to set the limits of the y-axis which we do with parameter ylim (see Figure 4-6):

```
longley %$% plot(Unemployed ~ Year, type = 'l',
                 ylim = range(c(Unemployed, Armed.Forces)))
longley %>% lines(Armed.Forces ~ Year, data = ., col = "blue")
```

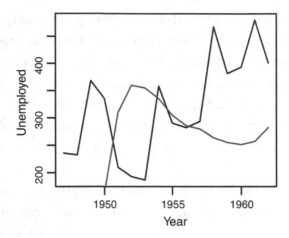

Figure 4-5. *Longley data showing Unemployed and Armed.Forces. The y-axis doesn't cover all of the Armed.Forces variable*

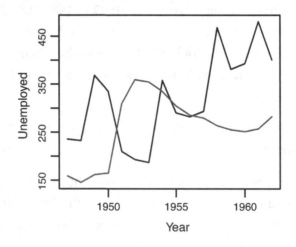

Figure 4-6. *Longley data showing Unemployed and Armed.Forces. The y-axis is wide enough to hold all the data*

Like plot(), the other plotting functions are usually generic. This means we can sometimes give them objects such as fitted models. The abline() function is one such case. It plots lines of the form $y = a + bx$, but there is a variant of it that takes a linear model as input and plots the best fitting line defined by the model. So we can plot the cars data together with the best-fitted line using the combination of the lm() and abline() functions (see Figure 4-7):

```
cars %>% plot(dist ~ speed, data = .)
cars %>% lm(dist ~ speed, data = .) %>% abline(col = "red")
```

Plotting using the basic graphics usually follows this pattern. First, there is a call to plot() that sets up the canvas to plot on—possibly adjusting the axes to make sure that later points will fit in on it. Then any additional data points are plotted—like the second time series we saw in the longley data. Finally, there might be some annotation like adding text labels or margin notes (see functions text() and mtext() for this).

If you want to select the shape of points or their color according to other data features, for example, plotting the iris data with data points in different colors according to the Species variable, then you need to map features to columns (see Figure 4-8):

```
color_map <- c("setosa" = "black",
               "versicolor" = "grey40",
               "virginica" = "grey75")
iris %$% plot(Petal.Length ~ Petal.Width,
              col = color_map[Species])
```

The basic graphics system has many functions for making publication-quality plots, but most of them work at a relatively low level. You have to map variables to colors or shapes explicitly if you want a variable to determine how points should be displayed. You have to set the xlim and ylim parameters to have the right *x*- and *y*-axes if the first points you plot do not cover the entire range of the data you want to plot. If you change an axis—say log-transform—or if you flip the *x*- and *y*-axes, then you will usually need to update several function calls. If you want to have different subplots—so-called facets— for different subsets of your data, then you have to subset and plot this explicitly.

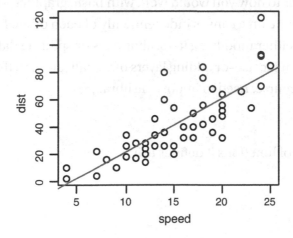

Figure 4-7. *The cars data points annotated with the best fitting line*

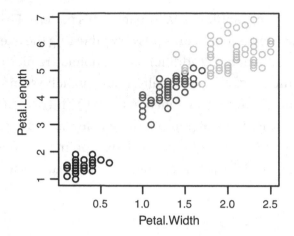

Figure 4-8. *Iris data plotted with different colors for different species*

So while the basic graphics system is powerful for making good-looking final plots, it is not necessarily optimal for exploring data where you often want to try different ways of visualizing it.

The Grammar of Graphics and the `ggplot2` Package

The `ggplot2` package provides an alternative to the basic graphic that is based on what is called the "grammar of graphics." The idea here is that the system gives you a small domain-specific language for creating plots (similar to how `dplyr` provides a domain-specific language for manipulating data frames). You construct plots through a list of function calls—similar to how you would work with basic graphics—but these function calls do not directly write on a canvas independently of each other. Rather, they all manipulate a plot by either modifying it—scaling axes or splitting data into subsets that are plotted on different facets—or adding layers of visualization to the plot.

To use it, you, of course, need to import the library:

```
library(ggplot2)
```

and you can get a list of functions it defines using

```
library(help = "ggplot2")
```

I can only give a very brief tutorial-like introduction to the package here. There are full books written about ggplot2 if you want to learn more details. After reading this chapter, you should be able to construct basic plots, and you should be able to find information about how to make more intricate plots by searching online.

We ease into ggplot2 by first introducing the qplot() function (it stands for quick plot). This function works similar to plot()—although it handles things a little differently—but creates the same kind of objects that the other ggplot2 functions operate on, and so it can be combined with those.

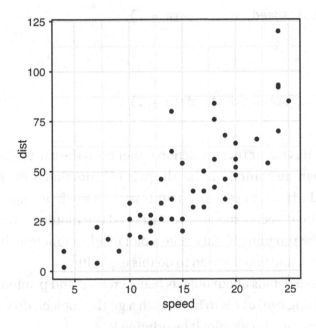

Figure 4-9. *Plot of the cars data using qplot (ggplot2)*

Using qplot()

The qplot() function can be used to plot simple scatter plots the same way as the plot() function. To plot the cars data (see Figure 4-9), we can use

```
cars %>% qplot(speed, dist, data = .)
```

What happens is slightly different, though. The qplot() function creates a ggplot object rather than directly plotting. It is just that when such objects are printed, which happened at the end of the statement, the effect of printing is that they are plotted.

That sounds a bit confusing, but it is what happens. The function used for printing R objects is a generic function, so the effect of printing an object depends on what the object implements for the `print()` function. For `ggplot` objects, this function plots the object. It works well with the kind of code we write, though, because in the preceding code the result of the entire expression is the return value of `qplot()`. When this is evaluated at the outermost level in the R prompt, the result is printed. So the `ggplot` object is plotted.

The preceding code is equivalent to

```
p <- cars %>% qplot(speed, dist, data = .)
p
```

which is equivalent to

```
p <- cars %>% qplot(speed, dist, data = .)
print(p)
```

The reason that it is the `print()` function rather than the `plot()` function—which would otherwise be more natural—is that the `print()` function is the function that is automatically called when we evaluate an expression at the R prompt. By using `print()`, we don't need to print objects explicitly, we just need the plotting code to be at the outermost level of the program. If you create a plot inside a function, however, it isn't automatically printed, and you do need to do this explicitly.

I mention all these details about objects being created and printed because the typical pattern for using `ggplot2` is to build such a `ggplot` object, do various operations on it to modify it, and then finally plot it by printing it.

When using `qplot()`, some transformations of the plotting object are done before `qplot()` returns the object. The quick in quick plot consists of `qplot()` guessing at what kind of plot you are likely to want and then doing transformations on a plot to get there. To get the full control of the final plot, we skip `qplot()` and do all the transformations explicitly—I personally never use `qplot()` anymore myself—but to get started and getting familiar with `ggplot2`, it is not a bad function to use.

With `qplot()`, we can make the visualization of data points depend on data variables in a more straightforward way than we can with `plot()`. To color the `iris` data according to `Species` in `plot()`, we needed to code up a mapping and then transform the `Species` column to get the colors. With `qplot()`, we just specify that we want the colors to depend on the `Species` variable (see Figure 4-10):

```
iris %>% qplot(Petal.Width, Petal.Length ,
               color = Species, data = .)
```

We get the legend for free when we are mapping the color like this, but we can modify it by doing operations on the ggplot object that qplot() returns, should we want to.

You can also use qplot() for other types of plots than scatter plots. If you give it a single variable to plot, it will assume that you want a histogram instead of a scatter plot and give you that (see Figure 4-11):

```
cars %>% qplot(speed, data = ., bins = 10)
```

If you want a density plot instead, you simply ask for it (see Figure 4-12):

```
cars %>% qplot(speed, data = ., geom = "density")
```

Similarly, you can get lines, box plots, violin plots, etc. by specifying a geometry. Geometries determine how the underlying data should be visualized. They might involve calculating some summary statistics, which they do when we create a histogram or a density plot, or they might just visualize the raw data, as we do with scatter plots. Still, they all describe how data should be visualized. Building a plot with ggplot2 involves adding geometries to your data, typically more than one geometry. To see how this is done, though, we leave qplot() and look at how we can create the plots we made earlier with qplot() using geometries instead.

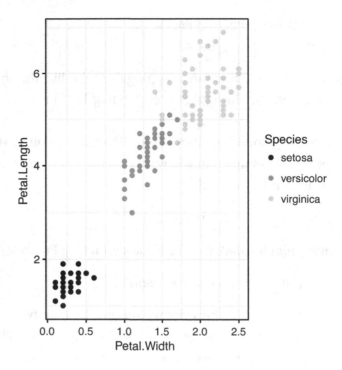

Figure 4-10. *Plot of iris data with colors determined by the species. Plotted with qplot (ggplot2)*

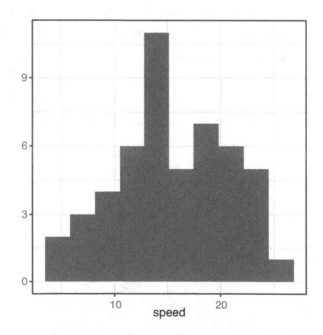

Figure 4-11. *Histogram of car speed created using qplot (ggplot2)*

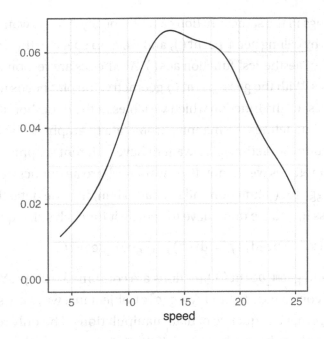

Figure 4-12. *Density of car speed created using qplot (ggplot2)*

Using Geometries

By stringing together several geometry commands, we can either display the same
data in different ways—for example, scatter plots combined with smoothed lines—or
put several data sources on the same plot. Before we see more complex constructions,
though, we can see how the preceding qplot() plots could be made by explicitly calling
geometry functions.

We start with the scatter plot for cars where we used

```
cars %>% qplot(speed, dist, data = .)
```

To create this plot using explicit geometries, we want a ggplot object, we need to
map the speed parameter from the data frame to the x-axis and the dist parameter to
the y-axis, and we need to plot the data as points:

```
ggplot(cars) + geom_point(aes(x = speed, y = dist))
```

We create an object using the ggplot() function. We give it the cars data as input.
When we give this object the data frame, the following operations can access the data. It is
possible to override which data frame the data we plot comes from, but unless otherwise
specified, we have access to the data we gave ggplot() when we created the initial object.

Next, we do two things in the same function call. We specify that we want x and y values to be plotted as points by calling geom_point(), and we map speed to the x values and dist to the y values using the "aesthetics" function aes(). Aesthetics are responsible for mapping from data to graphics. With the geom_point() geometry, the plot needs to have x and y values. The aesthetics tell the function which variables in the data should be used for these.

The aes() function defines the mapping from data to graphics just for the geom_point() function. Sometimes, we want to have different mappings for different geometries, and sometimes we do not. If we want to share aesthetics between functions, we can set it in the ggplot() function call instead. Then, like the data, the following functions can access it, and we don't have to specify it for each subsequent function call:

```
ggplot(cars, aes(x = speed, y = dist)) + geom_point()
```

The ggplot() and geom_point() functions are combined using +. You use + to string together a series of commands to modify a ggplot object in a way very similar to how we use %>% to string together a sequence of data manipulations. The only reason that these are two different operators here is historical; if the %>% operator had been in common use when ggplot2 was developed, it would most likely have used that. As it is, you use +. Because + works slightly different in ggplot2 than %>% does in magrittr, you cannot just use a function name when the function doesn't take any arguments, so you need to include the parentheses in geom_point().

Since ggplot()takes a data frame as its first argument, a typical pattern is first to modify data in a string of %>% or |> operations and then give it to ggplot() and follow that with a series of + operations. Doing that with cars would provide us with this simple pipeline—in larger applications, more steps are included in both the %>% pipeline and the + plot composition:

```
cars %>% ggplot(aes(x = speed, y = dist)) + geom_point()
```

For the iris data, we used the following qplot() call to create a scatter plot with colors determined by the Species variable:

```
iris %>% qplot(Petal.Width, Petal.Length ,
               color = Species, data = .)
```

The corresponding code using ggplot() and geom_point() looks like this:

```
iris %>% ggplot() +
  geom_point(aes(x = Petal.Width, y = Petal.Length,
                 color = Species))
```

Here, we could also have put the aesthetics in the `ggplot()` call instead of the `geom_point()` call.

When you specify the color as an aesthetic, you let it depend on another variable in the data. If you instead want to hardwire a color—or any graphics parameter in general— you simply have to move the parameter assignment outside the `aes()` call. If `geom_point()` gets assigned a color parameter, it will use that color for the points; if it doesn't, it will get the color from the aesthetics (see Figure 4-13):

```
iris |> ggplot() +
  geom_point(aes(x = Petal.Width, y = Petal.Length),
             color = "grey50")
```

The `qplot()` code for plotting a histogram and a density plot

```
cars %>% qplot(speed, data = ., bins = 10)
cars %>% qplot(speed, data = ., geom = "density")
```

can be constructed using `geom_histogram()` and `geom_density()`, respectively:

```
cars |> ggplot() + geom_histogram(aes(x = speed), bins = 10)
cars |> ggplot() + geom_density(aes(x = speed))
```

You can combine more geometries to display the data in more than one way. Doing this isn't always meaningful depending on how data is summarized—combining scatter plots and histograms might not be so useful. However, we can, for example, make a plot showing the car speed both as a histogram and a density (see Figure 4-14):

```
cars |> ggplot(aes(x = speed, y = ..count..)) +
  geom_histogram(bins = 10) +
  geom_density()
```

It just requires us to call both `geom_histogram()` and `geom_density()`. We do also need to add an extra aesthetics option for the y value. The reason is that histograms by default will show the counts of how many observations fall within a bin on the y-axis, while densities integrate to one. By setting `y = ..count..`, you tell both geometries to use counts as the y-axis. To get densities instead, you can use `y = ..density...`

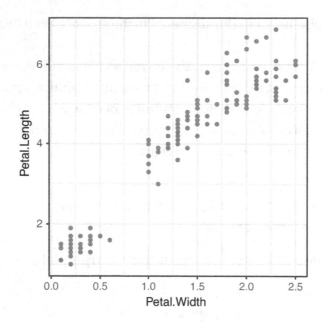

Figure 4-13. *Iris data where the color of the points is hardwired*

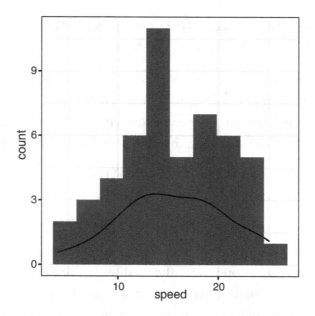

Figure 4-14. *Combined histogram and density plot for speed from the cars data*

We can also use combinations of geometries to show summary statistics of data together with a scatter plot. We added the result of a linear fit of the data to the scatter plot we did for the `cars` data with `plot()`. To do the same with `ggplot2`, we add a `geom_smooth()` call (see Figure 4-15):

```
cars %>% ggplot(aes(x = speed, y = dist)) +
  geom_point() + geom_smooth(method = "lm")

## `geom_smooth()` using formula 'y ~ x'
```

The message we get from `geom_smooth` is that it used the formula y ~ x in the linear model to smooth the data. It will let us know when we use a default instead of explicitly providing a formula for what we want smoothed. Here, it just means that it is finding the best line between the x and y values, which is exactly what we want. You could make the formula explicit by writing `geom_smooth(formula = y ~ x, method = "lm")`, or you could use a different formula, for example, `geom_smooth(formula = y ~ 1, method = "lm")`, to fit the y values to a constant, getting a horizontal line to fit the mean y value (you can try it out). The default is usually what we want.

Earlier, we told the `geom_smooth()` call to use the linear model method. If we didn't do this, it would instead plot a loess smoothing (see Figure 4-16):

```
cars %>% ggplot(aes(x = speed, y = dist)) +
  geom_point() + geom_smooth()

## `geom_smooth()` using method = 'loess' and formula 'y ~ x'
```

We can also use more than one geometry to plot more than one variable. For the `longley` data, we could use two different `geom_line()` to plot the `Unemployed` and the `Armed.Forces` data (see Figure 4-17):

```
longley %>% ggplot(aes(x = Year)) +
  geom_line(aes(y = Unemployed)) +
  geom_line(aes(y = Armed.Forces), color = "blue")
```

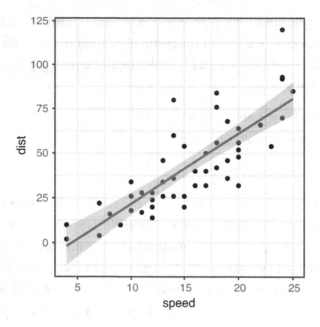

Figure 4-15. *Cars data plotted with a linear model smoothing*

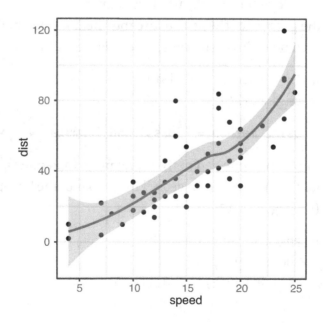

Figure 4-16. *Cars data plotted with a loess smoothing*

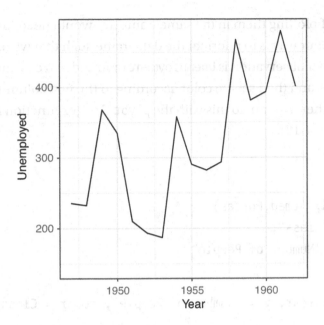

Figure 4-17. *Longley data plotted with ggplot2*

Here, we set the x value aesthetics in the `ggplot()` function since it is shared by the two `geom_line()` geometries, but we set the y value in the two calls, and we set the color for the `Armed.Forces` data, hardwiring it instead of setting it as an aesthetic. Because we are modifying a plot rather than just drawing on a canvas with the second `geom_line()` call, the y-axis is adjusted to fit both lines. We, therefore, do not need to set the y-axis limit anywhere.

We can also combine `geom_line()` and `geom_point()` to get both lines and points for our data (see Figure 4-18):

```
longley %>% ggplot(aes(x = Year)) +
  geom_point(aes(y = Unemployed)) +
  geom_line(aes(y = Unemployed)) +
  geom_point(aes(y = Armed.Forces), color = "blue") +
  geom_line(aes(y = Armed.Forces), color = "blue")
```

Plotting two variables using different aesthetics like this is fine for most applications, but it is not always the optimal way to do it. The problem is that we are representing that the two measures, `Unemployed` and `Armed.Forces`, are two different measures we have per year and that we can plot together in the plotting code. The data is not reflecting this as something we can compute on. Should we want to split the two measures into

139

subplots instead of plotting them in the same frame, we would need to write new plotting code. A better way is to reformat the data frame such that we have one column telling us whether an observation is Unemployment or Armed.Forces and another giving us the values and then set the color according to the first column and the y-axis according to the other. We can do this with the pivot_longer function from the tidyr package (see Figure 4-19):

```
longley %>%
  pivot_longer(
    c(Unemployed, Armed.Forces),
    names_to = "Class",
    values_to = "Number of People"
  ) %>%

ggplot(aes(x = Year, y = `Number of People`, color = Class)) +
geom_line()
```

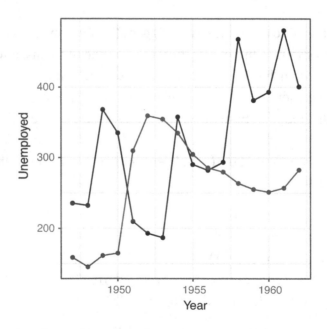

Figure 4-18. *Longley data plotted with ggplot2 using both points and lines*

In the `pivot_longer` expression, we are saying that we want to transform the `Unemployed` and the `Armed.Forces` columns. These are two different classes from the statistics, so we put the original column names into a new column called `Class`. The two columns count the number of people in the two classes, so the values from the two original classes will go into a new column that we name `Number of People`. The names in the `pivot_longer` expression are strings, and we can put anything there, but the y value in the `aes()` expression has to be a valid variable name, and those cannot have spaces, and we need to escape the string. We do that using backticks.

Once we have transformed the data, we can change the plot with little extra code. If, for instance, we want the two values on different facets, we can simply specify this (instead of setting the colors) (see Figure 4-20):

```
longley %>%
  pivot_longer(
    c(Unemployed, Armed.Forces),
    names_to = "Class",
    values_to = "Number of People"
  ) %>%
  ggplot(aes(x = Year, y = `Number of People`)) +
  geom_line() +
  facet_grid(Class ~ .)
```

Facets

Facets are subplots showing different subsets of the data. In the preceding example, we show the `Armed.Forces` variable in one subplot and the `Unemployed` variable in another. You can specify facets using one of two functions: `facet_grid()` creates facets from a formula `rows ~ columns`, and `facet_wrap()` creates facets from a formula `~ variables`. The former creates a row for the variables on the left-hand side of the formula and a column for the variables on the right-hand side and builds facets based on this. In the preceding example, we used "`key ~ .`", so we get a row per key. Had we used "`. ~ key`" instead, we would get a column per key. The `facet_wrap()` doesn't explicitly set up rows and columns, it simply makes a facet per combination of variables on the right-hand side of the formula and wraps the facets in a grid to display them.

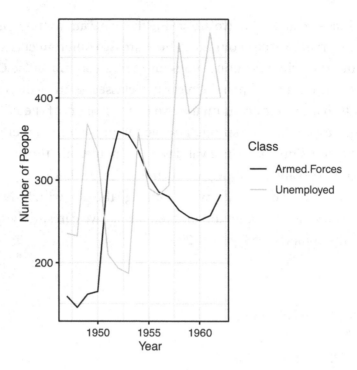

Figure 4-19. *Longley data plotted using tidy data*

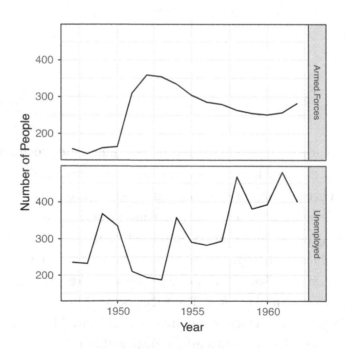

Figure 4-20. *Longley data plotted using facets*

By default, `ggplot2` will try to put values on the same axes when you create facets using `facet_grid()`. So in the preceding example, the `Armed.Forces` are shown on the same x- and y-axes as `Unemployment` even though the y values, as we have seen, are not covering the same range. We can use the `scales` parameter to change this. Facets within a column will always have the same x-axis, however, and facets within a row will have the same y-axis.

We can see this in action with the `iris` data. We can transform the `iris` data, so every column except `Species` gets squashed into two key-value columns using `pivot_longer`. We can select everything except selected columns by putting a - in front of their name when we select them. Then we can plot the measurements for each separate species like this:

```
iris %>%
  pivot_longer(
    -Species,
    names_to = "Measurement",
    values_to = "Value"
  ) %>%
  ggplot(aes(x = Species, y = Value)) +
  geom_boxplot() +
  facet_grid(Measurement ~ .)
```

We plot the four measurements for each species in different facets, but they are on slightly different scales, so we will only get a good look at the range of values for the largest range. We can fix this by setting the y-axis free; contrast Figures 4-21 and 4-22.

```
iris %>%
  pivot_longer(
    -Species,
    names_to = "Measurement",
    values_to = "Value"
  ) %>%
  ggplot(aes(x = Species, y = Value)) +
  geom_boxplot() +
  facet_grid(Measurement ~ ., scale = "free_y")
```

143

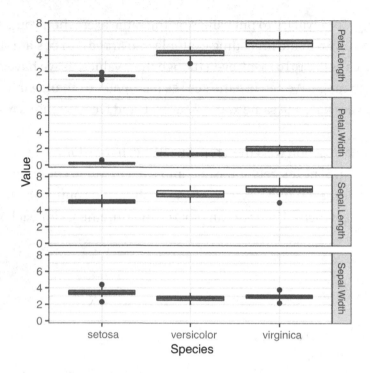

Figure 4-21. *Iris measures plotted on the same y-axis*

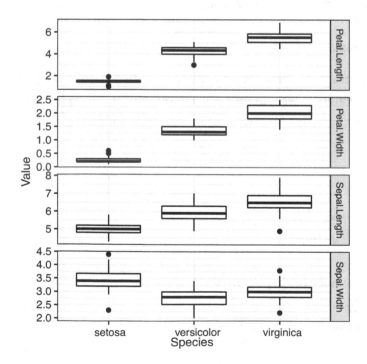

Figure 4-22. *Iris measures plotted on different y-axes*

By default, all the facets will have the same size. You can modify this using the `space` variable. This is mainly useful for categorical values if one facet has many more of the levels than another.

The labels used for facets are taken from the factors in the variables used to construct the facet. This is a good default, but for print quality plots, you often want to modify the labels a little. You can do this using the `labeller` parameter to `facet_grid()`. This parameter takes a function as an argument that is responsible for constructing labels. The easiest way to construct this function is by using another function, `labeller()`. You can give `labeller()` a named argument specifying a factor to make labels for with lookup tables for mapping levels to labels. For the `iris` data, we can use this to remove the dots in the measurement names (see Figure 4-23):

```
label_map <- c(Petal.Width = "Petal Width",
               Petal.Length = "Petal Length",
               Sepal.Width = "Sepal Width",
               Sepal.Length = "Sepal Length")

iris %>%
  pivot_longer(
    -Species,
    names_to = "Measurement",
    values_to = "Value"
  ) %>%
  ggplot(aes(x = Species, y = Value)) +
  geom_boxplot() +
  facet_grid(Measurement ~ ., scale = "free_y",
             labeller = labeller(Measurement = label_map))
```

Scaling

Geometries specify part of how data should be visualized and scales another. The geometries tell `ggplot2` how you want your data mapped to visual components, like points or densities, and scales tell `ggplot2` how dimensions should be visualized. The simplest scales to think about are the x- and y-axes, where values are mapped to positions on the plot as you are familiar with, but scales also apply to visual properties such as colors.

The simplest use we can make of scales is just to put labels on the axes. We can also do this using the xlab() and ylab() functions, and if setting labels were all we were interested in, we would, but as an example, we can see this use of scales. To set the labels in the cars scatter plot, we can write

```
cars %>%
  ggplot(aes(x = speed, y = dist)) +
  geom_point() + geom_smooth(method = "lm") +
  scale_x_continuous("Speed") +
  scale_y_continuous("Stopping Distance")
```

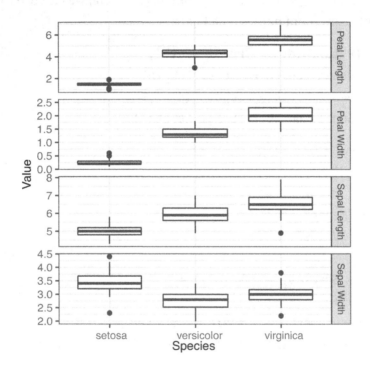

Figure 4-23. *Iris measures with measure labels adjusted*

Both the x- and y-axes are showing a continuous value, so we scale like that and give the scale a name as the parameter. This will then be the names put on the axis labels. In general, we can use the scale_x/y_continuous() functions to control the axis graphics, for instance, to set the breakpoints shown. If we want to plot the longley data with a tick mark for every year instead of every five years, we can set the breakpoints to every year:

```
longley %>%
  pivot_longer(
    c(Unemployed, Armed.Forces),
    names_to = "Class",
    values_to = "Number of People"
  ) %>%
  ggplot(aes(x = Year, y = `Number of People`)) +
  geom_line() +
  scale_x_continuous(breaks = 1947:1962) +
  facet_grid(Class ~ .)
```

You can also use the scale to modify the labels shown at tick marks or set limits on the values displayed.

Scales are also the way to transform data shown on an axis. If you want to log-transform the x- or y-axis, you can use the scale_x/y_log10() functions, for instance. This usually leads to a nicer plot compared to plotting data you log-transform yourself since the plotting code then knows that you want to show data on a log scale rather than showing transformed data on a linear scale.

To reverse an axis, you use scale_x/y_reverse(). This is better than reversing the data mapped in the aesthetic since all the plotting code will just be updated to the reversed axis; you don't need to update x or y values in all the function geometry calls. For instance, to show the speed in the cars data in decreasing instead of increasing order, we could write

```
cars %>%
  ggplot(aes(x = speed, y = dist)) +
  geom_point() +
  geom_smooth(method = "lm") +
  scale_x_reverse("Speed") +
  scale_y_continuous("Stopping Distance")
```

Neither axis has to be continuous. If you map a factor to x or y in the aesthetics, you get a discrete axis; see Figure 4-24 for the iris data plotted with the factor Species on the x-axis.

```
iris %>%
  ggplot(aes(x = Species, y = Petal.Length)) +
  geom_boxplot() +
  geom_jitter(width = 0.1, height = 0.1)
```

Since Species is a factor, the x-axis will be discrete, and we can show the data as a box plot and the individual data points using the jitter geometry. If we want to modify the x-axis, we need to use scale_x_discrete() instead of scale_x_continuous().

We can, for instance, use this to modify the labels on the axis to put the species in capital letters:

```
iris %>%
  ggplot(aes(x = Species, y = Petal.Length)) +
  geom_boxplot() +
  geom_jitter(width = 0.1, height = 0.1) +
  scale_x_discrete(labels = c("setosa" = "Setosa",
                              "versicolor" = "Versicolor",
                              "virginica" = "Virginica"))
```

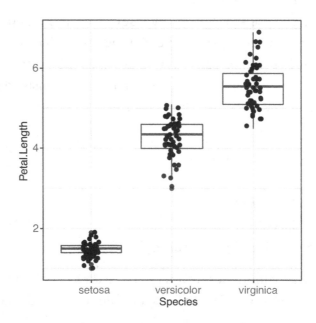

Figure 4-24. *Iris data plotted with a factor on the x-axis*

We provide a map from the data levels to labels. There is more than one way to set the labels, but this is by far the easiest.

Scales are also used to control colors. You use the various `scale_color_` functions to control the color of lines and points, and you use the `scale_fill_` functions to control the color of filled areas.

We can plot the `iris` measurements per species and give them a different color for each species. Since it is the boxes we want to color, we need to use the `fill` aesthetics. Otherwise, we would color the lines around the boxes. See Figure 4-25.

```
iris %>%
  pivot_longer(
    -Species,
    names_to = "Measurement",
    values_to = "Value"
  ) %>%
  ggplot(aes(x = Species, y = Value, fill = Species)) +
  geom_boxplot() +
  facet_grid(Measurement ~ ., scale = "free_y",
             labeller = labeller(Measurement = label_map))
```

There are different ways to modify color scales. There are two classes, as there are for axes, discrete and continuous. The `Species` variable in `iris` is discrete, so to modify the fill color, we need one of the functions for that. The simplest is just to give a color per species explicitly. We can do that with the `scale_fill_manual()` function (see Figure 4-26):

```
iris %>%
  pivot_longer(
    -Species,
    names_to = "Measurement",
    values_to = "Value"
  ) %>%
  ggplot(aes(x = Species, y = Value, fill = Species)) +
  geom_boxplot() +
  scale_fill_manual(values = c("black", "grey40", "grey60")) +

  facet_grid(Measurement ~ ., scale = "free_y",
             labeller = labeller(Measurement = label_map))
```

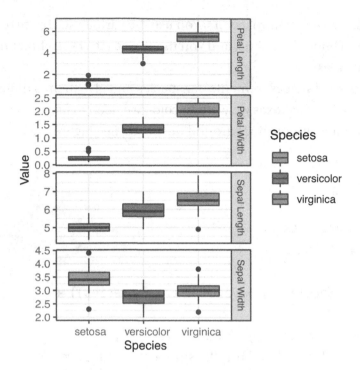

Figure 4-25. *Iris data plotted with default fill colors*

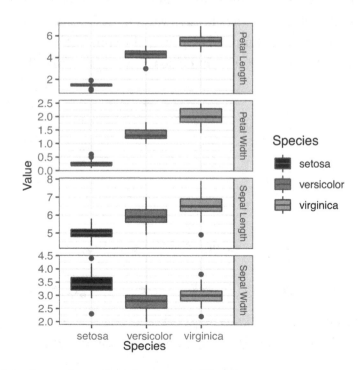

Figure 4-26. *Iris data plotted with custom fill colors*

Explicitly setting colors is a risky business, though, unless you have a good feeling for how colors work together and which combinations can be problematic for color blind people. It is better to use one of the "brewer" choices. These are methods for constructing good combinations of colors (see http://colorbrewer2.org), and you can use them with the scale_fill_brewer() function (see Figure 4-27):

```
iris %>%
  pivot_longer(
    -Species,
    names_to = "Measurement",
    values_to = "Value"
  ) %>%
  ggplot(aes(x = Species, y = Value, fill = Species)) +
  geom_boxplot() +
  scale_fill_brewer(palette = "Greens") +
  facet_grid(Measurement ~ ., scale = "free_y",
             labeller = labeller(Measurement = label_map))
```

Themes and Other Graphics Transformations

Most of using ggplot2 consist of specifying geometries and scales to control how data is mapped to visual components, but you also have much control over how the final plot will look through functions that only concern the final result.

Most of this is done by modifying the so-called theme. If you have tried the examples I have given in this chapter yourself, the results might look different from the figures in this book. This is because I have set up a default theme for the book using the command

```
theme_set(theme_bw())
```

The theme_bw() sets up the final visual appearance of the figures you see here. You can add a theme to a plot using + as you would any other ggplot2 modification or set it as default as I have done here. There are several themes you can use; you can look for functions that start with theme_, but all of them can be modified to get more control over a plot.

Besides themes, various other functions also affect the way a plot looks. There is far too much to cover here on all the things you can do with themes and graphics transformations, but I can show you an example that should give you an idea of what can be achieved.

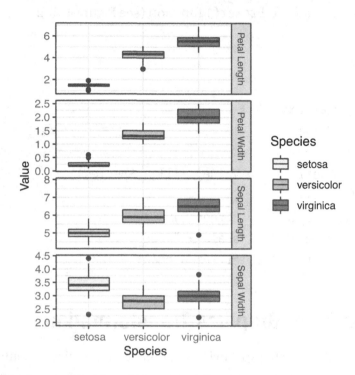

Figure 4-27. *Iris data plotted with brewer fill colors*

You can, for instance, change coordinate systems using various `coord_` functions—the simplest is just flipping x and y with `coord_flip()`. This can, of course, also be achieved by changing the aesthetics, but flipping the coordinates of a complex plot can be easier than updating aesthetics several places. For the `iris` plot we have looked at before, I might want to change the axes.

I also want to put the measurement labels on the left instead of on the right. You can control the placement of facet labels using the `switch` option to `facet_grid()`, and giving the `switch` parameter the value `y` will switch the location of that label:

```
iris %>%
  pivot_longer(
    -Species,
    names_to = "Measurement",
```

```
      values_to = "Value"
) %>%
ggplot(aes(x = Species, y = Value, fill = Species)) +
geom_boxplot() +
scale_x_discrete(labels = c("setosa" = "Setosa",
                            "versicolor" = "Versicolor",
                            "virginica" = "Virginica")) +
scale_fill_brewer(palette = "Greens") +
facet_grid(Measurement ~ ., switch = "y",
           labeller = labeller(Measurement = label_map)) +
coord_flip()
```

If I just flip the coordinates, the axis labels on the new x-axis will be wrong if I tell the facet_grid() function to have a free y-axis. With a free y-axis, it would have different ranges for the y values, which is what we want, but after flipping the coordinates, we will only see the values for one of the y-axes. The other values will be plotted as if they were on the same axis, but they won't be. So I have removed the scale parameter to facet_grid(). Try to put it back and see what happens.

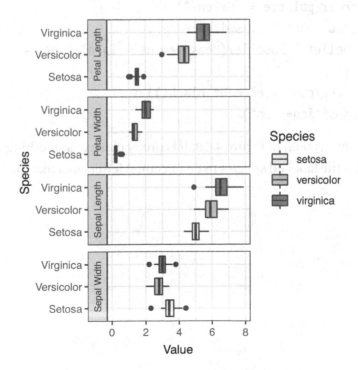

Figure 4-28. *Iris with flipped coordinates and switched facet labels*

The result so far is shown in Figure 4-28. We have flipped coordinates and moved labels, but the labels look ugly with the color background. We can remove it by modifying the theme using theme(`strip.background = element_blank()`). It just sets the strip. background, which is the graphical property of facet labels, to a blank element, so in effect it removes the background color. We can also move the legend label using a theme modification: theme(`legend.position="top"`).

```
iris %>%
  pivot_longer(
    -Species,
    names_to = "Measurement",
    values_to = "Value"
  ) %>%
  ggplot(aes(x = Species, y = Value, fill = Species)) +
  geom_boxplot() +
  scale_x_discrete(labels = c("setosa" = "Setosa",
                              "versicolor" = "Versicolor",
                              "virginica" = "Virginica")) +
  scale_fill_brewer(palette = "Greens") +
  facet_grid(Measurement ~ ., switch = "y",
             labeller = labeller(Measurement = label_map)) +
  coord_flip() +
  theme(strip.background = element_blank()) +
  theme(legend.position="top")
```

The result is now as seen in Figure 4-29. It is pretty close to something we could print. We just want the labelled species to be in capital letters just like the axis labels.

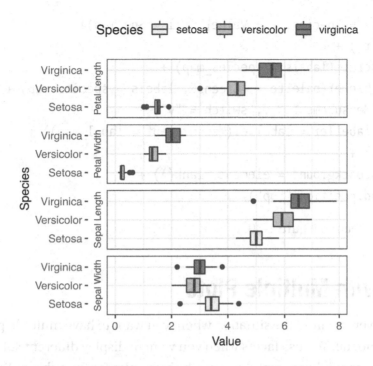

Figure 4-29. *Iris data with theme modifications*

Well, we know how to do that using the labels parameter to a scale so the final plotting code could look like this:

```
label_map <- c(Petal.Width = "Petal Width",
               Petal.Length = "Petal Length",
               Sepal.Width = "Sepal Width",
               Sepal.Length = "Sepal Length")

species_map <- c(setosa = "Setosa",
               versicolor = "Versicolor",
               virginica = "Virginica")

iris %>%
  pivot_longer(
    -Species,
    names_to = "Measurement",
    values_to = "Value"
  ) %>%
```

```
ggplot(aes(x = Species, y = Value, fill = Species)) +
geom_boxplot() +
scale_x_discrete(labels = species_map) +
scale_fill_brewer(palette = "Greens", labels = species_map) +
facet_grid(Measurement ~ ., switch = "y",
            labeller = labeller(Measurement = label_map)) +
coord_flip() +
theme(strip.background = element_blank()) +
theme(legend.position="top")
```

and the result is seen in Figure 4-30.

Figures with Multiple Plots

Using facets covers many of the situation where you want to have multiple panels in the same plot, but not all. You use facets when you want to display different subsets of the data in separate panels but essentially have the same plot for the subsets. Sometimes, you want to combine different types of plots, or plots of different data sets, as subplots in different panels. For that, you need to combine otherwise independent plots.

The `ggplot2` package doesn't directly support combining multiple plots, but it can be achieved using the underlying graphics system, `grid`. Working with basic `grid`, you have many low-level tools for modifying graphics, but for just combining plots, you want more high-level functions, and you can get that from the `gridExtra` package.

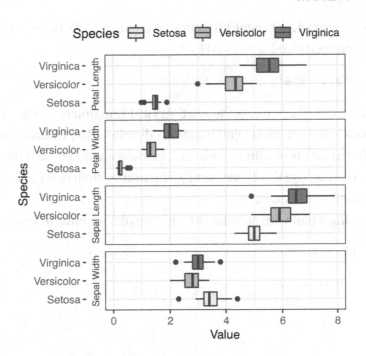

Figure 4-30. *Final version of the iris plot*

To combine plots, you first create them as you normally would. So, for example, we could make two plots of the iris data like this:

```
petal <- iris %>% ggplot() +
  geom_point(aes(x = Petal.Width, y = Petal.Length,
                 color = Species)) +
  scale_color_grey() +
  theme(legend.position="none")

sepal <- iris %>% ggplot() +
  geom_point(aes(x = Sepal.Width, y = Sepal.Length,
                 color = Species)) +
  scale_color_grey() +
  theme(legend.position="none")
```

We then import the gridExtra package:

```
library(gridExtra)
```

and can then use the `grid.arrange()` function to create a grid of plots, putting in the two plots we just created (see Figure 4-31):

```
grid.arrange(petal, sepal, ncol = 2)
```

Another approach I like to use is the `plot_grid()` function from the `cowplot` package. This package contains several functions developed by Claus O. Wilke (where the `cow` comes from) for his plotting needs, and loading it will redefine the default `ggplot2` theme. You can use the `theme_set()` function to change it back if you don't like the theme that `cowplot` provides.

Anyway, creating a plot with subplots using `cowplot`, we have to import the package:

```
library(cowplot)
```

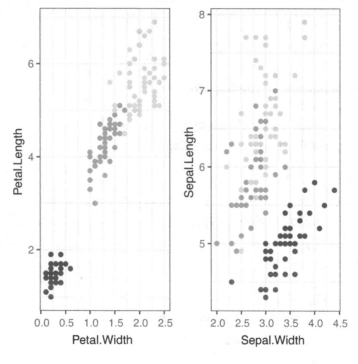

Figure 4-31. *Combining two plots of the iris data using grid.arrange*

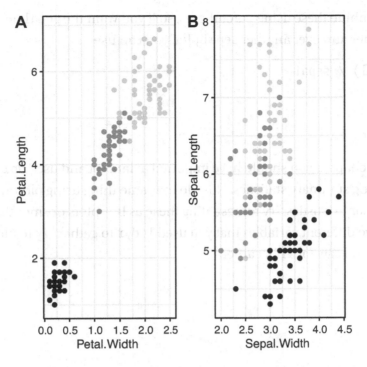

Figure 4-32. *Combining two plots of the iris data using cowplot*

If we don't want the theme it sets here, we need to change it again using theme_
set(), but otherwise we can combine the plots we have defined before using plot_
grid() (see Figure 4-32):

```
plot_grid(petal, sepal, labels = c("A", "B"))
```

With the patchwork package, combining plots is even easier. You can just add them
together to get them next to each other:

```
library(patchwork)
petal + sepal
```

The pipe operator does the same thing as the plus, so this composition does the
same thing:

```
petal | sepal
```

If you want to stack one plot over another, you use /:

```
petal / sepal
```

You can combine these using parentheses, so if you want the `petal` and `sepal` plots next to each other and over another `sepal` plot, you can use

```
(petal + sepal) / sepal
```

Exercises

In the previous chapter, you should have imported a data set and used `dplyr` and `tidyr` to explore it using summary statistics. Now do the same thing using plotting. If you looked at summary statistics, try representing these as box plots or smoothed scatter plots. If you have different variables that you used `tidyr` to gather, try to plot the data similar to what you saw for `iris` earlier.

## CHAPTER 5

Working with Large Data Sets

The concept of Big Data refers to enormous data sets, sets of sizes where you need data warehouses to store it, where you typically need sophisticated algorithms to handle the data and distributed computations to get anywhere with it. At the very least, we talk many gigabytes of data but also often terabytes or exabytes.

Dealing with Big Data is also part of data science, but it is beyond the scope of this book. This chapter is on large data sets and how to deal with data that slows down your analysis, but it is not about data sets so large that you cannot analyze it on your desktop computer.

If we ignore the Big Data issue, what a large data set is depends very much on what you want to do with the data. That comes down to the complexity of what you are trying to achieve. Some algorithms are fast and can scan through data in linear time—meaning that the time it takes to analyze the data is linear in the number of data points—while others take exponential time and cannot be applied to data sets with more than a few tens or hundreds of data points. The science of what you can do with data in a given amount of time, or a given amount of space (be it RAM or disk space or whatever you need), is called complexity theory and is one of the fundamental topics in computer science. In practical terms, though, it usually boils down to how long you are willing to wait for an analysis to finish, and it is a very subjective measure.

In this chapter, we will just consider a few cases where I have found in my own work that data gets a bit too large to do what I want, and I have had to deal with it in various ways. Your cases are likely to be different, but maybe you can get some inspiration, at least, from these cases.

© Thomas Mailund 2022
T. Mailund, *Beginning Data Science in R 4*, https://doi.org/10.1007/978-1-4842-8155-0_5

Subsample Your Data Before You Analyze the Full Data Set

The first point I want to make, though, is this: you very rarely need to analyze a complete data set to get at least an idea of how the data behaves. Unless you are looking for very rare events, you will get as much feeling for the data looking at a few thousands of data points as you would from looking at a few million.

Sometimes, you do need extensive data to find what you are looking for. This is the case, for example, when looking for associations between genetic variation and common diseases where the association can be very weak, and you need lots of data to distinguish between chance associations and genuine associations. But for most signals in data that are of practical importance, you will see the signals in smaller data sets. So before you throw the full power of all your data at an analysis, especially if that analysis turns out to be very slow, you should explore a smaller sample of your data.

Here, you must pick a random sample. There is often structure in data beyond the columns in a data frame. This could be a structure caused by when the data was collected. If the data is ordered by when the data was collected, then the first data points you have can be different from later data points. This isn't explicitly represented in the data, but the structure is there nevertheless. Randomizing your data alleviates problems that can arise from this. Randomizing might remove a subtle signal, but with the power of statistics, we can deal with random noise. It is much harder to deal with consistent biases we just don't know about.

If you have a large data set, and your analysis is being slowed down because of it, don't be afraid to pick a random subset and analyze that. You may see signals in the subsample that is not present in the full data set, but it is much less likely than you might fear. When you are looking for signals in your data, you always have to worry about false signals. But it is not more liable to pop up in a smaller data set than in a larger. And with a more extensive data set to check your results against later, you are less likely to stick with wrong results at the end of your analysis.

Getting spurious results is mostly a concern with traditional hypothesis testing. If you set a threshold for when a signal is significant at 5% for p-values, you will see spurious results one time out of twenty. If you don't correct for multiple testing, you will be almost guaranteed to see false results. These are unlikely to survive when you later throw the complete data at your models.

In any case, with large data sets, you are more likely to have statistically significant deviations from a null model, which are entirely irrelevant to your analysis. We usually use simple null models when analyzing data, and any complex data sets are not generated from a simple null model. With enough data, the chances are that anything you look at will have significant deviations from your simple null model. The real world does not draw samples from a simple linear model. There is always some extra complexity. You won't see it with a few data points, but with enough data, you can reject any null model. It doesn't mean that what you see has any practical importance.

If you have signals you can discover in a smaller subset of your data, and these signals persist when you look at the full data set, you can trust them that much more.

So, if the data size slows you down, downsample and analyze a subset.

You can use `dplyr` functions `sample_n()` and `sample_frac()` to sample from a data frame. Use `sample_n()` to get a fixed number of rows and `sample_frac()` to get a fraction of the data:

```
iris |> sample_n(size = 5)
```

```
##   Sepal.Length Sepal.Width Petal.Length
## 1          5.5         3.5          1.3
## 2          6.4         2.8          5.6
## 3          5.7         2.8          4.5
## 4          5.0         3.4          1.5
## 5          5.1         3.8          1.6
##   Petal.Width    Species
## 1         0.2     setosa
## 2         2.2  virginica
## 3         1.3 versicolor
## 4         0.2     setosa
## 5         0.2     setosa
```

```
iris |> sample_frac(size = 0.02)
```

```
##   Sepal.Length Sepal.Width Petal.Length
## 1          6.3         2.5          5.0
## 2          6.4         2.8          5.6
## 3          6.3         3.3          4.7
```

```
##    Petal.Width    Species
## 1           1.9 virginica
## 2           2.2 virginica
## 3           1.6 versicolor
```

(Your output will be different, since these are random functions, but it should look similar.)

Of course, to sample using `dplyr`, you need your data in a form that `dplyr` can manipulate, and if the data is too large even to load into R, then you cannot have it in a data frame to sample from, to begin with. Luckily, `dplyr` has support for using data that is stored on disk rather than in RAM, in various back-end formats, as we will see later. It is, for example, possible to connect a database to `dplyr` and sample from a large data set this way.

Running Out of Memory During an Analysis

R can be very wasteful of RAM. Even if your data set is small enough to fit in memory and small enough that the analysis time is not a substantial problem, it is easy to run out of memory because R remembers more than is immediately apparent.

In R, all objects are immutable,[1] so whenever you modify an object, you are actually creating a new object. The implementation of this is smart enough that you only have independent copies of data when it is different. Having two different variables to refer to the same data frame doesn't mean that the data frame is represented twice. Still, if you modify the data frame in one of the variables, then R will create a copy with the modifications, and you now have the data twice, accessible through the two variables. If you only refer to the data frame through one variable, then R is smart enough not to make a copy, though.

You can examine memory usage and memory changes using the `pryr` package:

```
library(pryr)
```

For example, we can see what the cost is of creating a new vector:

```
mem_change(x <- rnorm(10000))
```

---

[1] This is not entirely true; it is possible to make mutable objects, but it requires some work. Unless you go out of your way to create mutable objects, it is true.

```
## 83.9 kB
```

(The exact value you see here will depend on your computer and your installation, so don't be surprised if it differs from mine.)

R doesn't allow modification of data, so when you "modify" a vector, it makes a new copy that contains the changes. This doesn't significantly increase the memory usage because R is smart about only copying when more than one variable refers to an object:

```
mem_change(x[1] <- 0)
```

```
## 528 B
```

If we assign the vector to another variable, we do not use twice the memory, because both variables will just refer to the same object:

```
mem_change(y <- x)
```

```
## 584 B
```

but if we modify one of the vectors, we will have to make a copy, so the other vector remains the same:

```
mem_change(x[1] <- 0)
```

```
## 80.6 kB
```

This is another reason for using pipelines rather than assigning to many variables during an analysis. You are fine if you assign back to a variable, though, so the %<>% operator does not lead to a lot of copying.

Even using pipelines, you still have to be careful, though. Many functions in R will again copy data.

If a function does any modification to data, the data is copied to a local variable. There might be some sharing, so, for example, just referring to a data frame in a local variable does not create a copy. Still, if you, for example, split a data frame into training and test data in a function, then you will be copying and now represent all the data twice. This memory is freed after the function finishes its computations, so it is really only a problem if you are very close to the limit of RAM.

If such copied data is returned in some way from the function, it is not freed. It is, for example, not unusual that model fitting functions will save the entire fitting data in the returned object. If it is copied without modification, again we do not see a memory increase. Yet, if the function modifies it in any way, we are now using twice the memory as before.

When you have problems with running out of memory in data analysis in R, it is usually not that you cannot represent your data initially but that you end up having many copies. You can avoid this to some extent by not storing temporary data frames in variables and by not implicitly storing copies of data frames in the output of functions, or you can explicitly remove stored data using the rm() function to free up memory.

Too Large to Plot

The first point where I typically run into problems with large data sets is not that I run out of RAM, but when I am plotting, especially when making scatter plots; box plots or histograms summarize the data and are usually not a problem.

There are two problems when making scatter plots with a lot of data. The first is that if you create files from scatter plots, you will create a plot that contains every single individual point. That can be a huge file. Worse, it will take forever to plot, since a viewer will have to consider every single point. You can avoid this problem by creating raster graphics instead of PDFs, but that takes us to the second issue. With too many points, a scatter plot is just not informative any longer. Points will overlap, and you cannot see how many individual data points fall on the plot. This usually becomes a problem long before the computational time becomes an issue.

If, for example, we have a data frame with 10,000 points

```
d <- data.frame(x = rnorm(10000), y = rnorm(10000))
```

we can still make a scatter plot, and if the plot is saved as raster graphic instead of PDF, the file will not be too large to watch or print:

```
d |> ggplot(aes(x = x, y = y)) +
  geom_point()
```

The result will just not be all that informative; see Figure 5-1. The points are shown on top of each other, making it hard to see if the big black cloud of points has different densities in some places than others.

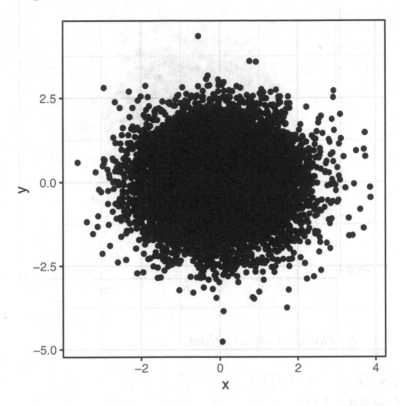

Figure 5-1. A scatter plot with too many points

The solution is to represent points such that they are still visible when there are many overlapping points. If the points are overlapping because they have the same x or y coordinates, you can jitter them; we saw that in the previous chapter. Another solution to the same problems is plotting the points with alpha levels, so each point is partly transparent. You can see the density of points because they are slightly transparent, but you still end up with a plot with very many points; see Figure 5-2.

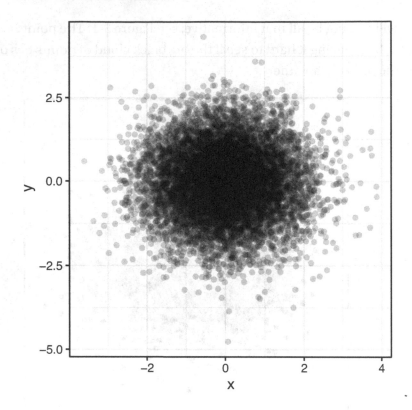

Figure 5-2. *A scatter plot with alpha values*

```
d |> ggplot(aes(x = x, y = y)) +
  geom_point(alpha = 0.2)
```

This, however, doesn't solve the problem that files will draw every single point and cause printing and file size problems. A scatter plot with transparency is just a way of showing the 2D density, though, and we can do that directly using the geom_density_2d() function; see Figure 5-3.

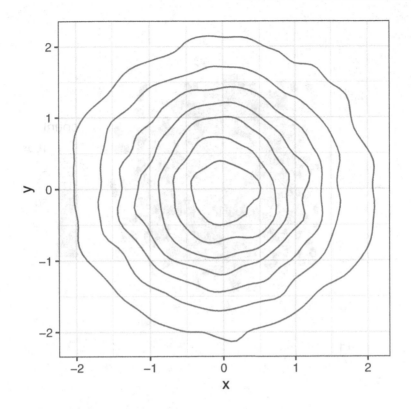

Figure 5-3. *A 2D density plot*

```
d |> ggplot(aes(x = x, y = y)) +
  geom_density_2d()
```

The plot shows the contour of the density.

An alternative way of showing a 2D density is using a so-called hex plot, the 2D equivalent of a histogram. The plot splits the 2D plane into hexagonal bins and displays the count of points falling into each bin.

To use it, you need to install the package hexbin and use the ggplot2 function geom_hex(); see Figure 5-4.

```
d |> ggplot(aes(x = x, y = y)) +
  geom_hex()
```

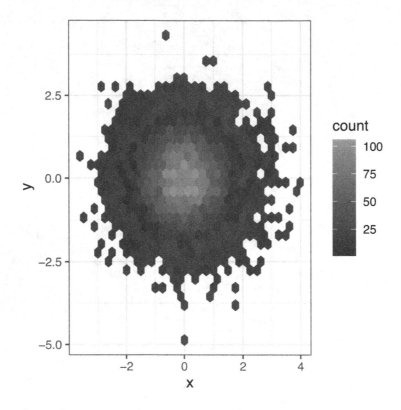

Figure 5-4. *A hex plot*

The colors used by geom_hex() are the fill colors, so you can change them using the scale_fill functions. You can also combine hex and 2D density plots to get both the bins and contours displayed; see Figure 5-5.

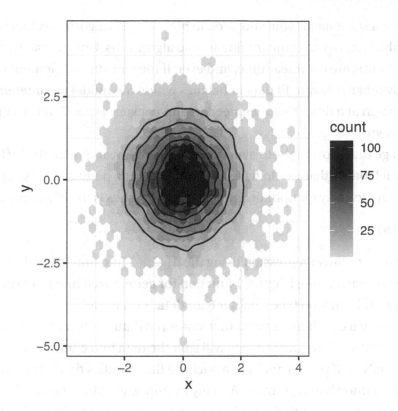

Figure 5-5. *A plot combining hex and 2D density*

```
d |> ggplot(aes(x = x, y = y)) +
  geom_hex() +
  scale_fill_gradient(low = "lightgray", high = "grey10") +
  geom_density2d(color = "black")
```

Too Slow to Analyze

When plotting data, the problem is usually only in scatter plots. Otherwise, you don't
have to worry about having too many points or too large plot files. Even when plotting
lots of points, the real problem doesn't show up until you create a plot and load it into
your viewer or send it to the printer.

With enough data points, though, most analyses will slow down, and that can be a
problem.

The easy solution is again to subsample your data and work with that. It will show
you the relevant signals in your data without slowing down your analysis.

171

If that is not a solution for you, you need to pick analysis algorithms that work more efficiently. That typically means linear time algorithms. Unfortunately, many standard algorithms are not linear time, and even if they are, the implementation does not necessarily make it easy to fit data in batches where the model parameters can be updated one batch at a time. You often need to find packages specifically written for that or make your own.

One package that provides both a memory-efficient linear model fitting (it avoids creating a model matrix that would have rows for each data point and solving equations for that) and functionality for updating the model in batches is the `biglm` package:

```
library(biglm)
```

You can use it for linear regression using the `biglm()` function instead of the `lm()` function, and you can use the `bigglm()` function for generalized linear regression instead of the `glm()` function (see Chapter 6 for details on these).

If you are using a data frame format that stores the data on disk and has support for `biglm` (see the next section), the package will split the data into chunks it can load into memory and analyze. If you do not have a package that handles this automatically, you can split the data into chunks yourself. As a toy example, we can consider the `cars` data set and try to fit a linear model of stopping distance as a function of speed but do this in batches of ten data points. Of course, we can easily fit such a small data set without splitting it into batches, we don't even need to use the `biglm()` function for it, but as an example, it will do.

Defining the slice indices requires some arithmetic, and after that, we can extract subsets of the data using the `slice()` function from `dplyr`. We can create a linear model from the first slice and then update using the following:

```
slice_size <- 10
n <- nrow(cars)
slice <- cars |> slice(1:slice_size)
model <- biglm(dist ~ speed, data = slice)
for (i in 1:(n/slice_size-1)) {
  slice <- cars |> slice((i*slice_size+1):((i+1)*slice_size))
  model <- update(model, moredata = slice)
}
```

```
Model

## Large data regression model: biglm(dist ~ speed, data = slice)
## Sample size = 50
```

Bayesian model fitting methods have a (somewhat justified) reputation for being slow, but Bayesian models based on conjugate priors are ideal for this. Having a conjugate prior means that the posterior distribution you get out of analyzing one data set can be used as the prior distribution for the next data set. This way, you can split the data into slices and fit the first slice with a real prior and the subsequent slices with the result of the previous model fits.

The Bayesian linear regression model in the second project, the last chapter of this book, is one such model. There, we implement an `update()` function that fits a new model based on a data set and a previously fitted model. Using it on the `cars` data, splitting the data into chunks of size 10, would look very similar to the `biglm` example.

Even better are models where you can analyze slices independently and then combine the results to get a model for the full data set. These can not only be analyzed in batches, but the slices can be handled in parallel, exploiting multiple cores or multiple computer nodes. For gradient descent optimization approaches, you can compute gradients for slices independently and then combine them to make a step in the optimization.

There are no general solutions for dealing with data that is too large to be efficiently analyzed, though. It requires thinking about the algorithms used and usually also some custom implementation of these unless you are lucky and can find a package that can handle data in batches.

Too Large to Load

R wants to keep the data it works on in memory. So if your computer doesn't have the RAM to hold it, you are out of luck. At least if you work with the default data representations like 'data.frame'. R usually also wants to use 32-bit integers for indices. Since it uses both positive and negative numbers for indices, you are limited to indexing around two billion data points even if you could hold more in memory.

There are different packages for dealing with this. One such is the `ff` package that works with the kind of tables we have used so far but uses memory-mapped files to represent the data and loads data chunks into memory as needed:

```
library(ff)
```

It essentially creates flat files and has functionality for mapping chunks of these into memory when analyzing them.

It represents data frames as objects of class `ffdf`. These behave just like data frames if you use them as such, and you can translate a data frame into an `ffdf` object using the `as.ffdf()` function.

You can, for example, convert the `cars` data into an `ffdf` object using

```
ffcars <- as.ffdf(cars)
summary(ffcars)
```

```
##         Length Class     Mode
## speed 50       ff_vector list
## dist  50       ff_vector list
```

Of course, if you can already represent a data frame in memory, there is no need for this translation, but `ff` also has functions for creating `ffdf` objects from files. If, for example, you have a large file as comma-separated values, you can use `read.csv.ffdf()`.

With `ff`, you get various functions for computing summary statistics efficiently from the memory-mapped flat files. These are implemented as generic functions (we will cover generic functions in Chapter 12), and this means that for most common summaries, we can work efficiently with `ffdf` objects. Not every function supports this, however, so sometimes functions will (implicitly) work on an `ffdf` object as if it was a plain `data.frame`. This means that the full data might be loaded into memory. This usually doesn't work if the data is too large to fit.

To deal with data that you cannot load into memory, you will have to analyze it in batches. This means that you need special functions for analyzing data, and, unfortunately, this quite often means that you have to implement analysis algorithms yourself.

You cannot use `ffdf` objects together with `dplyr`, which is a main drawback of using `ff` to represent data. However, the `dplyr` package itself provides support for different back ends, such as relational databases. If you can work with data as flat files, there is no benefit for putting it in databases, but large data sets are usually stored in databases that are accessed through the Structured Query Language (SQL). This is a language that is worth learning, but beyond the scope of this book. In any case, dplyr can be used to access such databases. This means that you can write dplyr pipelines of data manipulation function calls; these will be translated into SQL expressions that are then sent to the database system, and you can get the results back.

With dplyr, you can access commonly used database systems such as MySQL (www. mysql.com) or PostgreSQL (www.postgresql.org). These require that you set up a server for the data, though, so a simpler solution, if your data is not already stored in a database, is to use LiteSQL (https://en.wikipedia.org/wiki/LiteSQL).

LiteSQL databases sit on your filesystem and provide a file format and ways of accessing it using SQL. You can open or create a LiteSQL file using the src_sqlite() function:[2]

```
iris_db <- DBI::dbConnect(RSQLite::SQLite(),
                          path = "iris_db.sqlite3")
```

and load a data set into it using copy_to():

```
copy_to(iris_db, iris, temporary = FALSE)
```

Of course, if you can already represent a data frame in RAM, you wouldn't usually copy it to a database. It only slows down analysis to go through a database system compared to keeping the data in memory—but the point is, of course, that you can populate the database outside of R and then access it using dplyr.

Setting the temporary option to FALSE here ensures that the table you fill into the database survives between sessions. If you do not set temporary to FALSE, it will only exist as long as you have the database open; after you close it, it will be deleted. This is useful for many operations but not what we want here.

Once you have a connection to a database, you can pull out a table using tbl():

```
iris_db_tbl <- tbl(iris_db, "iris")
iris_db_tbl
```

```
## # Source:    table<iris> [?? x 5]
## # Database: sqlite 3.38.0 []
```

---

[2] You will have to install the package RSQLite to run this code, since that package implements the underlying functionality.

```
##     Sepal.Length Sepal.Width Petal.Length
##            <dbl>       <dbl>        <dbl>
## 1          5.1         3.5          1.4
## 2          4.9         3            1.4
## 3          4.7         3.2          1.3
## 4          4.6         3.1          1.5
## 5          5 3.        6            1.4
## 6          5.4         3.9          1.7
## 7          4.6         3.4          1.4
## 8          5 3.        4            1.5
## 9          4.4         2.9          1.4
## 10         4.9         3.1          1.5
## # … with more rows, and 2 more variables:
## #   Petal.Width <dbl>, Species <chr>
```

and use dplyr functions to make a query to it:

```
iris_db_tbl %>% group_by(Species) %>%
  summarise(mean.Petal.Length = mean(Petal.Length, na.rm = TRUE))
```

```
## # Source:   lazy query [?? x 2]
## # Database: sqlite 3.38.0 []
##   Species     mean.Petal.Length
##   <chr>                   <dbl>
## 1 setosa                   1.46
## 2 versicolor               4.26
## 3 virginica                5.55
```

Using dplyr with SQL databases is beyond the scope of this book, so I will just refer you to the documentation for the package.

Manipulating data using dplyr with a database back end is only useful for doing analysis exclusively using dplyr, of course. To fit models and such, you will still have to batch data, so some custom code is usually still required.

Exercises

Subsampling

Take the data set you worked on the last two chapters and pick a subset of the data. Summarize it and compare to the results you get for the full data. Plot the subsamples and compare that to the plots you created with the full data.

Hex and 2D Density Plots

If you have used any scatter plots to look at your data, translate them into hex or 2D density plots.

CHAPTER 6

Supervised Learning

This chapter and the next concern the mathematical modelling of data that is the essential core of data science. We can call this statistics, or we can call it machine learning. At its heart, it is the same thing. It is all about extracting information out of data.

Machine Learning

Machine learning is the discipline of developing and applying models and algorithms for learning from data. Traditional algorithms implement fixed rules for solving particular problems like sorting numbers or finding the shortest route between two cities. To develop algorithms like that, you need a deep understanding of the problem you are trying to solve—a thorough understanding that you can rarely obtain unless the problem is particularly simple or you have abstracted away all the unusual cases. Far more often, you can collect examples of good or bad solutions to the problem you want to solve without being able to explain precisely why a given solution is good or bad. Or you can obtain data that provides examples of relationships between data you are interested in without necessarily understanding the underlying reasons for these relationships.

This is where machine learning can help. Machine learning concerns learning from data. You do not explicitly develop an algorithm for solving a particular problem. Instead, you use a generic learning algorithm that you feed examples of solutions to and let it learn how to solve the problem from those examples.

This might sound very abstract, but most statistical modelling is indeed examples of this. Take, for example, a linear model $y = \alpha x + \beta + \epsilon$ where ϵ is the stochastic noise (usually assumed to be normal distributed). When you want to model a linear relationship between x and y, you don't figure out α and β from the first principle. You can write an algorithm for sorting numbers without having studied the numbers beforehand, but you cannot usually figure out what the linear relationship is between y and x without looking at data. When you fit the linear model, you are doing machine

179

© Thomas Mailund 2022
T. Mailund, *Beginning Data Science in R 4*, https://doi.org/10.1007/978-1-4842-8155-0_6

learning. (Well, I suppose if you do it by hand, it isn't machine learning, but you are not likely to fit linear models by hand that often.) People typically do not call simple models like linear regression machine learning, but that is mostly because the term "machine learning" is much younger than these models. Linear regression is as much machine learning as neural networks are.

Supervised Learning

Supervised learning is used when you have variables you want to predict using other variables—situations like linear regression where you have some input variables, for example, x, and you want a model that predicts output (or response) variables, $y = f(x)$.

Unsupervised learning, the topic for the next chapter, is instead concerned with discovering patterns in data when you don't necessarily know what kind of questions you are interested in learning–when you don't have x and y values and want to know how they are related, but instead have a collection of data, and you want to discover what patterns there are in the data.

For the simplest case of supervised learning, we have one response variable, y, and one input variable, x, and we want to figure out a function, f, mapping input to output, that is, such that $y = f(x)$. What we have to work with is example data of matching x and y. Let us write that as vectors $x = (x_1, \ldots, x_n)$ and $y = (y_1, \ldots, y_n)$ where we want to figure out a function f such that $y_i = f(x_i)$.

We will typically accept that there might be some noise in our observations, so f doesn't map perfectly from x to y. Therefore, we can change the setup slightly and assume that the data we have is $x = (x_1, \ldots, x_n)$ and $t = (t_1, \ldots, t_n)$, where t is target values and where $t_i = y_i + \epsilon_i$, $y_i = f(x_i)$, and ϵi is the error in the observation t_i.

How we model the error ϵ_i and the function f are choices that are up to us. It is only modelling, after all, and we can do whatever we want. Not all models are equally good, of course, so we need to be a little careful with what we choose and how we evaluate if the choice is good or bad, but in principle, we can do anything.

The way most machine learning works is that an algorithm, implicitly or explicitly, defines a class of parameterized functions $f(-; \theta)$, each mapping input to output $f(-; \theta)$: $x \mapsto f(x; \theta) = y(\theta)$ (now the value we get for the output depends on the parameters of the function, θ), and the learning consists of choosing parameters θ such that we minimize the errors, that is, such that $f(x_i; \theta)$ is as close to ti as we can get. We want to get close for all our data points, or at least get close on average, so if we let y(θ) denote the vector $(y(\theta)_1, \ldots, y(\theta)_n) = (f(x_1; \theta), \ldots, f(x_n; \theta))$, we want to minimize the distance from y(θ) to t, $\|y(\theta) - t\|$, for some distance measure $\|\cdot\|$.

Regression vs. Classification

There are two main types of supervised learning: regression and classification. Regression is used when the output variable we try to target is a number. Classification is used when we try to target some categorical variables.

Take linear regression, $y = \alpha x + \beta$ (or $t = \alpha x + \beta + \epsilon$). It is regression because the variable we are trying to target is a number. The parameterized class of functions, f_θ, are all lines. If we let $\theta = (\theta_1, \theta_0)$ and $\alpha = \theta_1$, $\beta = \theta_0$, then $y(\theta) = f(x; \theta) = \theta_1 x + \theta_0$. Fitting a linear model consists of finding the best θ, where best is defined as the θ that gets y(θ) closest to t. The distance measure used in linear regression is the squared Euclidean distance $|y^{(\theta)} - t|^2 = \sum_{i=1}^{n} \left(y_i(\theta) - t_i \right)^2$.

The reason it is the squared distance instead of just the distance is mostly mathematical convenience—it is easier to maximize θ that way—but also related to us interpreting the error term ϵ as normal distributed. Whenever you fit data in linear regression, you are minimizing this distance; you are finding the parameters θ that best fit the data in the sense of minimizing the distance from y(θ) to t:

$$\hat{\theta} = \arg\min_{\theta_1, \theta_0} \sum_{i=1}^{n} \left(\theta_1 x_i + \theta_0 - t_i \right)^2.$$

For an example of classification, let us assume that the targets t_i are binary, encoded as 0 and 1, but that the input variables x_i are still real numbers. A common way of defining the mapping function $f(-; \theta)$ is to let it map x to the unit interval [0, 1] and interpret the resulting y(θ) as the probability that t is 1. In a classification setting, you would then predict 0 if $f(x; \theta) < 0.5$ and predict 1 if $f(x; \theta) > 0.5$ (and have some strategy for dealing with $f(x; \theta) = 0.5$). In linear classification, the function f_θ could look like this:

$$f(x; \theta) = \sigma(\theta_1 x + \theta_0)$$

where σ is a sigmoid function (a function mapping $\mathbb{R} \mapsto [0, 1]$ that is "S-shaped"). A common choice of σ is the logistic function $\sigma : z \mapsto \dfrac{1}{1 + e^{-z}}$, in which case we call the fitting of $f(-; \theta)$ logistic regression.

Whether we are doing regression or classification, and whether we have linear models or not, we are simply trying to find parameters θ such that our predictions y(θ) are as close to our targets t as possible. The details that differ between different machine learning methods are how the class of prediction functions $f(-; \theta)$ is defined, what kind of parameters θ we have, and how we measure the distance between y(θ) and t. There are a lot of different choices here and a lot of different machine learning algorithms. Many of them are already implemented in R, however, so we rarely will have to implement our own. We just need to find the right package that implements the learning algorithms we need.

Inference vs. Prediction

A question always worth considering when we fit parameters of a model is this: Do we care about the model parameters or do we just want to make a function that is good at predicting?

If you were taught statistics the same way I was, your introduction to linear regression was mostly focused on the model parameters. You inferred the parameters θ_1 and θ_0 mostly to figure out if $\theta_1 \neq 0$, that is, to find out if there was a (linear) relationship between x and y or not. When we fit our function to data to learn about the parameters, we say we are doing inference, and we are inferring the parameters.

This focus on model parameters makes sense in many situations. In a linear model, the coefficient θ_1 tells us if there is a significant correlation between x and y, meaning we are statistically relatively certain that the correlation exists, and whether it is substantial, meaning that θ_1 is large enough to care about in practical situations.

When we care about model parameters, we usually want to know more than just the best-fitting parameters, $\hat{\theta}$. We want to know how certain we are that the "true parameters" are close to our estimated parameters. This usually means estimating not just the best parameters but also confidence intervals or posterior distributions of parameters. How easy it is to estimate these depends very much on the models and algorithms used.

I put "true parameters" in quotes earlier, where I talked about how close estimates were to the "true parameters," for a good reason. True parameters only exist if the data you are analyzing were simulated from a function f_θ where some true θ exist. When you are estimating parameters, $\hat{\theta}$, you are looking for the best choice of parameters assuming that the data were generated by a function f_θ. Outside of statistics textbooks, there is no reason to think that your data was generated from a function in the class of functions you consider. Unless we are trying to model causal relationships—modelling how we think the world actually works as forces of nature—that is usually not an underlying assumption of model fitting. A lot of the theory we have for doing statistics on inferred parameters does assume that we have the right class of functions, and that is where you get confidence intervals and such from. In practice, data does not come from these sorts of functions, so treat the results you get from theory with some skepticism.

We can get more empirical distributions of parameters directly from data if we have a lot of data—which we usually do have when doing data science—using sampling methods. I will briefly return to that later in this chapter.

We don't always care about the model parameters, though. For linear regression, it is easy to interpret what the parameters mean, but in many machine learning models, the parameters aren't that interpretable—and we don't really care about them. All we care about is if the model we have fitted is good at predicting the target values. Evaluating how well we expect a function to be able to predict is also something that we sometimes have theoretical results regarding, but as for parameter estimation, we shouldn't really trust these too much. It is much better to use the actual data to estimate this, and as for getting empirical distributions of model parameters, it is something we return to later.

Whether you care about model parameters or not depends on your application and quite often on how you think your model relates to reality.

Specifying Models

The general pattern for specifying models in R is using what is called a "formula," which is a type of objects built into the language. The simplest form is y ~ x which we should interpret as saying $y = f(x)$. Implicitly, there is assumed some class of functions indexed with model parameters, $f(-; \theta)$, and which class of functions we are working with depends on which R functions we use.

Linear Regression

If we take a simple linear regression, $f_\theta(x) = \theta_1 x + \theta_0$, we need the function lm().

For an example, we can use the built-in data set cars, which just contains two variables, speed and breaking distance, where we can consider speed the x value and breaking distance the y value:

```
cars |> head()

##    speed dist
## 1     4    2
## 2     4   10
## 3     7    4
## 4     7   22
## 5     8   16
## 6     9   10
```

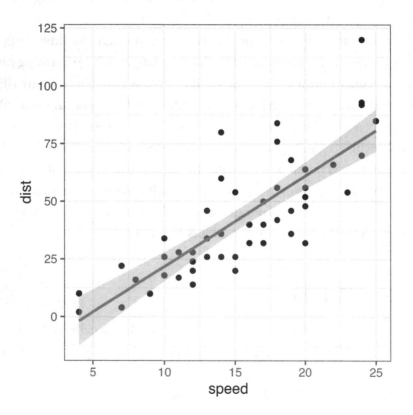

Figure 6-1. *Plot of breaking distance vs. speed for cars*

If we plot the data set (see Figure 6-1), we see that there is a very clear linear relationship between speed and distance:

```
cars |> ggplot(aes(x = speed, y = dist)) +
  geom_point() +
  geom_smooth(formula = y ~ x, method = "lm")
```

In this plot, I used the method "lm" for the smoothed statistics to see the fit. By default, the geom_smooth() function would have given us a loess curve, but since we are interested in linear fits, we tell it to use the lm method. By default, geom_smooth() will also plot the uncertainty of the fit. This is the gray area in the plot. This is the area where the line is likely to be (assuming that the data is generated by a linear model). Do not confuse this with where data points are likely to be, though. If target values are given by $t = \theta_1 x + \theta_0 + \epsilon$ where ϵ has a very large variance, then even if we knew θ_1 and θ_0 with high certainty, we still wouldn't be able to predict with high accuracy where any individual point would fall. There is a difference between prediction accuracy and inference accuracy. We might know model parameters with very high accuracy without being able to predict very well. We might also be able to predict very well without knowing all model parameters that well. If a given model parameter has little influence on where target variables fall, then the training data gives us little information about that parameter. This usually doesn't happen unless the model is more complicated than it needs to be, though, since we often want to remove parameters that do not affect the data.

To actually fit the data and get information about the fit, we use the lm() function with the model specification, dist ~ speed, and we can use the summary() function to see information about the fit:[1]

```
cars %>% lm(dist ~ speed, data = .) %>% summary()

##
## Call:
## lm(formula = dist ~ speed, data = .)
##
## Residuals:
##     Min      1Q  Median      3Q     Max
## -29.069  -9.525  -2.272   9.215  43.201
##
## Coefficients:
##             Estimate Std. Error t value Pr(>|t|)
## (Intercept) -17.5791     6.7584  -2.601   0.0123
## speed         3.9324     0.4155   9.464 1.49e-12
```

---

[1] We need the %>% operator here because of where we want the cars data to go in the call to lm; we can't do that this easily with |>.

```
##
## (Intercept) *
## speed          ***
## ---
## Signif. codes:
## 0 '***' 0.001 '**' 0.01 '*' 0.05 '.' 0.1 ' ' 1
##
## Residual standard error: 15.38 on 48 degrees of freedom
## Multiple R-squared:  0.6511, Adjusted R-squared:  0.6438
## F-statistic: 89.57 on 1 and 48 DF,  p-value: 1.49e-12
```

or we can use the coefficients() function to get the point estimates and the confint() function to get the confidence intervals for the parameters:

```
cars %>% lm(dist ~ speed, data = .) %>% coefficients()

## (Intercept)       speed
##   -17.579095    3.932409

cars %>% lm(dist ~ speed, data = .) %>% confint()

##                    2.5 %     97.5 %
## (Intercept) -31.167850 -3.990340
## speed          3.096964   4.767853
```

Here, (Intercept) is θ_0 and speed is θ_1.

To illustrate the fitting procedure and really drive the point home, we can explicitly draw models with different parameters, that is, draw lines with different choices of θ. To simplify matters, I am going to set $\theta_0 = 0$. Then I can plot the lines $y = \theta_1 x$ for different choices of θ_1 and visually see the fit; see Figure 6-2.

```
predict_dist <- function(speed, theta_1)
    data.frame(speed = speed,
    dist = theta_1 * speed,
    theta = as.factor(theta_1))

cars %>% ggplot(aes(x = speed, y = dist, colour = theta)) +
    geom_point(colour = "black") +
    geom_line(data = predict_dist(cars$speed, 2)) +
```

```
geom_line(data = predict_dist(cars$speed, 3)) +
geom_line(data = predict_dist(cars$speed, 4)) +
scale_color_discrete(name=expression(theta[1]))
```

In this plot, I want to color the lines according to their θ_1 parameter, but since the cars data frame doesn't have a theta column, I'm in a bit of a pickle. I specify that theta should determine the color anyway, but when I plot the points, I overwrite this and say that these should be plotted black; then it doesn't matter that I didn't have theta values. In the geom_line() calls, where I plot the lines, I do have a theta, and that will determine the colors for the lines. The lines are plotted according to their theta value which I set in the predict_dist() function.

Each of the lines shows a choice of model. Given an input value x, they all produce an output value $y(\theta) = f(x; \theta)$. So we can fix θ and consider the mapping $x \mapsto \theta_1 x$. This is the function we use when predicting the output for a given value of x. If we fix x instead, we can also see it as a function of θ: $\theta_1 \mapsto \theta_1 x$. This is what we use when we fit parameters to the data, because if we keep our data set fixed, this mapping defines an error function, that is, a function that given parameters gives us a measure of how far our predicted values are from our target values. If, as before, our input values and target values are vectors x and t, then the error function is

$$E_x, t\left(\theta_i\right) = \sum_{i=1}^{n} \left(\theta_1 x_i - t_i\right)^2$$

and we can plot the errors against different choices of θ_1 (Figure 6-3). Where this function is minimized, we find our best estimate for θ_1:

```
# Get the error value for the specific theta
fitting_error <- Vectorize(function(theta)
   sum((theta * cars$speed - cars$dist)**2)
)
# Plot the errors for a range of thetas
tibble(theta = seq(0, 5, length.out = 50)) |> # set the theta values
   mutate(errors = fitting_error(theta)) |> # add the errors
   ggplot(aes(x = theta, y = errors)) +
   geom_line() +
   xlab(expression(theta[1])) + ylab(expression(E(theta[1])))
```

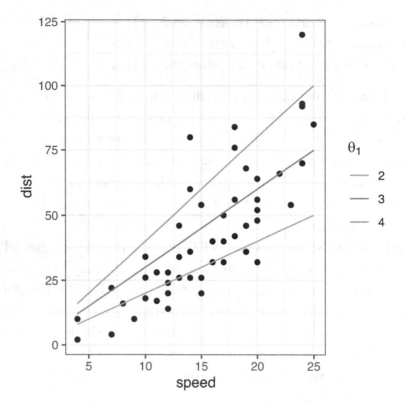

Figure 6-2. *Prediction lines for different choices of parameters*

To wrap up this example, we can also plot and fit the best model where $\theta_0 = 0$. The formula needed to remove the intercept is of the form "y ~ x - 1". It is the "- 1" that removes the intercept:

```
cars %>% lm(dist ~ speed - 1, data = .) %>% coefficients()
```

```
##     speed
## 2.909132
```

We can also plot this regression line, together with the confidence interval for where it lies, using geom_smooth(). See Figure 6-4. Here, though, we need to use the formula y ~ x - 1 rather than dist ~ speed - 1. This is because the geom_smooth() function works on the ggplot2 layers that have x and y coordinates and not the data in the data frame as such. We map the speed variable to the x-axis and the dist variable to the y variable in the aesthetics, but it is x and y that geom_smooth() works on:

```
cars |> ggplot(aes(x = speed, y = dist)) +
    geom_point() +
    geom_smooth(method = "lm", formula = y ~ x - 1)
```

Logistic Regression (Classification, Really)

Using other statistical models works the same way. We specify the class of functions, f_θ, using a formula and use a function to fit its parameters. Consider binary classification and logistic regression.

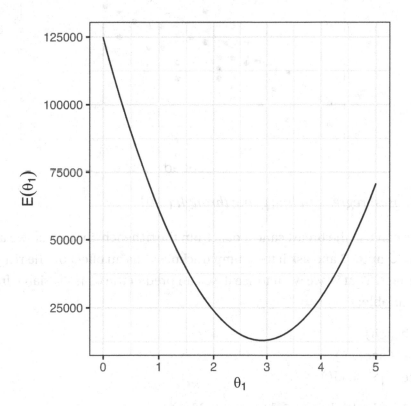

Figure 6-3. *Error values for different choices of parameters*

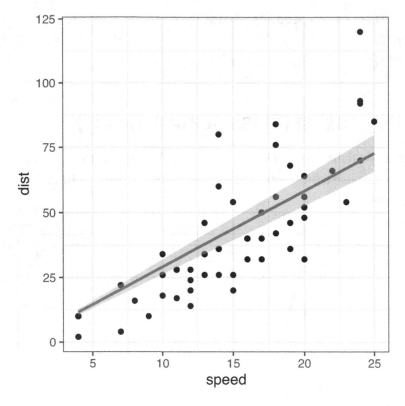

Figure 6-4. *Best regression line going through (0,0)*

Here, we can use the breast cancer data from the `mlbench` library that we also discussed in Chapter 3 and ask if the clump thickness has an effect on the risk of a tumor being malignant. That is, we want to see if we can predict the `Class` variable from the `Cl.thickness` variable:

```
library(mlbench)
data("BreastCancer")
BreastCancer |> head()
```

```
##         Id Cl.thickness Cell.size Cell.shape
## 1 1000025            5         1          1
## 2 1002945            5         4          4
## 3 1015425            3         1          1
## 4 1016277            6         8          8
## 5 1017023            4         1          1
## 6 1017122            8        10         10
```

```
##    Marg.adhesion Epith.c.size Bare.nuclei
## 1              1            2           1
## 2              5            7          10
## 3              1            2           2
## 4              1            3           4
## 5              3            2           1
## 6              8            7          10
##    Bl.cromatin Normal.nucleoli Mitoses      Class
## 1            3               1       1     benign
## 2            3               2       1     benign
## 3            3               1       1     benign
## 4            3               7       1     benign
## 5            3               1       1     benign
## 6            9               7       1  malignant
```

We can plot the data against the fit; see Figure 6-5. Since the malignant status is either 0 or 1, the points would overlap, but if we add a little jitter to the plot, we can still see them, and if we make them slightly transparent, we can see the density of the points.

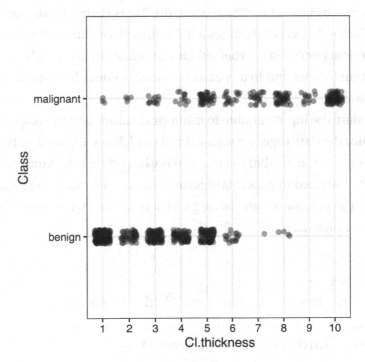

Figure 6-5. *Breast cancer class vs. clump thickness*

```
BreastCancer |>
    ggplot(aes(x = Cl.thickness, y = Class)) +
    geom_jitter(height = 0.05, width = 0.3, alpha = 0.4)
```

For classification, we still specify the prediction function $y = f(x)$ using the formula
y ~ x. The outcome parameter for y ~ x is just binary now. To fit a logistic regression,
we need to use the glm() function (generalized linear model) with the family set to
"binomial". This specifies that we use the logistic function to map from the linear space
of x and θ to the unit interval. Aside from that, fitting and getting results are very similar.

We cannot directly fit the breast cancer data with logistic regression, though. There
are two problems. The first is that the breast cancer data set considers the clump
thickness ordered factors, but for logistic regression, we need the input variable to be
numeric. While, generally, it is not advisable to directly translate categorical data into
numeric data, judging from the plot it seems okay in this case.

Using the function as.numeric() will do this, but remember that this is a risky
approach when working with factors! It actually would work for this data set, but we will
use the safer approach of first translating the factor into strings and then into numbers.

The second problem is that the glm() function expects the response variable to be
numerical, coding the classes like 0 or 1, while the BreastCancer data again encodes
the classes as a factor. Generally, it varies a little from algorithm to algorithm whether a
factor or a numerical encoding is expected for classification, so you always need to check
the documentation for that, but in any case, it is simple enough to translate between the
two representations.

We can translate the input variable to numerical values and the response variable
to 0 and 1 and plot the data together with a fitted model; see Figure 6-6. For the geom_
smooth() function, we specify that the method is glm and that the family is binomial. To
specify the family, we need to pass this argument on to the smoothing method, and that
is done by giving the parameter method.args a list of named parameters; here, we give it
list(family = "binomial"):

```
BreastCancer |>
    mutate(Thickness =
                as.numeric(as.character(Cl.thickness))) |>
    mutate(Malignant = ifelse(Class != "benign", 1, 0)) |>
    ggplot(aes(x = Thickness, y = Malignant)) +
```

```
geom_jitter(height = 0.05, width = 0.3, alpha = 0.4) +
geom_smooth(method = "glm", formula = y ~ x,
            method.args = list(family = "binomial"))
```

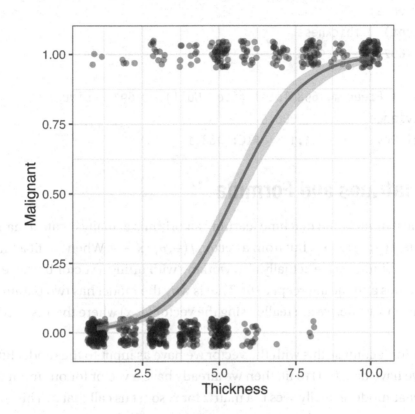

Figure 6-6. *Logistic regression fit to breast cancer data*

To get the fitted object, we use glm() like we used lm() for the linear regression:[2]

```
BreastCancer %>%
  mutate(Thickness =
           as.numeric(as.character(Cl.thickness))) %>%
  mutate(Malignant = ifelse(Class != "benign", 1, 0)) %>%
  glm(Malignant ~ Thickness,
      family = "binomial",
    data = .)
```

---

[2] In this pipeline, we have switched to %>% again, because we need the left-hand side to go into the data argument in glm.

```
##
## Call: glm(formula = Malignant ~ Thickness, family = "binomial", data = .)
##
## Coefficients:
## (Intercept)    Thickness
##      -5.1602       0.9355
##
## Degrees of Freedom: 698 Total (i.e. Null);    697 Residual
## Null Deviance:           900.5
## Residual Deviance: 464.1     AIC: 468.1
```

Model Matrices and Formula

Most statistical models and machine learning algorithms actually create a map not from a single value, $f(-; \theta) : x \mapsto y$, but from a vector, $f(-; \theta) : \mathrm{x} \mapsto y$. When we fit a line for single x and y values, we are actually also working with fitting a vector because we have both the x values and the intercept to fit. That is why the model has two parameters, θ_0 and θ_1. For each x value, we are really using the vector $(1, x)$ where the 1 is used to fit the intercept.

We shouldn't confuse this with the vector we have as input to the model fitting, though. If we have data (x, t) to fit, then we already have a vector for our input data. But what the linear model actually sees is a matrix for x, so let us call that X. This matrix, known as the model matrix, has a row per value in x, and it has two columns, one for the intercept and one for the x values:

$$X = \begin{bmatrix} 1 & x_1 \\ 1 & x_2 \\ 1 & x_3 \\ \vdots & \vdots \\ 1 & x_n \end{bmatrix}$$

We can see what model matrix R generates for a given data set and formula using the model.matrix() function. For the cars data, if we wish to fit dist vs. speed, we get this:

```
cars %>%
    model.matrix(dist ~ speed, data = .) %>%
```

```
head(5)
```

```
## (Intercept) speed
## 1            1     4
## 2            1     4
## 3            1     7
## 4            1     7
## 5            1     8
```

If we remove the intercept, we simply get this:

```
cars %>%
    model.matrix(dist ~ speed - 1, data = .) %>%
    head(5)
```

```
##    speed
## 1      4
## 2      4
## 3      7
## 4      7
## 5      8
```

Pretty much all learning algorithms work on a model matrix, so, in R, they are implemented to take a formula for specifying the model and then building the model matrix from that and the input data.

For linear regression, the map is a pretty simple one. If we let the parameters $\theta^T = (\theta_0, \theta_1)$, then it is just multiplying that with the model matrix, X:

$$X \cdot \theta = \begin{bmatrix} 1 & x_1 \\ 1 & x_2 \\ 1 & x_3 \\ \vdots & \vdots \\ 1 & x_n \end{bmatrix} \cdot \begin{bmatrix} \theta_0 \\ \theta_1 \end{bmatrix} = \begin{bmatrix} \theta_0 & + & \theta_1 x_1 \\ \theta_0 & + & \theta_1 x_2 \\ \theta_0 & + & \theta_1 x_3 \\ & \vdots & \\ \theta_0 & + & \theta_1 x_n \end{bmatrix}$$

This combination of formulas and model matrices is a powerful tool for specifying models. Since all the algorithms we use for fitting data work on model matrices anyway, there is no reason to hold back on how complex formulas to give them. The formulas will just be translated into model matrices anyhow, and they can all deal with them.

If you want to fit more than one parameter, no problem. You just write y ~ x + z, and the model matrix will have three columns:

$$X = \begin{bmatrix} 1 & x_1 & z_1 \\ 1 & x_2 & z_2 \\ 1 & x_3 & z_3 \\ \vdots & \vdots & \vdots \\ 1 & x_n & z_n \end{bmatrix}$$

Our model fitting functions are just as happy to fit this model matrix as the one we get from just a single variable.

So if we wanted to fit the breast cancer data to both cell thickness and cell size, we can do that just by adding both explanatory variables in the formula:

```
BreastCancer %>%
    mutate(Thickness =
                as.numeric(as.character(Cl.thickness))),
           CellSize =
                as.numeric(as.character(Cell.size))) %>%
    mutate(Malignant = ifelse(Class != "benign", 1, 0)) %>%
    model.matrix(Malignant ~ Thickness + CellSize,
                data = .) %>%
    head(5)
```

```
##   (Intercept) Thickness CellSize
## 1           1         5        1
## 2           1         5        4
## 3           1         3        1
## 4           1         6        8
## 5           1         4        1
```

The generalized linear model fitting function will happily work with that:

```
BreastCancer %>%
    mutate(Thickness =
                as.numeric(as.character(Cl.thickness)),
```

```
        CellSize =
              as.numeric(as.character(Cell.size))) %>%
    mutate(Malignant = ifelse(Class != "benign", 1, 0)) %>%
    glm(Malignant ~ Thickness + CellSize, family = "binomial", data = .)
```

```
##
## Call: glm(formula = Malignant ~ Thickness + CellSize, family = "binomial",
##     data = .)
##
## Coefficients:
## (Intercept)      Thickness     CellSize
##     -7.1517        0.6174        1.1751
##
## Degrees of Freedom: 698 Total (i.e. Null);    696 Residual
## Null Deviance:          900.5
## Residual Deviance: 212.3     AIC: 218.3
```

Translating data into model matrices also works for factors; they are just represented as a binary vector for each level:

```
BreastCancer %>%
    mutate(Malignant = ifelse(Class != "benign", 1, 0)) %>%
    model.matrix(Malignant ~ Bare.nuclei, data = .) %>%
    head(5)
```

```
##   (Intercept) Bare.nuclei2 Bare.nuclei3
## 1           1            0            0
## 2           1            0            0
## 3           1            1            0
## 4           1            0            0
## 5           1            0            0
##   Bare.nuclei4 Bare.nuclei5 Bare.nuclei6
## 1            0            0            0
## 2            0            0            0
## 3            0            0            0
```

```
## 4                    1                0                 0
## 5                    0                0                 0
##   Bare.nuclei7 Bare.nuclei8 Bare.nuclei9
## 1              0            0                  0
## 2              0            0                  0
## 3              0            0                  0
## 4              0            0                  0
## 5              0            0                  0
##   Bare.nuclei10
## 1                0
## 2                1
## 3                0
## 4                0
## 5                0
```

The translation for ordered factors gets a little more complicated, but R will happily do it for you:

```
BreastCancer %>%
    mutate(Malignant = ifelse(Class != "benign", 1, 0)) %>%
    model.matrix(Malignant ~ Cl.thickness, data = .) %>%
    head(5)
```

```
##   (Intercept) Cl.thickness.L Cl.thickness.Q
## 1             1     -0.05504819     -0.34815531
## 2             1     -0.05504819     -0.34815531
## 3             1     -0.27524094     -0.08703883
## 4             1      0.05504819     -0.34815531
## 5             1     -0.16514456     -0.26111648
##   Cl.thickness.C Cl.thickness^4 Cl.thickness^5
## 1      0.1295501      0.33658092     -0.21483446
## 2      0.1295501      0.33658092     -0.21483446
## 3      0.3778543     -0.31788198     -0.03580574
## 4     -0.1295501      0.33658092      0.21483446
## 5      0.3346710      0.05609682     -0.39386318
##   Cl.thickness^6 Cl.thickness^7 Cl.thickness^8
## 1     -0.3113996      0.3278724      0.2617852
```

```
## 2      -0.3113996       0.3278724       0.2617852
## 3       0.3892495      -0.5035184       0.3739788
## 4      -0.3113996      -0.3278724       0.2617852
## 5       0.2335497       0.2459043      -0.5235703
##    Cl.thickness^9
## 1      -0.5714300
## 2      -0.5714300
## 3      -0.1632657
## 4       0.5714300
## 5       0.3809534
```

If you want to include interactions between your parameters, you specify that using * instead of +:

```
BreastCancer %>%
    mutate(Thickness =
                as.numeric(as.character(Cl.thickness)),
            CellSize =
                as.numeric(as.character(Cell.size))) %>%
    mutate(Malignant = ifelse(Class != "benign", 1, 0)) %>%
    model.matrix(Malignant ~ Thickness * CellSize,
                data = .) %>%
    head(5)
```

```
##    (Intercept) Thickness CellSize
## 1            1         5        1
## 2            1         5        4
## 3            1         3        1
## 4            1         6        8
## 5            1         4        1
##    Thickness:CellSize
## 1                   5
## 2                  20
## 3                   3
```

```
## 4                         48
## 5                          4
```

How interactions are modelled depends a little bit on whether your parameters are factors or numeric, but for numeric values, the model matrix will just contain a new column with the two values multiplied. For factors, you will get a new column for each level of the factor:

```
BreastCancer %>%
    mutate(Thickness =
             as.numeric(as.character(Cl.thickness))) %>%
    mutate(Malignant = ifelse(Class != "benign", 1, 0)) %>%
    model.matrix(Malignant ~ Thickness * Bare.nuclei, data = .) %>%
    head(3)
```

```
##   (Intercept) Thickness Bare.nuclei2 Bare.nuclei3
## 1           1         5            0            0
## 2           1         5            0            0
## 3           1         3            1            0
##   Bare.nuclei4 Bare.nuclei5 Bare.nuclei6
## 1            0            0            0
## 2            0            0            0
## 3            0            0            0
##   Bare.nuclei7 Bare.nuclei8 Bare.nuclei9
## 1            0            0            0
## 2            0            0            0
## 3            0            0            0
##   Bare.nuclei10 Thickness:Bare.nuclei2
## 1             0                      0
## 2             1                      0
## 3             0                      3
##   Thickness:Bare.nuclei3 Thickness:Bare.nuclei4
## 1                      0                      0
## 2                      0                      0
## 3                      0                      0
##   Thickness:Bare.nuclei5 Thickness:Bare.nuclei6
## 1                      0                      0
```

```
## 2                      0                    0
## 3                      0                    0
##    Thickness:Bare.nuclei7 Thickness:Bare.nuclei8
## 1                      0                    0
## 2                      0                    0
## 3                      0                    0
##    Thickness:Bare.nuclei9 Thickness:Bare.nuclei10
## 1                      0                    0
## 2                      0                    5
## 3                      0                    0
```

The interaction columns all have : in their name, and you can specify an interaction term directly by writing that in the model formula as well:

```
BreastCancer %>%
    mutate(Thickness =
              as.numeric(as.character(Cl.thickness))) %>%
    mutate(Malignant = ifelse(Class != "benign", 1, 0)) %>%
    model.matrix(Malignant ~ Thickness : Bare.nuclei, data = .) %>%
    head(3)
```

```
##    (Intercept) Thickness:Bare.nuclei1
## 1            1                      5
## 2            1                      0
## 3            1                      0
##    Thickness:Bare.nuclei2 Thickness:Bare.nuclei3
## 1                      0                    0
## 2                      0                    0
## 3                      3                    0
##    Thickness:Bare.nuclei4 Thickness:Bare.nuclei5
## 1                      0                    0
## 2                      0                    0
## 3                      0                    0
##    Thickness:Bare.nuclei6 Thickness:Bare.nuclei7
## 1                      0                    0
```

```
## 2                              0                              0
## 3                              0                              0
##    Thickness:Bare.nuclei8 Thickness:Bare.nuclei9
## 1                              0                              0
## 2                              0                              0
## 3                              0                              0
##    Thickness:Bare.nuclei10
## 1                           0
## 2                           5
## 3                           0
```

If you want to use all the variables in your data except the response variable, you can even use the formula y ~ . where the . will give you all parameters in your data except y.

Using formulas and model matrices also means that we do not have to use our data raw. We can transform it before we give it to our learning algorithms. In general, we can transform our data using any function ϕ. It is traditionally called phi because we call what it produces features of our data, and the point of it is to pull out the relevant features of the data to give to the learning algorithm. It usually maps from vectors to vectors, so you can use it to transform each row in your raw data into the rows of the model matrix which we will then call Φ instead of X:

$$\Phi = \begin{bmatrix} -\phi(x_1)- \\ -\phi(x_2)- \\ -\phi(x_3)- \\ \cdots \\ -\phi(x_n)- \end{bmatrix}$$

If this sounds very abstract, perhaps it will help to see some examples. We go back to the cars data, but this time, we want to fit a polynomial to the data instead of a line. If d denotes breaking distance and s the speed, then we want to fit $d = \theta_0 + \theta_1 s + \theta_1 s^2 + \cdots + \theta_n s^n$. Let us just do $n = 2$, so we want to fit a second-degree polynomial. Don't be confused about the higher degrees of the polynomial, it is still a linear model. The linear in linear model refers to the θ parameters, not the data. We just need to map the single s parameter into a vector with the different polynomial degrees, so 1 for the intercept, s for the linear component, and s^2 for the squared component. So $\phi(s) = (1, s, s^2)$.

We can write that as a formula. There, we don't need to specify the intercept term explicitly—it will be included by default, and if we don't want it, we have to remove it with -1 in the formula—but we need speed, and we need speed^2:

```
cars %>%
    model.matrix(dist ~ speed + speed^2, data = .) %>%
    head()
```

```
##   (Intercept) speed
## 1           1     4
## 2           1     4
## 3           1     7
## 4           1     7
## 5           1     8
## 6           1     9
```

Now this doesn't quite work—you can see that we only got the intercept and speed—and the reason is that multiplication is interpreted as interaction terms even if it is interaction with the parameter itself. And interaction with itself doesn't go into the model matrix because that would just be silly.

To avoid that problem, we need to tell R that the speed^2 term should be interpreted just the way it is. We do that using the identity function, I():

```
cars %>%
    model.matrix(dist ~ speed + I(speed^2), data = .) %>%
    head()
```

```
##   (Intercept) speed I(speed^2)
## 1           1     4         16
## 2           1     4         16
## 3           1     7         49
## 4           1     7         49
## 5           1     8         64
## 6           1     9         81
```

Now our model matrix has three columns, which is precisely what we want.

We can fit the polynomial using the linear model function like this:

```
cars %>% lm(dist ~ speed + I(speed^2), data = .) %>%
    summary()
##
## Call:
## lm(formula = dist ~ speed + I(speed^2), data = .)
##
## Residuals:
##      Min      1Q   Median      3Q      Max
## -28.720  -9.184   -3.188   4.628   45.152
##
## Coefficients:
##                Estimate  Std. Error  t value  Pr(>|t|)
## (Intercept)    2.47014    14.81716    0.167     0.868
## speed          0.91329     2.03422    0.449     0.656
## I(speed^2)     0.09996     0.06597    1.515     0.136
##
## Residual standard error: 15.18 on 47 degrees of freedom
## Multiple R-squared: 0.6673, Adjusted R-squared: 0.6532
## F-statistic: 47.14 on 2 and 47 DF, p-value: 5.852e-12
```

or we can plot it like this (see Figure 6-7):

```
cars %>% ggplot(aes(x = speed, y = dist)) +
    geom_point() +
    geom_smooth(method = "lm", formula = y ~ x + I(x^2))
```

This is a slightly better fitting model, but that wasn't the point. You can see how you can transform data in a formula to have different features to give to your fitting algorithms.

Validating Models

How did I know the polynomial fit was better than the linear fit? Well, theoretically a second-degree polynomial should always be a better fit than a line since a line is a special case of a polynomial. We just set θ_2 to zero. If the best-fitted polynomial doesn't have $\theta_2 = 0$, then that is because we can fit the data better if it is not.

The result of fitting the polynomial tells me, in the output from the `summary()` function, that the variables are not significant. It tells me that both from the linear and the squared component, though, so it isn't that useful. Clearly, the points are on a line, so it cannot be correct that there isn't a linear component. I cannot use the summary that much because it is only telling me that when I have both components, then neither of them is statistically significant. That doesn't mean much.

But should I even care, though? If I know that the more complex model always fits better, then shouldn't I just always use it? The problem with that idea is that while the most complex model will always fit the training data—the data I use for fitting the model—better, it will not necessarily generalize better. If I use a high enough degree polynomial—if I have a degree that is the same as the number of data points—I can fit the data perfectly. But it will be fitting both the systematic relationship between x and y and also the statistical errors in our targets t. It might be utterly useless for predicting point number $n + 1$.

What I really need to know is whether one or the other model is better at predicting the distance from the speed.

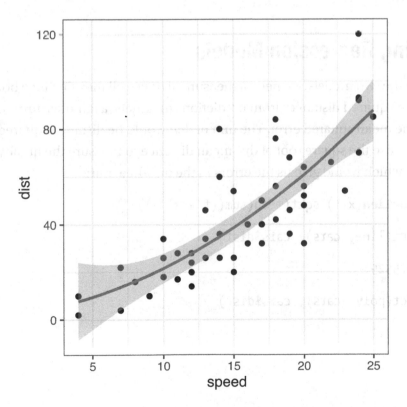

Figure 6-7. *The cars data fitted to a second-degree polynomial*

We can fit the two models and get their predictions using the `predict()` function. It takes the fitted model as the first argument and data to predict on as the second:

```
line <- cars %>% lm(dist ~ speed, data = .)
poly <- cars %>% lm(dist ~ speed + I(speed^2), data = .)

predict(line, cars) |> head()

##          1         2        3        4         5
## -1.849460 -1.849460 9.947766 9.947766 13.880175
##          6
## 17.812584

predict(poly, cars) |> head()

##         1        2         3         4         5
##  7.722637 7.722637 13.761157 13.761157 16.173834
##          6
## 18.786430
```

Evaluating Regression Models

To compare the two models, we need a measure of how well they fit. Since both models are fitting the squared distances from predictions to targets, a fair measure would be looking at the mean squared error. The unit of that would be distance squared, though, so we usually use the square root of this mean distance to measure the quality of the predictions, which would give us the errors in the distance unit:

```
rmse <- function(x,t) sqrt(mean(sum((t - x)^2)))

rmse(predict(line, cars), cars$dist)

## [1] 106.5529

rmse(predict(poly, cars), cars$dist)

## [1] 104.0419
```

Now clearly the polynomial fits slightly better, which it should, based on theory, but there is a bit of a cheat here. We are looking at how the models work on the data we used to fit them. The more complex model will always be better at this. That is the problem we are dealing with. The more complex model might be overfitting the data and capturing the statistical noise we don't want it to capture. What we really want to know is how well the models generalize; how well do they work on data they haven't already seen and used to fit their parameters?

We have used all the data we have to fit the models. That is generally a good idea. You want to use all the data available to get the best-fitted model. But to compare models, we need to have data that isn't used in the fitting.

We can split the data into two sets, one we use for training and the other we use to test the models. There are 50 data points, so I can take the first 25 to train my models on and the next 25 to test them on:

```
training_data <- cars[1:25,]
test_data <- cars[26:50,]

line <- training_data %>% lm(dist ~ speed, data = .)
poly <- training_data %>% lm(dist ~ speed + I(speed^2), data = .)

rmse(predict(line, test_data), test_data$dist)

## [1] 88.89189

rmse(predict(poly, test_data), test_data$dist)

## [1] 83.84263
```

The second-degree polynomial is still better, but I am also still cheating. There is more structure in my data set than just the speed and distances. The data frame is sorted according to the distance, so the training set has all the short distances and the test data all the long distances. They are not similar. That is not good.

In general, you cannot know if there is such structure in your data. In this particular case, it is easy to see because the structure is that obvious, but sometimes it is more subtle. So when you split your data into training and test data, you will want to sample data points randomly. That gets rid of the structure that is in the order of the data points.

We can use the sample() function to sample randomly zeros and ones:

```
sampled_cars <- cars |>
    mutate(training = sample(0:1, nrow(cars), replace = TRUE))

sampled_cars |> head()
```

```
##    speed dist training
## 1     4    2        1
## 2     4   10        0
## 3     7    4        1
## 4     7   22        1
## 5     8   16        0
## 6     9   10        1
```

This doesn't give us 50/50 training and test data since which data point gets into each category will depend on the random samples, but it will be roughly half the data we get for training:

```
training_data <- sampled_cars |> filter(training == 1)
test_data <- sampled_cars |> filter(training == 0)

training_data |> head()
```

```
##    speed dist training
## 1     4    2        1
## 2     7    4        1
## 3     7   22        1
## 4     9   10        1
## 5    10   18        1
## 6    10   34        1
```

```
test_data |> head()
```

```
##    speed dist training
## 1     4   10        0
## 2     8   16        0
## 3    10   26        0
## 4    12   14        0
## 5    12   20        0
## 6    12   24        0
```

Now we can get a better estimate of how the functions are working:

```
line <- training_data %>% lm(dist ~ speed, data = .)
poly <- training_data %>% lm(dist ~ speed + I(speed^2), data = .)

rmse(predict(line, test_data), test_data$dist)

## [1] 66.06671

rmse(predict(poly, test_data), test_data$dist)

## [1] 65.8426
```

Now, of course, the accuracy scores depend on the random sampling when we create the training and test data, so you might want to use more samples. We will return to that in the next section.

Once you have figured out what the best model is, you will still want to train it on all the data you have. Splitting the data is just a tool for evaluating how well different models work. For the final model you choose to work with, you will always want to fit it with all the data you have.

Evaluating Classification Models

If you want to do classification rather than regression, then the root mean square error is not the function to use to evaluate your model. With classification, you want to know how many data points are classified correctly and how many are not.

As an example, we can take the breast cancer data and fit a model:

```
formatted_data <- BreastCancer |>
    mutate(Thickness =
                as.numeric(as.character(Cl.thickness)),
           CellSize =
                as.numeric(as.character(Cell.size))) %>%
    mutate(Malignant = ifelse(Class != "benign", 1, 0))

fitted_model <- formatted_data %>%
    glm(Malignant ~ Thickness + CellSize,
        family = "binomial",
        data = .)
```

To get its prediction, we can again use `predict()`, but we will see that for this particular model, the predictions are probabilities of a tumor being malignant:

```
predict(fitted_model, formatted_data, type = "response") |> head()
```

```
##          1          2          3          4
## 0.05266571 0.65374326 0.01591478 0.99740926
##          5          6
## 0.02911157 0.99992795
```

We would need to translate that into actual predictions. The natural choice here is to split the probabilities at 50%. If we are more certain that a tumor is malignant than benign, we will classify it as malignant:

```
classify <- function(probability) ifelse(probability < 0.5, 0, 1)
classified_malignant <- classify(predict(fitted_model, formatted_data))
```

Where you want to put the threshold of how to classify depends on your data and the consequences of the classification. In a clinical situation, maybe you want to examine further a tumor with less than 50% probability that it is malignant, or maybe you don't want to tell patients that a tumor might be malignant if it is only 50% probable. The classification should take into account how sure you are about the classification, and that depends a lot on the situation you are in. Of course, you don't want to bet against the best knowledge you have, so I am not suggesting that you should classify everything below probability 75% as the "false" class, for instance. The only thing you gain from this is making worse predictions than you could. But sometimes you want to leave some data unpredicted. So here you can use the probabilities the model predicts to leave some data points as NA. How you want to use that your prediction gives you probabilities instead of just classes—assuming it does, it depends on the algorithm used for classifying—is up to you and the situation you are analyzing.

Confusion Matrix

In any case, if we just put the classification threshold at 50/50, then we can compare the predicted classification against the actual classification using the `table()` function:

```
table(formatted_data$Malignant, classified_malignant)
```

```
##      classified_malignant
##         0   1
##   0   447  11
##   1    31 210
```

This table, contrasting predictions against true classes, is known as the confusion matrix. The rows count how many zeros and ones we see in the `formatted_data$Malignant` argument and the columns how many zeros and ones we see in the `classified_malignant` argument. So the first row is where the data says the tumors are not malignant, and the second row is where the data says that the tumors are malignant. The first column is where the predictions say the tumors are not malignant, while the second column is where the predictions say that they are.

This, of course, depends on the order of the arguments to `table()`; it doesn't know which argument contains the data classes and which contains the model predictions. It can be a little hard to remember which dimension, rows or columns, are the predictions, but you can provide a parameter, dnn (dimnames names), to make the table remember it for you:

```
table(formatted_data$Malignant, classified_malignant,
      dnn = c("Data", "Predictions"))
```

```
##        Predictions
## Data    0   1
##    0  447  11
##    1   31 210
```

The correct predictions are on the diagonal, and the off-diagonal values are where our model predicts incorrectly.

The first row is where the data says that tumors are not malignant. The first element, where the model predicts that the tumor is benign, and the data agrees, is called the true negatives. The element to the right of it, where the model says a tumor is malignant but the data says it is not, is called the false positives.

The second row is where the data says that tumors are malignant. The first column is where the prediction says that it isn't a malignant tumor, and this is called the false negatives. The second column is the cases where both the model and the data say that the tumor is malignant. That is the true positives.

The terms positives and negatives are a bit tricky here. I managed to sneak them past you by having the classes called zeros and ones which you already associate with true and false and positive and negative and by having a data set where it was more natural to think of malignant tumors as being the ones we want to predict.

The classes do not have to be zeros and ones. That was just easier in this particular model where I had to translate the classes into zeros and ones for the logistic classification anyway. But really, the classes are "benign" and "malignant":

```
classify <- function(probability)
    ifelse(probability < 0.5, "benign", "malignant")
classified <- classify(predict(fitted_model, formatted_data))

table(formatted_data$Class, classified,
      dnn=c("Data", "Predictions"))

##           Predictions
## Data       benign malignant
## benign       447        11
## malignant     31       210
```

What is positive and what is negative now depends on whether we want to predict malignant or benign tumors. Of course, we really want to predict both well, but the terminology considers one class true and the other false.

The terms carry over into several of the terms used in classification described in the following where the classes and predictions are not so explicitly stated. In the confusion matrix, we can always see exactly what the true classes are and what the predicted classes are, but once we start summarizing it in various ways, this information is no longer explicitly available. The summaries still will often depend on which class we consider "positive" and which we consider "negative," though.

Since which class is which really is arbitrary, so it is always worth a thought deciding which you want to call which and definitely something you want to make explicit in any documentation of your analysis.

Accuracy

The simplest measure of how well a classification is doing is the accuracy. It measures how many classes it gets right out of the total, so it is the diagonal values of the confusion matrix divided by the total:

```
confusion_matrix <- table(formatted_data$Class, classified,
                          dnn=c("Data", "Predictions"))
accuracy <- sum(diag(confusion_matrix)) / sum(confusion_matrix)
accuracy
```

```
## [1] 0.9399142
```

This measure of the classification accuracy is pretty simple to understand, but you have to be careful in what you consider a good accuracy. Of course, "good" is a subjective term, so let us get technical and think in terms of "better than chance." That means that your baseline for what you consider "good" is randomly guessing. This, at least, is not subjective.

It is still something you have to consider a bit carefully, though. Because what does randomly guessing mean? We naturally think of a random guess as one that chooses either class with the same 50/50 probability. If the data has the same number of observations for each of the two classes, then that would be a good strategy and would get the average accuracy of 0.5. So better than chance would, in that case, be better than 0.5. The data doesn't have to have the same number of instances for each class. The breast cancer data does not. The breast cancer data has more benign tumors than malignant tumors:

```
table(BreastCancer$Class)
```

```
##
##    benign malignant
##       458       241
```

Here, you would be better off guessing more benign than malignant. If you had to guess and already knew that you were more than twice as likely to have a benign than a malignant tumor, you would always guess benign:

```
tbl <- table(BreastCancer$Class)
tbl["benign"] / sum(tbl)
```

```
##      benign
## 0.6552217
```

Always guessing "benign" is a lot better than 50/50. Of course, it is arguable whether this is guessing, but it is a strategy for guessing, and you want your model to do better than this simple strategy.

Always guessing the most frequent class—assuming that the frequency of the classes in the data set is a representative for the frequency in new data as well (which is a strong assumption)—is the best strategy for guessing.

If you actually want to see "random" guessing, you can get an estimate of this by simply permuting the classes in the data. The function sample() can do this:

```
table(BreastCancer$Class, sample(BreastCancer$Class))
```

```
##
##             benign malignant
## benign         291       167
## malignant      167        74
```

This gives you an estimate for random guessing, but since it is random, you would want to get more than one to get a feeling for how much it varies with the guess:

```
accuracy <- function(confusion_matrix)
    sum(diag(confusion_matrix))/sum(confusion_matrix)
sample_table <- function()
    table(BreastCancer$Class, sample(BreastCancer$Class))

replicate(8, sample_table() |> accuracy())
```

```
## [1] 0.5450644 0.5336195 0.5879828 0.5565093
## [5] 0.5622318 0.5565093 0.5364807 0.5278970
```

As you can see, even random permutations do better than 50/50—but the better guess is still just the most frequent class, and at the very least, you would want to beat that.

Sensitivity and Specificity

We want a classifier to have a high accuracy, but accuracy isn't everything. The costs in real life of misclassifying often have different consequences when you classify something like a benign tumor as malignant from when you classify a malignant tumor as benign. In a clinical setting, you have to weight the false positives against the false negatives and the consequences they have. You are interested in more than pure accuracy.

We usually use two measures of the predictions of a classifier that takes that into account: the specificity and the sensitivity of the model. The first measure captures how often the model predicts a negative case correctly. In the breast cancer data, this is how often, when the model predicts a tumor as benign, it actually is:

```
(specificity <- confusion_matrix[1,1] /
    (confusion_matrix[1,1] + confusion_matrix[1,2]))
```

```
## [1] 0.9759825
```

The sensitivity does the same thing but for the positives. It captures how well, when the data has the positive class, your model predicts this correctly:

```
(sensitivity <- confusion_matrix[2,2]/
    (confusion_matrix[2,1] + confusion_matrix[2,2]))
```

```
## [1] 0.8713693
```

If your accuracy is 100%, then both of these will also be 100%. But there is usually a trade-off between the two. Using the "best guessing" strategy of always picking the most frequent class will set one of the two to 100% but at the cost of the other. In the breast cancer data, the best guess is always benign, the negative case, and always guessing benign will give us a specificity of 100%.

This strategy can always achieve 100% for one of the two measures but at the cost of setting the other to 0%. If you only ever guess at one class, you are perfect when the data is actually from that class, but you are always wrong when the data is from the other class.

Because of this, we are never interested in optimizing either measure alone. That is trivial. We want to optimize both. We might consider specificity more important than sensitivity or vice versa, but even if we want one to be 100%, we also want the other to be as good as we can get it.

215

To evaluate how much better than chance we are doing, we can again compare to random permutations. This tells us how well we are doing compared to random guesses for both:

```
specificity <- function(confusion_matrix)
    confusion_matrix[1,1] /
    (confusion_matrix[1,1]+confusion_matrix[1,2])

sensitivity <- function(confusion_matrix)
    confusion_matrix[2,2] /
    (confusion_matrix[2,1]+confusion_matrix[2,2])

prediction_summary <- function(confusion_matrix)
    c("accuracy" = accuracy(confusion_matrix),
      "specificity" = specificity(confusion_matrix),
      "sensitivity" = sensitivity(confusion_matrix))

random_prediction_summary <- function()
    prediction_summary(
      table(BreastCancer$Class, sample(BreastCancer$Class))
)

replicate(3, random_prediction_summary())

##                   [,1]      [,2]      [,3]
## accuracy    0.5536481 0.5536481 0.5278970
## specificity 0.6593886 0.6593886 0.6397380
## sensitivity 0.3526971 0.3526971 0.3153527
```

Other Measures

The specificity is also known as the true negative rate since it measures how many of the negative classifications are true. Similarly, the sensitivity is known as the true positive rate. There are analogue measures for getting things wrong. The false negative rate is the analogue of the true negative rate, but instead of dividing the true negatives by all the negatives, it divides the false negatives by all the negatives. The false positive rate similarly divides the false positives by all the positives. Having these two measures together with sensitivity and specificity is not really adding much. The true negative rate is just one minus the false negative rate and similar for the true positive rate and false positive rate. They just focus on when the model gets things wrong instead of when it gets things right.

All four measures split the confusing matrix into the two rows. They look at when the data says the class is true and when the data says the class is false. We can also look at the columns instead and consider when the predictions are true and when the predictions are false.

When we look at the column where the predictions are false—for the breast cancer when the tumors are predicted as benign—we have the false omission rate, which is the false negatives divided by all the predicted negatives:

```
confusion_matrix[2,1] / sum(confusion_matrix[,1])
```

```
## [1] 0.06485356
```

The negative predictive value is instead the true negatives divided by the predicted negatives:

```
confusion_matrix[1,1] / sum(confusion_matrix[,1])
```

```
## [1] 0.9351464
```

These two will always sum to one, so we are really only interested in one of them, but which we choose is determined by which we find more important.

For the predicted positives, we have the positive predictive values and false discovery rate:

```
confusion_matrix[2,2] / sum(confusion_matrix[,2])
```

```
## [1] 0.9502262
```

```
confusion_matrix[1,2] / sum(confusion_matrix[,2])
```

```
## [1] 0.04977376
```

The false discovery rate, usually abbreviated FDR, is the one most frequently used. It is closely related to the threshold used on p-values (the significance thresholds) in classical hypothesis testing. Remember that if you have a 5% significance threshold in classical hypothesis testing, it means that when the null hypothesis is true, you will predict it is false 5% of the time. This means that your false discovery rate is 5%.

The classical approach is to pick an acceptable false discovery rate; by convention, this is 5%, but there is nothing magical about that number—it is simply convention—and then that threshold determines how extreme a test statistic has to be before we switch from predicting a negative to predicting a positive. This approach entirely ignores the cases where the data is from the positive class. It has its uses, but not for classification where you have data from both the positive class and the negative class, so we will not consider it more here. You will have seen it in statistics classes, and you can learn more about it in any statistics textbook.

More Than Two Classes

All of the above considers a situation where we have two classes, one we call positive and one we call negative. This is a common case, which is the reason we have so many measures for dealing with it, but it is not the only case. Quite often, we need to classify data into more than two classes.

The only measure you can reuse there is the accuracy. The accuracy is always the sum along the diagonal divided by the total number of observations. Accuracy still isn't everything in those cases. Some classes are perhaps more important to get right than others—or just harder to get right than others—so you have to use a lot of sound judgment when evaluating a classification. There are just fewer rules of thumb to use here, so you are more left to your own judgment.

Sampling Approaches

To validate classifiers, I suggested splitting the data into a training data set and a test data set. I also mentioned that there might be hidden structures in your data set, so you always want to make this split a random split of the data.

Generally, there are a lot of benefits you can get out of randomly splitting your data or randomly subsampling from your data. We have mostly considered prediction in this chapter, where splitting the data into training and a test data lets us evaluate how well a model does at predicting on unseen data. But randomly splitting or subsampling from data is also very useful for inference. When we do inference, we can typically get confidence intervals for model parameters, but these are based on theoretical results that assume that the data is from some (usually) simple distribution. Data is generally not. If you want to know how a parameter is distributed from the empirical distribution of the data, you will want to subsample and see what distribution you get.

Random Permutations of Your Data

With the `cars` data, we split the observations into two equally sized data sets. Since this data is ordered by the stopping distance, splitting it into the first half and the second half makes the data sets different in distributions.

The simplest approach to avoiding this problem is to reorder your data randomly before you split it. Using the `sample()` function, we can get a random permutation of any input vector—we saw that earlier—and we can exploit this to get a random order of your data set.

Using `sample(1:n)`, we get a random permutation of the numbers from 1 to n. We can select rows in a data frame by giving it a vector of indices for the rows. Combining these two observations, we can get a random order of `cars` observations this way:

```
permuted_cars <- cars[sample(1:nrow(cars)),]
permuted_cars |> head(3)
```

```
##    speed dist
## 9     10   34
## 1      4    2
## 48    24   93
```

The numbers to the left of the data frame are the original row numbers (it really is the row names, but it is the same in this case).

We can write a simple function for doing this for general data frames:

```
permute_rows <- function(df) df[sample(1:nrow(df)),]
```

Using this, we can add it to a data analysis pipeline where we would write

```
permuted_cars <- cars |> permute_rows()
```

Splitting the data into two sets, training and testing, is one approach to subsampling, but a general version of this is used in something called cross-validation. Here, the idea is to get more than one result out of the random permutation we use. If we use a single training/test split, we only get one estimate of how a model performs on a data set. Using more gives us an idea about the variance of this.

We can split a data set into *n* groups like this:

```
group_data <- function(df, n) {
    groups <- rep(1:n, each = nrow(df)/n)
    split(df, groups)
}
```

You don't need to understand the details of this function for now, but it is a good exercise to try to figure it out, so you are welcome to hit the documentation and see if you can work it out.

The result is a `list`, a data structure we haven't explored yet (but we will later in the book, when we do some more serious programming). It is necessary to use a list here since vectors or data frames cannot hold complex data, so if we combined the result in one of those data structures, they would just be merged back into a single data frame here.

As it is, we get something that contains *n* data structures that each have a data frame of the same form as the `cars` data:

```
grouped_cars <- cars |> permute_rows() |> group_data(5)
grouped_cars |> str()

## List of 5
##  $ 1:'data.frame':    10 obs. of  2 variables:
##   ..$ speed: num [1:10] 12 4 14 13 18 19 14 4 ...
##   ..$ dist : num [1:10] 28 2 60 26 76 46 26 10 ...
##  $ 2:'data.frame':    10 obs. of  2 variables:
##   ..$ speed: num [1:10] 15 13 12 18 20 18 12 15 ..
##   ..$ dist : num [1:10] 54 34 20 84 32 42 24 26 ..
##  $ 3:'data.frame':    10 obs. of  2 variables:
##   ..$ speed: num [1:10] 11 24 22 9 14 13 24 20 ...
##   ..$ dist : num [1:10] 28 70 66 10 36 46 120 48..
##  $ 4:'data.frame':    10 obs. of  2 variables:
##   ..$ speed: num [1:10] 12 10 25 17 17 23 19 11 ..
##   ..$ dist : num [1:10] 14 26 85 40 32 54 68 17 ..
##  $ 5:'data.frame':    10 obs. of  2 variables:
##   ..$ speed: num [1:10] 10 20 15 16 13 7 19 24 ...
##   ..$ dist : num [1:10] 18 52 20 32 34 22 36 92 ..
```

```
grouped_cars[[1]] # First sample
```

```
##    speed dist
## 15    12   28
## 1      4    2
## 22    14   60
## 16    13   26
## 34    18   76
## 37    19   46
## 20    14   26
## 2      4   10
## 5      8   16
## 42    20   56
```

All you really need to know for now is that to get an entry in a list, you need to use [[ ]] indexing instead of [ ] indexing.

If you use [ ], you will also get the data, but the result will be a list with one element, which is not what you want:

```
grouped_cars[1]
```

```
## $`1`
##    speed dist
## 15    12   28
## 1      4    2
## 22    14   60
## 16    13   26
## 34    18   76
## 37    19   46
## 20    14   26
## 2      4   10
## 5      8   16
## 42    20   56
```

We can use the different groups to get estimates of the model parameters in the linear model for cars:

```
lm(dist ~ speed, data = grouped_cars[[1]])$coefficients
```

```
##  (Intercept)      speed
##   -10.006702  3.540214
```

With a bit of programming, we can get the estimates for each group:

```
get_coef <- function(df)
    lm(dist ~ speed, data = df)$coefficients

# Get estimates from first group
estimates <- get_coef(grouped_cars[[1]])
for (i in 2:length(grouped_cars)) {
    # Append the next group
    estimates <- rbind(estimates, get_coef(grouped_cars[[i]]))
}

Estimates
```

```
##            (Intercept)      speed
## estimates   -10.00670  3.540214
##             -33.89655  4.862069
##             -29.25116  4.744186
##             -11.82554  3.555755
##             -14.87639  3.494779
```

Right away, I will stress that this is not the best way to do this, but it shows you how it could be done. We will get to better approaches shortly. Still, you can see how splitting the data this way lets us get distributions for model parameters.

There are several reasons why this isn't the optimal way of coding this. The row names are ugly, but that is easy to fix. The way we combine the estimates in the data frame is inefficient—although it doesn't matter much with such a small data set—and later in the book, we will see why. The main reason, though, is that explicit loops like this make it hard to follow the data transformations since it isn't a pipeline of processing.

The package purrr lets us work on lists using pipelines. You import the package:

```
library(purrr)
```

and then you have access to the function map_df() that lets you apply a function to each element of the list:

```
estimates <- grouped_cars |> map_df(get_coef)
```

The map_df function maps (as the name suggests) across its input, applying the function to each element in the input list. The results of each function call are turned into rows in a data frame (where the _df part of the name comes from). This pipeline is essentially doing the same as the more explicit loop we wrote before; there is just less code to write. If you are used to imperative programming languages, this will look very succinct, but if you have experience in functional programming languages, it should look familiar.

Cross-Validation

A problem with splitting the data into many small groups is that we get a large variance in estimates. Instead of working with each little data set independently, we can remove one of the data sets and work on all the others. This will mean that our estimates are no longer independent, but the variance goes down. The idea of removing a subset of the data and then cycling through the groups evaluating a function for each group that is left out is called cross-validation. Well, it is called cross-validation when we use it to validate prediction, but it works equally well for inferring parameters.

If we already have the grouped data frames in a list, we can remove one element from the list using [-i] indexing—just as we can for vectors—and the result is a list containing all the other elements. We can then combine the elements in the list into a single data frame using the do.call("rbind",.) magical invocation.

So we can write a function that takes the grouped data frames and gives us another list of data frames that contains data where a single group is left out. One way to do this is listed as follows; that implementation uses the bind_rows function from the dplyr package (get it using library(dplyr)):

```
cross_validation_groups <- function(grouped_df) {
    remove_group <- function(group)
        # remove group "group" from the list
        grouped_df[-group] |>
        # merge the remaining groups into one data frame
        bind_rows()
```

```
# Iterate over indices from 1 to number of groups
seq_along(grouped_df) |>
    # get the data frame with this group removed
    map(remove_group)
}
```

This function is a little more spicy than those we have written before, but it doesn't use anything we haven't seen already. In the function, we write another helper function, remove_group. You can write functions inside other functions, and if you do, then the inner function can see the variables in the outer function. Our remove_group function can see the grouped_df data frame. We give it an argument, group, and it removes the group with that index using -group in the subscript grouped_df[-group]. Since grouped_df is a list of data frames, removing one of them still leaves us with a list of data frames, but we would rather have a single data frame. The bind_rows function merges the list into a single data frame containing all the data points we didn't remove.

With the function written, we can create the cross-validation groups. We use seq_along(grouped_df) to create all the numbers from one to the length of grouped_df—using this function is slightly safer than writing 1:len(grouped_df) because it correctly handles empty lists, so you should get used to using it. We loop over all these numbers with a map function—this function is also from the purrr package and behaves like map_df except that it returns a list and not a data frame—and apply remove_group to each index. This results in a list of data frames, which is exactly what we want.

We could have combined this with the group_data() function, but I prefer to write functions that do one simple thing and combine them instead using pipelines. We can use this function and all the stuff we did earlier to get estimates using cross-validation:

```
cars |>
    permute_rows() |> # randomize for safety...
    group_data(5) |> # get us five groups
    cross_validation_groups() |> # then make five cross-validation groups
    # For each cross-validation group, estimate the cofficients and put
    # the results in a data frame
    map_df(
        # We need a lambda expression here because lm doesn't take
        # the data frame as its first argument
        \(df) lm(dist ~ speed, data = df)$coefficients
```

```
)
```

```
## # A tibble: 5 × 2
##    `(Intercept)` speed
##            <dbl> <dbl>
## 1          -18.0  3.92
## 2          -14.4  3.86
## 3          -20.1  4.15
## 4          -16.2  3.79
## 5          -19.5  3.96
```

Where cross-validation is typically used is when leaving out a subset of the data for testing and using the rest for training.

We can write a simple function for splitting the data this way, similar to the cross_validation_groups() function. It cannot return a list of data frames but needs to return a list of lists, each list containing a training data frame and a test data frame. It looks like this:

```
cross_validation_split <- function(grouped_df) {
  seq_along(grouped_df) |> map(
    \(group) list(
      # Test is the current group
      test = grouped_df[[group]],
      # Training is all the others
      training = grouped_df[-group] |> bind_rows()
    ))
}
```

The function follows the same pattern as the previous, I just haven't bothered with writing an inner function; instead I use a lambda expression (\(group) ...). It creates a list with two elements, test and training. In test, we put the current group—we subscript with [[group]] to get the actual data frame instead of a list that holds it—and in training, we put all the other groups. Here, we use [-group] to get a list of all the other elements—[[-group]] would not work for us—and then we use the bind_rows() function we saw earlier to merge the list into a single data frame.

Don't worry if you don't understand all the details of it. After reading later programming chapters, you will. Right now, I hope you just get the gist of it.

I will not show you the result. It is just long and not that pretty, but if you want to see it, you can type in

```
cars |>
    permute_rows() |>
    group_data(5)   |>
    cross_validation_split()
```

As we have seen, we can index into a list using [[]]. We can also use the $name indexing like we can for data frames, so if we have a list lst with a training data set and a test data set, we can get them as lst$training and lst$test.

```
prediction_accuracy <- function(test_and_training) {
    test_and_training |>
        map_dbl(
            \(tt) {
            # Fit the model using training data
            fit <- lm(dist ~ speed, data = tt$training)
            # Then make predictions on the test data
            predictions <- predict(fit, newdata = tt$test)
            # Get root mean square error of result
            rmse(predictions, tt$test$dist)
            }
        )
}
```

You should be able to understand most of this function even though we haven't covered much R programming yet, but if you do not, then don't worry.

You can then add this function to your data analysis pipeline to get the cross-validation accuracy for your different groups:

```
cars |>
    permute_rows() |>
    group_data(5) |>
    cross_validation_split() |>
    prediction_accuracy()
## [1] 56.62113 38.55348 33.52728 59.27442 48.77524
```

The prediction accuracy function isn't general. It is hardwired to use a linear model and the model `dist ~ speed`. It is possible to make a more general function, but that requires a lot more R programming skills, so we will leave the example here.

Selecting Random Training and Testing Data

In the example earlier where I split the data `cars` into training and test data using `sample(0:1, n, replacement = TRUE)`, I didn't permute the data and then deterministically split it afterward. Instead, I sampled training and test based on probabilities of picking any given row as training and test.

What I did was adding a column to the data frame where I randomly picked whether an observation should be used for the training or for the test data. Since it required first adding a new column and then selecting rows based on it, it doesn't work well as part of a data analysis pipeline. We can do better and slightly generalize the approach at the same time.

To do this, I shamelessly steal two functions from the documentation of the `purrr` package. They do the same thing as the grouping function I wrote earlier. If you do not quite follow the example, do not worry. But I suggest you try to read the documentation for any function you do not understand and at least try to work out what is going on. Follow it as far as you can, but don't sweat it if there are things you do not fully understand. After finishing the entire book, you can always return to the example.

The grouping function earlier defined groups by splitting the data into *n* equally sized groups. The first function here instead samples from groups specified by probabilities. It creates a vector naming the groups, just as I did before. It just names the groups based on named values in a probability vector and creates a group vector based on probabilities given by this vector:

```
random_group <- function(n, probs) {
  probs <- probs / sum(probs)
  g <- findInterval(seq(0, 1, length = n), c(0, cumsum(probs)),
                    rightmost.closed = TRUE)
  names(probs)[sample(g)]
}
```

If we pull the function apart, we see that it first normalizes a probability vector. This just means that if we give it a vector that doesn't sum to one, it will still work. To use it, it makes the code easier to read if it already sums to one, but the function can deal with it, even if it doesn't.

The second line, which is where it is hardest to read, just splits the unit interval into n subintervals and assigns a group to each subinterval based on the probability vector. This means that the first chunk of the n intervals is assigned to the first group, the second chunk to the second group, and so on. It is not doing any sampling yet, it just partitions the unit interval into n subintervals and assigns each subinterval to a group.

The third line is where it is sampling. It now takes the n subintervals, permutes them, and returns the names of the probability vector each one falls into.

We can see it in action by calling it a few times. We give it a probability vector where we call the first probability "training" and the second "test":

```
random_group(8, c(training = 0.5, test = 0.5))
```

```
## [1] "training" "training" "training" "test"
## [5] "test"     "training" "test"     "test"
```

```
random_group(8, c(training = 0.5, test = 0.5))
```

```
## [1] "training" "test"     "training" "test"
## [5] "training" "test"     "training" "test"
```

We get different classes out when we sample, but each class is picked with 0.5 probability. We don't have to pick them 50/50, though; we can choose more training than test data, for example:

```
random_group(8, c(training = 0.8, test = 0.2))
```

```
## [1] "training" "training" "training" "training"
## [5] "test"     "training" "test"     "training"
```

The second function just uses this random grouping to split the data set. It works exactly like the cross-validation splitting we saw earlier:

```
partition <- function(df, n, probs) {
    replicate(n, split(df, random_group(nrow(df), probs)), FALSE)
}
```

The function replicates the subsampling *n* times. Here, *n* is not the number of observations you have in the data frame, but a parameter to the function. It lets you pick how many subsamples of the data you want.

We can use it to pick four random partitions. Here, with training and test, select with 50/50 probability:

```
random_cars <- cars |> partition(4, c(training = 0.5, test = 0.5))
```

If you evaluate it on your computer and look at `random_cars`, you will see that resulting values are a lot longer now. This is because we are not looking at smaller data sets this time; we have as many observations as we did before (which is 50), but we have randomly partitioned them.

We can combine this `partition()` function with the accuracy prediction from before:

```
random_cars |> prediction_accuracy()
## [1] 62.76781 87.52504 92.00689 84.99749
```

Examples of Supervised Learning Packages

So far in this chapter, we have looked at classical statistical methods for regression (linear models) and classification (logistic regression), but there are many machine learning algorithms for both, and many are available as R packages.

They all work similarly to the classical algorithms. You give the algorithms a data set and a formula specifying the model matrix. From this, they do their magic. All the ideas presented in this chapter can be used together with them.

In the following, I go through a few packages, but there are many more. A Google search should help you find a package if there is a particular algorithm you are interested in applying.

I present their use with the same two data sets we have used earlier, the `cars` data where we aim at predicting the stopping distance from the speed and the `BreastCancer` where we try to predict the class from the cell thickness. For both these cases, the classical models—a linear model and a logistic regression—are more ideal solutions, and these models will not outcompete them, but for more complex data sets, they can usually be quite powerful.

Decision Trees

Decision trees work by building a tree from the input data, splitting on a parameter in each inner node according to a variable value. This can be splitting on whether a numerical value is above or below a certain threshold or which level a factor has.

Decision trees are implemented in the rpart package, and models are fitted just as linear models are:

```
library(rpart)

## Warning: package 'rpart' was built under R version
## 4.1.2

model <- cars %>% rpart(dist ~ speed, data = .)
rmse(predict(model, cars), cars$dist)

## [1] 117.1626
```

Building a classifying model works very similar. We do not need to translate the cell thickness into a numerical value, though; we can use the data frame as it is (but you can experiment with translating factors into numbers if you are interested in exploring this):

```
model <- BreastCancer %>%
   rpart(Class ~ Cl.thickness, data = .)
```

The predictions when we used the glm() function were probabilities for the tumor being malignant. The predictions made using the decision tree give you the probabilities both for being benign and malignant:

```
predict(model, BreastCancer) |> head()

##        benign malignant
## 1 0.82815356 0.1718464
## 2 0.82815356 0.1718464
## 3 0.82815356 0.1718464
## 4 0.82815356 0.1718464
## 5 0.82815356 0.1718464
## 6 0.03289474 0.9671053
```

To get a confusion matrix, we need to translate these probabilities into the corresponding classes. The output of predict() is not a data frame but a matrix, so we first convert it into a data frame using the function as.data.frame(), and then we use the %$% operator in the pipeline to get access to the columns by name in the next step:

```
predicted_class <-
    predict(model, BreastCancer) %>%
    as.data.frame() %$%
    ifelse(benign > 0.5, "benign", "malignant")

table(BreastCancer$Class, predicted_class)

##           predicted_class
##            benign malignant
## benign        453         5
## malignant      94       147
```

Another implementation of decision trees is the ctree() function from the party package:

```
library(party)

model <- cars %>% ctree(dist ~ speed, data = .)
rmse(predict(model, cars), cars$dist)

## [1] 117.1626

model <- BreastCancer %>%
    ctree(Class ~ Cl.thickness, data = .)

predict(model, BreastCancer) %>% head()

## [1] benign    benign    benign    benign
## [5] benign    malignant
## Levels: benign malignant

table(BreastCancer$Class, predict(model, BreastCancer))

##
##           benign malignant
##  benign      453         5
##  malignant    94       147
```

I like this package slightly more since it can make plots of the fitted models (see Figure 6-8):

```
cars %>% ctree(dist ~ speed, data = .) %>% plot()
```

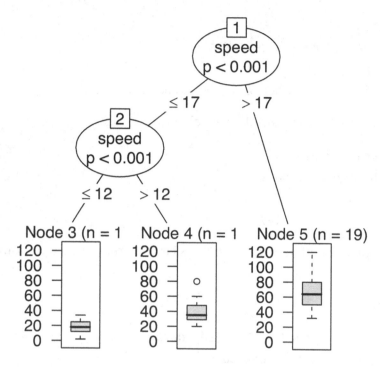

Figure 6-8. *Plot of the cars decision tree*

Random Forests

Random forests generalize decision trees by building several of them and combining them. They are implemented in the randomForest package:

```
library(randomForest)

model <- cars %>% randomForest(dist ~ speed, data = .)
rmse(predict(model, cars), cars$dist)

## [1] 83.7541
```

For classification, the predictions are the actual classes as a factor, so no translation is needed to get a confusion matrix:

```
model <- BreastCancer %>%
    randomForest(Class ~ Cl.thickness, data = .)

predict(model, BreastCancer) %>% head()

##           1       2       3          4        5
##      benign  benign  benign malignant  benign
##           6
##   malignant
##   Levels: benign malignant

table(BreastCancer$Class, predict(model, BreastCancer))

##
##           benign malignant
##   benign     437        21
##   malignant   76       165
```

Neural Networks

You can use a package called nnet to construct neural networks:

```
library(nnet)
```

You can use it for both classification and regression. We can see it in action on the cars data set:

```
model <- cars %>% nnet(dist ~ speed, data = ., size = 5)

## # weights:    16
## initial    value 123632.602158
## final    value 120655.000000
## converged

rmse(predict(model, cars), cars$dist)

## [1] 347.3543
```

The neural networks require a `size` parameter specifying how many nodes you want in the inner layer of the network. Here, I have just used five.

For classification, you use a similar call:

```
model <- BreastCancer %>%
    nnet(Class ~ Cl.thickness, data = ., size = 5)
```

```
## # weights: 56
## initial value 453.502123
## iter 10 value 226.317196
## iter 20 value 225.125028
## iter 30 value 225.099296
## iter 40 value 225.098355
## final value 225.098268
## converged
```

The output of the `predict()` function is probabilities for the tumor being malignant:

```
predict(model, BreastCancer) %>% head()
```

```
##           [,1]
## 1 0.3461460
## 2 0.3461460
## 3 0.1111139
## 4 0.5294021
## 5 0.1499858
## 6 0.9130386
```

We need to translate it into classes, and for this, we can use a lambda expression:

```
predicted_class <- predict(model, BreastCancer) %>%
    { ifelse(. < 0.5, "benign", "malignant") }
```

```
table(BreastCancer$Class, predicted_class)
```

```
##            predicted_class
##             benign malignant
## benign        437        21
## malignant      76       165
```

Support Vector Machines

Another popular method is support vector machines. These are implemented in the ksvm() function in the kernlab package:

```
library(kernlab)
```

```
model <- cars %>% ksvm(dist ~ speed, data = .)
rmse(predict(model, cars), cars$dist)
```

```
## [1] 92.41686
```

For classification, the output is again a factor we can use directly to get a confusion matrix:

```
model <- BreastCancer %>%
    ksvm(Class ~ Cl.thickness, data = .)
```

```
predict(model, BreastCancer) %>% head()
```

```
## [1] benign    benign    benign    malignant
## [5] benign    malignant
## Levels: benign malignant
```

```
table(BreastCancer$Class, predict(model, BreastCancer))
```

```
##
##           benign malignant
## benign       437        21
## malignant     76       165
```

Naive Bayes

Naive Bayes essentially assumes that each explanatory variable is independent of the others and uses the distribution of these for each category of data to construct the distribution of the response variable given the explanatory variables.

Naive Bayes is implemented in the e1071 package:

```
library(e1071)
```

The package doesn't support regression analysis—after all, it needs to look at conditional distributions for each output variable value—but we can use it for classification. The function we need is `naiveBayes()`, and we can use the `predict()` output directly to get a confusion matrix:

```
model <- BreastCancer %>%
    naiveBayes(Class ~ Cl.thickness, data = .)

predict(model, BreastCancer) %>% head

## [1] benign     benign     benign     malignant
## [5] benign     malignant
## Levels: benign malignant

table(BreastCancer$Class, predict(model, BreastCancer))

##
##             benign malignant
## benign         437        21
## malignant       76       165
```

Exercises

Fitting Polynomials

Use the `cars` data to fit higher degree polynomials and use training and test data to explore how they generalize. At which degree do you get the better generalization?

Evaluating Different Classification Measures

Earlier, I wrote functions for computing the accuracy, specificity (true negative rate), and sensitivity (true positive rate) of a classification. Write similar functions for the other measures described before. Combine them in a `prediction_summary()` function like I did earlier.

Breast Cancer Classification

You have seen how to use the `glm()` function to predict the classes for the breast cancer data. Use it to make predictions for training and test data, randomly splitting the data in these two classes, and evaluate all the measures with your `prediction_summary()` function.

If you can, then try to make functions similar to the ones I used to split data and evaluate models for the `cars` data.

Leave-One-Out Cross-Validation (Slightly More Difficult)

The code I wrote earlier splits the data into n groups and constructs training and test data based on that. This is called n-fold cross-validation. There is another common approach to cross-validation called leave-one-out cross-validation. The idea here is to remove a single data observation and use that for testing and all the rest of the data for training.

This isn't used that much if you have a lot of data—leaving out a single data point will not change the trained model much if you have lots of data points anyway—but for smaller data sets, it can be useful.

Try to program a function for constructing subsampled training and test data for this strategy.

Decision Trees

Use the `BreastCancer` data to predict the tumor class, but try including more of the explanatory variables. Use cross-validation or sampling of training/test data to explore how it affects the prediction accuracy.

Random Forests

Use the `BreastCancer` data to predict the tumor class, but try including more of the explanatory variables. Use cross-validation or sampling of training/test data to explore how it affects the prediction accuracy.

Neural Networks

The `size` parameter for the `nnet` function specifies the complexity of the model. Test how the accuracy depends on this variable for classification on the `BreastCancer` data.

Earlier, we only used the cell thickness variable to predict the tumor class. Include the other explanatory variables and explore if having more information improves the prediction power.

Support Vector Machines

Use the `BreastCancer` data to predict the tumor class, but try including more of the explanatory variables. Use cross-validation or sampling of training/test data to explore how it affects the prediction accuracy.

Compare Classification Algorithms

Compare the logistic regression, the neural networks, the decision trees, the random forests, and the support vector machines in how well they classify tumors in the `BreastCancer` data. For each, take the best model you obtained in your experiments.

CHAPTER 7

Unsupervised Learning

For supervised learning, we have one or more targets we want to predict using a set of explanatory variables. But not all data analysis consists of making prediction models. Sometimes, we are just trying to find out what structure is actually in the data we analyze. There can be several reasons for this. Sometimes, unknown structures can tell us more about the data. Sometimes, we want to explicitly avoid unknown structures (if we have data sets that are supposed to be similar, we don't want to discover later that there are systematic differences). Whatever the reason, unsupervised learning concerns finding unknown structures in data.

Dimensionality Reduction

Dimensionality reduction, as the name hints at, is a method used when you have high-dimensional data and want to map it down into fewer dimensions. The purpose here is usually to visualize data to try and spot patterns from plots. The analysis usually just transforms the data and doesn't add anything to it. It possibly removes some information, but by reducing the number of dimensions, it can be easier to analyze.

The type of data where this is necessary is when the data has lots of columns. Not necessarily many observations, but each observation has very many variables, and there is often little information in any single column. One example is genetic data where there is often hundreds of thousands, if not millions, of genetic positions observed in each individual, and at each of these positions, we have a count of how many of a given genetic variant is present at these markers, a number from zero to two. There is little information in any single marker, but combined they can be used to tell a lot about an individual. The first example we shall see in this chapter, principal component analysis, is frequently used to map thousands of genetic markers into a few more informative dimensions to reveal relationships between different individuals.

I will not use data with very high dimensionality but illustrate them with smaller data sets where the methods can still be useful.

239

© Thomas Mailund 2022
T. Mailund, *Beginning Data Science in R 4*, https://doi.org/10.1007/978-1-4842-8155-0_7

Principal Component Analysis

Principal component analysis (PCA) maps your data from one vector space to another of the same dimensionality as the first. So it doesn't reduce the number of dimensions as such. However, it chooses the coordinate system of the new space such that the most information is in the first coordinate, the second most information in the second coordinate, and so on.

In its simplest form, it is just a linear transformation. It changes the basis of your vector space such that the most variance in the data is along the first basis vector, and each basis vector then has increasingly less of the variance. The basis of the new vector space is called the components, and the name "principal component" refers to looking at the first few, the most important, the principal components.

There might be some transformations of the data first to normalize it, but the final step of the transformation is always such a linear map. Hence, after the transformation, there is exactly the same amount of information in your data; it is just represented along different dimensions.

Because the PCA just transforms your data, your data has to be numerical vectors, to begin with. For categorical data, you will need to modify the data first. One approach is to represent factors as a binary vector for each level, as is done with model matrices in supervised learning. If you have a lot of factors in your data, though, PCA might not be the right tool.

It is beyond the scope of this book to cover the theory of PCA in any detail—but many other textbooks will—so let us just dig into how it is used in R.

To illustrate this, I will use the `iris` data set. It is not high-dimensional, but it will do as a first example.

Remember that this data contains four measurements, sepal length and width and petal length and width, for flowers from three different species:

```
iris |> head()
```

```
##   Sepal.Length Sepal.Width Petal.Length
## 1          5.1         3.5          1.4
## 2          4.9         3.0          1.4
## 3          4.7         3.2          1.3
## 4          4.6         3.1          1.5
## 5          5.0         3.6          1.4
## 6          5.4         3.9          1.7
```

```
##   Petal.Width Species
## 1         0.2 setosa
## 2         0.2 setosa
## 3         0.2 setosa
## 4         0.2 setosa
## 5         0.2 setosa
## 6         0.4 setosa
```

To see if there is information in the data that would let us distinguish between the three species based on the measurements, we could try to plot some of the measurements against each other. See Figures 7-1 and 7-2.

```
iris |> ggplot() +
  geom_point(aes(x = Sepal.Length, y = Sepal.Width, colour = Species))

iris |> ggplot() +
  geom_point(aes(x = Petal.Length, y = Petal.Width, colour = Species))
```

It does look as if we should be able to distinguish the species. Setosa stands out on both plots, but Versicolor and Virginia overlap on the first.

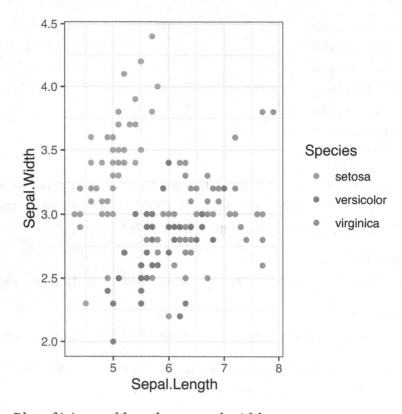

Figure 7-1. *Plot of iris sepal length vs. sepal width*

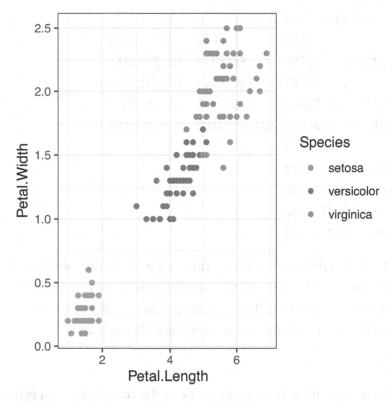

Figure 7-2. *Plot of iris petal length vs. petal width*

Since this is such a simple data set, and since there is obviously a structure if we just plot a few dimensions against each other, this is not a case where we would usually pull out the cannon that is PCA, but this is a section on PCA, so we will.

Since PCA only works on numerical data, we need to remove the Species parameter, but after that, we can do the transformation using the function prcomp:

```
pca <- iris |> select(-Species) |> prcomp()
pca

## Standard deviations (1, .., p=4):
## [1] 2.0562689 0.4926162 0.2796596 0.1543862
##
```

```
## Rotation (n x k) = (4 x 4):
##                         PC1          PC2          PC3
## Sepal.Length  0.36138659  -0.65658877   0.58202985
## Sepal.Width  -0.08452251  -0.73016143  -0.59791083
## Petal.Length  0.85667061   0.17337266  -0.07623608
## Petal.Width   0.35828920   0.07548102  -0.54583143
##                         PC4
## Sepal.Length   0.3154872
## Sepal.Width   -0.3197231
## Petal.Length  -0.4798390
## Petal.Width    0.7536574
```

The object that this produces contains different information about the result. The standard deviations tell us how much variance is in each component and the rotation what the linear transformation is. If we plot the pca object, we will see how much of the variance in the data that is on each component (see Figure 7-3):

```
pca |> plot()
```

The first thing you want to look at after making the transformation is how the variance is distributed along the components. If the first few components do not contain most of the variance, the transformation has done little for you. When it does, there is some hope that plotting the first few components will tell you about the data.

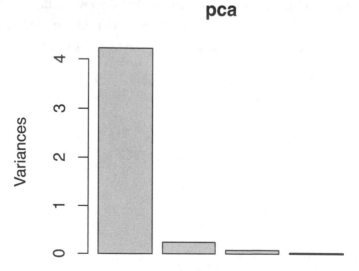

Figure 7-3. *Plot of the variance on each principal component for the iris data set*

To map the data to the new space spanned by the principal components, we use the `predict()` function:

```
mapped_iris <- pca |> predict(iris)
mapped_iris |> head()
```

```
##              PC1        PC2         PC3
## [1,] -2.684126 -0.3193972  0.02791483
## [2,] -2.714142  0.1770012  0.21046427
## [3,] -2.888991  0.1449494 -0.01790026
## [4,] -2.745343  0.3182990 -0.03155937
## [5,] -2.728717 -0.3267545 -0.09007924
## [6,] -2.280860 -0.7413304 -0.16867766
##              PC4
## [1,]  0.002262437
## [2,]  0.099026550
## [3,]  0.019968390
## [4,] -0.075575817
## [5,] -0.061258593
## [6,] -0.024200858
```

This can also be used with new data that wasn't used to create the `pca` object. Here, we just give it the same data we used before. We don't actually have to remove the `Species` variable; it will figure out which of the columns to use based on their names. We can now plot the first two components against each other (see Figure 7-4):

```
mapped_iris |>
    as_tibble() |>
    cbind(Species = iris$Species) |>
    ggplot() +
    geom_point(aes(x = PC1, y = PC2, colour = Species))
```

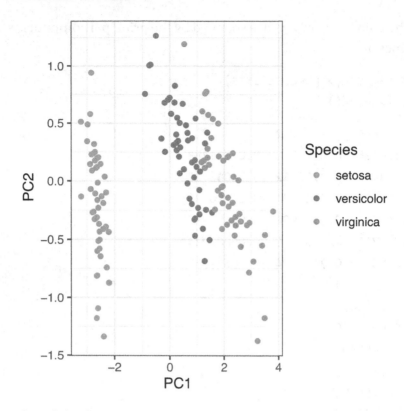

Figure 7-4. *Plot of the first two principal components for the iris data set*

The `mapped_iris` object returned from the `predict()` function is not a data frame but a matrix. That won't work with `ggplot()`, so we need to transform it back into a data frame, and we do that with `as_tibble`. This gives us the tibble variant of data frames, but we could also have used `as.data.frame` to give the classical data structure. Since we want to color the plot according to species, we need to add that information again—remember the `pca` object does not know about this factor data—so we do that with `cbind()`. After that, we plot.

We didn't gain much from this. There was about as much information in the original columns as there is in the transformed data. But now that we have seen PCA in action, we can try it out on a little more interesting example.

We will look at the `HouseVotes84` data from the `mlbench` package:

```
library(mlbench)
data(HouseVotes84)
HouseVotes84 |> head()
```

```
##           Class  V1 V2 V3   V4   V5 V6 V7 V8 V9 V10
## 1 republican      n  y  n    y    y  y  n  n  n   y
## 2 republican      n  y  n    y    y  y  n  n  n   n
## 3   democrat <NA>  y  y <NA>      y  y  n  n  n   n
## 4   democrat      n  y  y    n <NA>  y  n  n  n   n
## 5   democrat      y  y  y    n    y  y  n  n  n   n
## 6   democrat      n  y  y    n    y  y  n  n  n   n
##      V11 V12 V13 V14 V15 V16
## 1 <NA>   y   y   y   n   y
## 2    n   y   y   y   n <NA>
## 3    y   n   y   y   n   n
## 4    y   n   y   n   n   y
## 5    y <NA>  y   y   y   y
## 6    n   n   y   y   y   y
```

The data contains the votes cast for both republicans and democrats on 16 different proposals. The types of votes are yea, nay, and missing/unknown. Now, since votes are unlikely to be accidentally lost, missing data here means someone actively decided not to vote, so it isn't really missing. There is probably some information in that as well.

Now an interesting question we could ask is whether there are differences in voting patterns between republicans and democrats. We would expect that, but can we see it from the data?

The individual columns are binary (well, trinary if we consider the missing data as actually informative) and do not look very different between the two groups, so there is little information in each individual column. We can try doing a PCA on the data:

```
HouseVotes84 |> select(-Class) |> prcomp()
```

```
## Error in colMeans(x, na.rm = TRUE): 'x' must be numeric
```

Okay, R is complaining that the data isn't numeric. We know that PCA needs numeric data, but we are giving it factors. We need to change that, so we can try to map the votes into zeros and ones.

We want to mutate every column in the data frame, since all the columns are encoded as factors and we need them as numerical values. However, doing `mutate` (V1 = ..., V2 = ..., ...) for all the columns Vi is tedious, so let's not do it that way.

We can map over the columns of a data frame using the mapping functions from the package `purrr` that we have used earlier. Since we want to process all the columns (after removing `Class`), this looks like a good alternative. We want the result to be a data frame, so the `map_df` variant must be the one we want. For the function to apply to each column, we want each `vote` to map to 0 or 1, depending on whether it is n or y, but missing data should remain missing. Most R functions leave missing data alone, however, so if we do `ifelse(vote == "n", 0, 1)`, the NA values will remain NA. For `map_df`, then, we need a function `\(vote) ifelse(vote == "n", 0, 1)`. Here we go:

```
HouseVotes84 |>
    select(-Class) |>
    map_df(\(vote) ifelse(vote == "n", 0, 1)) |>
    prcomp()
```

```
## Error in svd(x, nu = 0, nv = k): infinite or missing values in 'x'
```

That doesn't work either, but now the problem is the missing data. We have mapped nay to 0 and yea to 1, but missing data remains missing data.

We should always think carefully about how we deal with missing data, especially in a case like this where it might actually be informative. One approach we could take is to translate each column into three binary columns indicating if a vote was cast as yea, nay, or not cast.

I have left that as an exercise. Here, I will just pretend that if someone abstained from voting, then they are equally likely to have voted yea or nay, and translate missing data into 0.5. This isn't true, and you shouldn't do that in a real data analysis, but we are interested in PCA and not truth here, so it is time to move on.

Since I want to map the data onto the principal components afterward, and since I don't want to write the data transformations twice, I save it in a variable and then perform the PCA:

```
vote_patterns <- HouseVotes84 |>
    select(-Class) |>
    map_df(\(vote) ifelse(vote == "n", 0, 1)) |>
    map_df(\(vote) ifelse(is.na(vote), 0.5, vote))

pca <- vote_patterns |> prcomp()
```

Now we can map the vote patterns onto the principal components and plot the first against the second (see Figure 7-5):

```
mapped_votes <- pca |> predict(vote_patterns)
mapped_votes |>
    as_tibble() |>
    cbind(Class = HouseVotes84$Class) |>
    ggplot() +
    geom_point(aes(x = PC1, y = PC2, colour = Class))
```

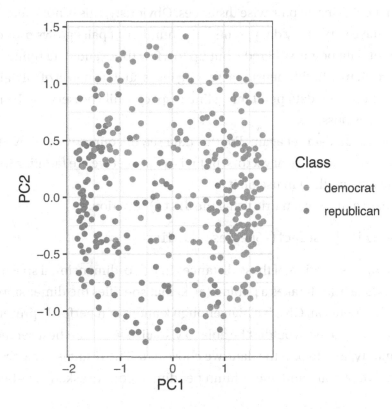

Figure 7-5. *Plot of the first two principal components for the house votes data set*

It looks like there is a clear separation in the voting patterns, at least on the first principal component. This is not something we could immediately see from the original data.

Multidimensional Scaling

Sometimes, it is easier to have a measure of distance between objects than representing them as numerical vectors. Consider, for example, strings. You could translate them into numbers based on their encoding, but the space of possible strings is vast—infinite if you do not restrict their length—so it is not a practical approach. However, there are many measures of how different two strings are. For strings, at least, it is easier to define a distance measure than a mapping into numeric values.

When what we have is a distance measure, we can represent our data as a distance matrix, one that contains all pair-wise distances. Obviously, this is not a feasible solution if you have very many data points—the number of pairs grows proportionally to the number of data points squared—but up to a few thousand data points, it is not a significant problem. Multidimensional scaling takes such a matrix of all pair-wise distances and maps each data point into a linear space while preserving the pair-wise distances as well as possible.

Consider the `iris` data set again. For this data set, of course, we do have the data points represented as numerical vectors, but it is a data set we are familiar with, so it is good to see the new method in use on it.

We can create a distance matrix using the `dist()` function:

```
iris_dist <- iris |> select(-Species) |> dist()
```

To create a representation of these distances in a two-dimensional space we use the function `cmdscale()`. It takes a parameter, k, that specifies the dimensionality we want to place the points in. Give it a high enough k and it can perfectly preserve all pair-wise distances, but we wouldn't be able to visualize it. We are best served with low dimensionality, and to plot the data, we chose two. The result is a matrix with one row per original data point and one column per dimension we asked for—here, of course, two:

```
mds_iris <- iris_dist |> cmdscale(k = 2)
mds_iris |> head()

##              [,1]        [,2]
## [1,] -2.684126   0.3193972
## [2,] -2.714142  -0.1770012
## [3,] -2.888991  -0.1449494
## [4,] -2.745343  -0.3182990
```

```
## [5,] -2.728717  0.3267545
## [6,] -2.280860  0.7413304
```

We can translate this matrix into a data frame and plot it (see Figure 7-6):

```
mds_iris |>
    as_tibble(.name_repair = ~ c("x", "y")) |>
    cbind(Species = iris$Species) |>
    ggplot() +
    geom_point(aes(x = x, y = y, colour = Species))
```

The as_tibble(.name_repair = ~ c("x", "y")) part translates the matrix we have into a data frame, as the other times where we have used as_tibble, but we need to do something special this time. The matrix doesn't have column names, and we need that for data frames. The .name_repair argument to as_tibble is a powerful tool for working with column names when you create a data frame, and all it can do is beyond the scope of this book. Here, we just set the column names to "x" and "y." You do need the tilde, ~, before the string vector, though, for this to work with as_tibble, but the details of what is going on here are more technical than we care to delve into at this point.

Because we gave the matrix column names, we can use them in aes(x = x, y = y, ...). If we hadn't given them names, we couldn't refer to them here.

The plot looks essentially the same as the PCA plot earlier, which is not a coincidence, except that it is upside down.

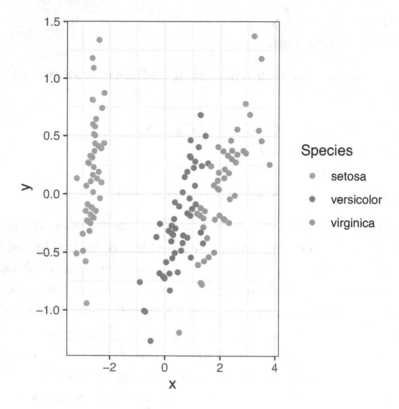

Figure 7-6. *Multidimensional scaling plot for iris data*

We can do exactly the same thing with the voting data—here, we can reuse the cleaned data that has translated the factors into numbers—and the result is shown in Figure 7-7.

```
mds_votes <- vote_patterns |> dist() |> cmdscale(k = 2)

mds_votes |>
  as_tibble(.name_repair = ~ c("x", "y")) |>
  cbind(Class = HouseVotes84$Class) |>
  ggplot() +
  geom_point(aes(x = x, y = y, colour = Class))
```

Should you ever have the need for computing a distance matrix between strings, by the way, you might want to look at the `stringdist` package. As an example illustrating this, we can simulate some strings. The following code first has a function for simulating random strings over the letters "A," "C," "G," and "T," and the second function then adds a random length to that. We then create ten strings using these functions:

```
random_ngram <- function(n)
  sample(c('A','C','G','T'), size = n, replace = TRUE) |>
  paste0(collapse = "")

random_string <- function(m) {
  n <- max(1, m + sample(c(-1,1), size = 1) * rgeom(1, 1/2))
  random_ngram(n)
}

strings <- replicate(10, random_string(5))
```

Using the stringdist package, we can compute the all-pairs distance matrix:

```
library(stringdist)
```

```
string_dist <- stringdistmatrix(strings)
```

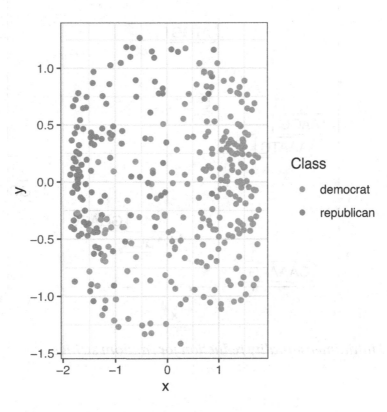

Figure 7-7. *Multidimensional scaling plot for house voting data*

We can now plot the strings in two-dimensional space, roughly preserving their distances (see Figure 7-8):

```
string_dist |>
  cmdscale(k = 2) |>
  as_tibble(.name_repair = ~ c("x", "y")) |>
  cbind(String = strings) |>
  ggplot(aes(x = x, y = y)) +
  geom_point() +
  geom_label(aes(label = String),
             hjust = 0, nudge_y = -0.1)
```

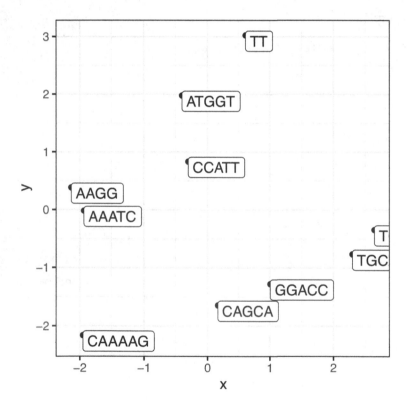

Figure 7-8. *Multidimensionality reduction for random strings*

Clustering

Clustering methods seek to find similarities between data points and group data according to these similarities. Such clusters can either have a hierarchical structure or not; when the structure is hierarchical, each data point will be associated with several clusters, ordered from the more specific to the more general, and when the structure is not hierarchical, any data point is typically only assigned a single cluster. The following are two of the most popular clustering algorithms, one of each kind of clustering.

k-means Clustering

In k-means clustering, you attempt to separate the data into k clusters, where the number k is determined by you. The data usually has to be in the form of numeric vectors. Strictly speaking, the method will work as long as you have a way of computing the mean of a set of data points and the distance between pairs of data points. The R function for k-means clustering, kmeans, wants numerical data.

The algorithm essentially works by first guessing at k "centers" of proposed clusters. Then each data point is assigned to the center it is closest to, creating a grouping of the data, and then all centers are moved to the mean position of their clusters. This is repeated until an equilibrium is reached. Because the initial centers are randomly chosen, different calls to the function will not necessarily lead to the same result. At the very least, expect the labelling of clusters to be different between the various calls.

Let us see it in action. We use the iris data set, and we remove the Species column to get a numerical matrix to give to the function:

```
clusters <- iris |>
select(-Species) |>
kmeans(centers = 3)
```

We need to specify k, the number of centers, in the parameters to kmeans(), and we choose three. We know that there are three species, so this is a natural choice. Life isn't always that simple, but here it is the obvious choice.

The function returns an object with information about the clustering. The two most interesting pieces of information are the centers, the variable centers, and the cluster assignment, the variable cluster.

Let us first have a look at the centers:

```
clusters$centers
```

```
##     Sepal.Length Sepal.Width Petal.Length
## 1       5.901613    2.748387     4.393548
## 2       6.850000    3.073684     5.742105
## 3       5.006000    3.428000     1.462000
##     Petal.Width
## 1      1.433871
## 2      2.071053
## 3      0.246000
```

These are simply vectors of the same form as the input data points. They are the center of mass for each of the three clusters we have computed.

The cluster assignment is simply an integer vector with a number for each data point specifying which cluster that data point is assigned to:

```
clusters$cluster |> head()
```

```
## [1] 3 3 3 3 3 3
```

```
clusters$cluster |> table()
```

```
##
##  1  2  3
## 62 38 50
```

There are 50 data points for each species, so if the clustering perfectly matched the species, we should see 50 points for each cluster as well. The clustering is not perfect, but we can try plotting the data and see how well the clustering matches up with the species class.

We can first plot how many data points from each species are assigned to each cluster (see Figure 7-9):

```
iris |>
    cbind(Cluster = clusters$cluster) |>
    ggplot() +
    geom_bar(aes(x = Species, fill = as.factor(Cluster)),
             position = "dodge") +
    scale_fill_discrete("Cluster")
```

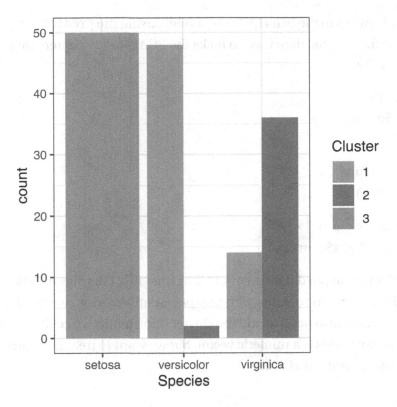

Figure 7-9. *Cluster assignments for the three iris species*

We first combine the `iris` data set with the cluster association from `clusters` and then make a bar plot. The `position` argument is `"dodge"`, so the cluster assignments are plotted next to each other instead of stacked on top of each other.

Not unexpectedly, from what we have learned of the data by plotting it earlier, Setosa seems clearly distinct from the other two species, which, according to the four measurements we have available at least, overlap in features.

There is a bit of luck involved here as well. A different starting point for where `kmeans()` placed the first centers will affect the final result, and had it put two clusters inside the cloud of the Setosa data points, it would have split those points into two clusters and merged the Versicolor and Virginia points into a single cluster, for instance.

It is always a good idea to visually examine how the clustering result matches where the actual data points fall. We can do this by plotting the individual data points and see how the classification and clustering looks. We could plot the points for any pair of features, but we have seen how to map the data onto principal components, so we could try to plot the data on the first two of these. As you remember, we can map data points from the four features to the principal components using the predict() function. This works both for the original data used to make the PCA and the centers we get from the *k*-means clustering:

```
pca <- iris |>
   select(-Species) |>
   prcomp()

mapped_iris <- pca |>
   predict(iris)

mapped_centers <- pca |>
   predict(clusters$centers)
```

We can plot the mapped data points, PC1 against PC2 (see Figure 7-10). To display the principal components together with the species information, we need to add a Species column. We also need to add the cluster information since that isn't included in the mapped vectors. This is a numeric vector, but we want to treat it as categorical, so we need to translate it using as.factor():

```
mapped_iris |>
   as_tibble() |>
   cbind(Species = iris$Species,
         Clusters = as.factor(clusters$cluster)) |>
   ggplot() +
   geom_point(aes(x = PC1, y = PC2,
           colour = Species, shape = Clusters)) +
geom_point(aes(x = PC1, y = PC2), size = 5, shape = "X",
      data = as_tibble(mapped_centers))
```

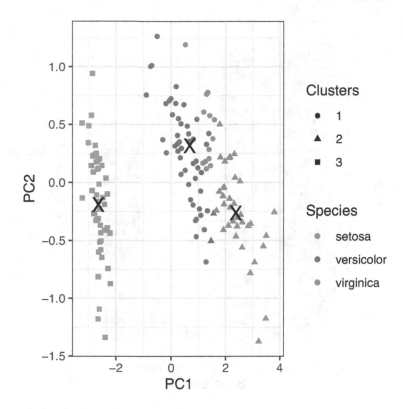

Figure 7-10. *Clusters and species for iris*

In the plot, I also show the centers. I use the `data` argument to `geom_point()` to give it this data, and I set the size to 5 and set the shape to "X".

As mentioned, there is some luck involved in getting a good clustering like this. The result of a second run of the `kmeans()` function is shown in Figures 7-11 and 7-12.

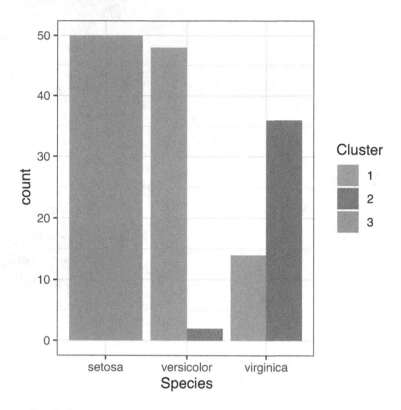

Figure 7-11. *A bad cluster assignment for the three iris species*

If you go back and look at Figure 7-10 and think that some of the square points are closer to the center of the "triangular cluster" than the center of the "square cluster," or vice versa, you are right. Don't be too disturbed by this; two things are deceptive here. One is that the axes are not on the same scale, so distances along the x-axis are farther than distances along the y-axis. A second is that the distances used to group data points are in the four-dimensional space of the original features, while the plot is a projection onto the two-dimensional plane of the first two principal components.

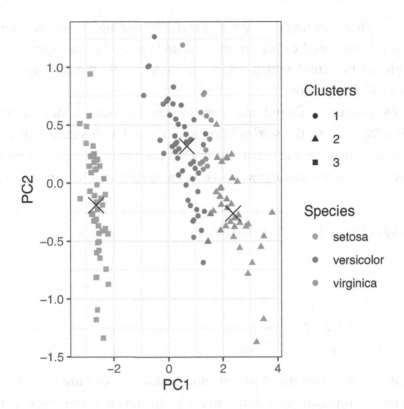

Figure 7-12. *Clusters and species for iris for a bad clustering*

There is something to worry about, though, concerning distances. The algorithm is based on the distance from cluster centers to data points, but if you have one axis in centimeters and another in meters, a distance along one axis is numerically a hundred times farther than along the other. This is not merely solved by representing all features in the same unit. First of all, that isn't always possible. There is no meaningful way of translating time or weight into a distance. Even if it was, what is being measured is also relevant for the unit we consider. The height of a person is meaningfully measured in meters, but you do not want something like cell size to be measured in meters.

This is also an issue for principal component analysis. Obviously, a method that tries to create a vector space basis based on the variance in the data is going to be affected by the units used in the input data. The usual solution is to rescale all input features so they are centered at zero and have variance one. You subtract from each data point the mean of the feature and divide by the standard deviation. This means that measured in standard deviations, all dimensions have the same variation.

The prcomp() function takes parameters to do the scaling. Parameter center, which defaults to TRUE, translates the data points to mean zero, and parameter scale. (notice the "." at the end of the name), which defaults to FALSE, scales the data points to have variance one at all dimensions.

The kmeans() functions do not take these parameters, but you can explicitly rescale a numerical data frame using the scale() function. I have left this as an exercise.

Now let us consider how the clustering does at predicting the species more formally. This returns us to familiar territory: we can build a confusion matrix between species and clusters.

```
table(iris$Species, clusters$cluster)
```

```
##
##               1  2  3
##   setosa      0  0 50
##   versicolor 48  2  0
##   virginica  14 36  0
```

One problem here is that the clustering doesn't know about the species, so even if there were a one-to-one correspondence between clusters and species, the confusion matrix would only be diagonal if the clusters and species were in the same order.

We can associate each species to the cluster most of its members are assigned to. This isn't a perfect solution—two species could be assigned to the same cluster this way, and we still wouldn't be able to construct a confusion matrix—but it will work for us in the case we consider here. We can count how many observations from each cluster are seen in each species like this:

```
tbl <- table(iris$Species, clusters$cluster)
(counts <- apply(tbl, 1, which.max))
```

```
## setosa versicolor virginica
##      3          1          2
```

and build a table mapping species to clusters to get the confusion matrix like this:

```
map <- rep(NA, each = 3)
map[counts] <- names(counts)
table(iris$Species, map[clusters$cluster])
```

```
##
##            setosa versicolor virginica
##   setosa       50          0         0
##   versicolor    0         48         2
##   virginica     0         14        36
```

A final word on k-means is this: Since k is a parameter that needs to be specified, how do you pick it? Here, we knew that there were three species, so we picked three for k as well. But when we don't know if there is any clustering in the data, to begin with, or if there is a lot, how do we choose k? Unfortunately, there isn't a general answer to this. There are several rules of thumb, but no perfect solution you can always apply. Try some, see what happens, and then use your understanding of the data to interpret what you see.

Hierarchical Clustering

Hierarchical clustering is a technique you can use when you have a distance matrix of your data. Here, the idea is that you build up a tree structure of nested clusters by iteratively merging clusters. You start with putting each data point in their own singleton clusters. Then iteratively you find two clusters that are close together and merge them into a new cluster. You continue this until all data points are in the same large cluster. Different algorithms exist, and they mainly vary in how they choose which cluster to merge next and how they compute the distance between clusters. In R, the function hclust() implements several algorithms—the parameter method determines which is used—and we can see it in use with the iris data set. We first need a distance matrix. This time, I first scale the data:

```
iris_dist <- iris |> select(-Species) |> scale() |> dist()
```

Now the clustering is constructed by calling hclust() on the distance matrix:

```
clustering <- hclust(iris_dist)
```

We can plot the result using the generic plot() function; see Figure 7-13. There is not much control over how the clustering is displayed using this function, but if you are interested in plotting trees, you should have a look at the ape package:

```
plot(clustering)
```

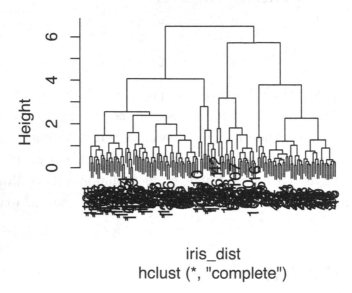

Figure 7-13. *Hierarchical clustering of iris data*

To create plots that work well with `ggplot2` graphics, you want the package `ggdendro`, see Figure 7-14:

```
library(ggdendro)

clustering |>
  ggdendrogram() + theme_dendro()
```

Using `ggdendro`, you can get access to the raw plotting segments which gives you control over much of the visualization of the tree.

Only visualizing the clustering is rarely enough, so to work with the result we need to be able to extract the actual groupings. The `cutree()` function—it stands for cut tree, but there is only one t in the name—lets us do this. You can give it a parameter h to cut the tree into clusters by splitting the tree at height h, or you can give it parameter k to cut the tree at the level where there is exactly k clusters.

Since we are working with the `iris` data, it is natural for us to want to split the data into three clusters:

```
clusters <- clustering |> cutree(k = 3)
```

The result is in the same format as we had for *k*-means clustering, that is, a vector with integers specifying which cluster each data point belongs to. Since we have the information in the familiar format, we can try plotting the clustering information as a bar plot (Figure 7-15):

```
iris |>
    cbind(Cluster = clusters) |>
    ggplot() +
    geom_bar(aes(x = Species, fill = as.factor(Cluster)),
             position = "dodge") +
    scale_fill_discrete("Cluster")
```

or plot the individual plots together with species and cluster information (Figure 7-16).

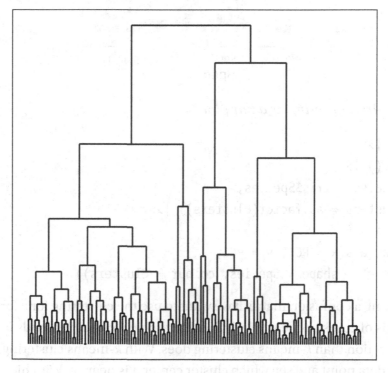

Figure 7-14. *Hierarchical clustering of iris data plotted with ggdendro*

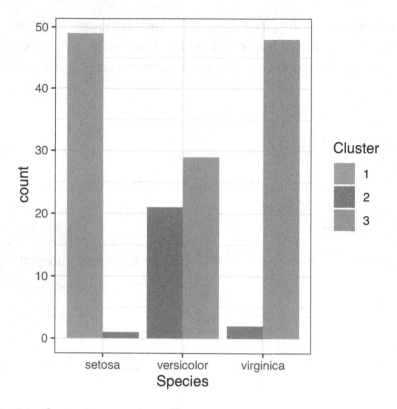

Figure 7-15. *Iris clustering as a bar plot*

```
mapped_iris |>
  as_tibble() |>
  cbind(Species = iris$Species,
        Clusters = as.factor(clusters)) |>
  ggplot() +
  geom_point(aes(x = PC1, y = PC2,
                 shape = Species, colour = Clusters))
```

Constructing a confusion matrix if we want to use the clustering for a form of classification is of course done similarly, but hierarchical clustering lends itself much less to classification than k-means clustering does. With k-means clustering, it is simple to take a new data point and see which cluster center it is nearest. With hierarchical clustering, you would need to rebuild the entire tree to see where it falls.

Association Rules

The last unsupervised learning method we will see is aimed at categorical data, ordered or unordered. Just like you have to translate factors into numerical data to use methods such as PCA, you will need to translate numerical data into factors to use association rules. This typically isn't a problem, and you can use the function cut() to split a numerical vector into a factor and combine it with ordered() if you want it ordered.

Association rules search for patterns in your data by picking out subsets of the data, X and Y, based on predicates on the input variables and evaluate rules $X \Rightarrow Y$. Picking X and Y is a brute-force choice (which is why you need to break the numerical vectors into discrete classes).[1]

Any statement $X \Rightarrow Y$ is called a rule, and the algorithm evaluates all rules (at least up to a certain size) to figure out how good each rule is.

The association rules algorithm is implemented in the arules package:

```
library(arules)
```

---

[1] The algorithm could do it for you by considering each point between two input values, but it doesn't, so you have to break the data.

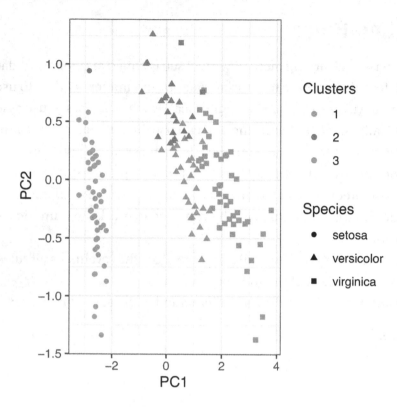

Figure 7-16. *Iris points plotted with species and hierarchical clustering information*

To see it in action, we use the income data set from the kernlab package:

```
library(kernlab)

data(income)
income |> head()
```

```
##          INCOME     SEX MARITAL.STATUS    AGE
## 1    [75.000-        F        Married 45-54
## 2    [75.000-        M        Married 45-54
## 3    [75.000-        F        Married 25-34
## 4    -10.000)        F         Single 14-17
## 5    -10.000)        F         Single 14-17
## 6 [50.000-75.000)    M        Married 55-64
```

```
##                    EDUCATION                    OCCUPATION
## 1 1 to 3 years of college               Homemaker
## 2          College graduate               Homemaker
## 3          College graduate  Professional/Managerial
## 4            Grades 9 to 11    Student, HS or College
## 5            Grades 9 to 11    Student, HS or College
## 6 1 to 3 years of college                 Retired
##          AREA   DUAL.INCOMES HOUSEHOLD.SIZE UNDER18
## 1 10+ years              No          Three   None
## 2 10+ years              No           Five    Two
## 3 10+ years             Yes          Three    One
## 4 10+ years     Not Married           Four    Two
## 5 4-6 years     Not Married           Four    Two
## 6 10+ years              No            Two   None
##    HOUSEHOLDER  HOME.TYPE ETHNIC.CLASS  LANGUAGE
## 1         Own      House         White      <NA>
## 2         Own      House         White   English
## 3        Rent  Apartment         White   English
## 4      Family      House         White   English
## 5      Family      House         White   English
## 6         Own      House         White   English
```

This data contains income information together with several explanatory variables and is already in a form the arules can deal with: all columns are factorial.

The same data is actually also available in the arules package as the Income data set, but here it is representing in a different format than a data frame, so we will use this version of the data:

```
data(Income)
Income |> head()
```

```
## transactions in sparse format with
##  6 transactions (rows) and
##  50 items (columns)
```

To construct the rules, we use the `apriori()` function. It takes various arguments for controlling which rules the function will return, but we can use it with all default parameters:

```
rules <- income |> apriori()

## Apriori
##
## Parameter specification:
##  confidence minval smax arem  aval
##         0.8    0.1    1 none FALSE
##  originalSupport maxtime support minlen maxlen
##             TRUE       5     0.1      1     10
##  target  ext
##   rules TRUE
##
## Algorithmic control:
##  filter tree heap memopt load sort verbose
##     0.1 TRUE TRUE  FALSE TRUE    2    TRUE
##
## Absolute minimum support count: 899
##
## set item appearances ...[0 item(s)] done [0.00s].
## set transactions ...[84 item(s), 8993 transaction(s)] done [0.01s].
## sorting and recoding items ... [42 item(s)] done [0.00s].
## creating transaction tree ... done [0.00s].
## checking subsets of size 1 2 3 4 5 6 done [0.02s].
## writing ... [785 rule(s)] done [0.00s].
## creating S4 object  ... done [0.00s].
```

The `rules` object we create this way is not a simple object like a data frame, but it will let us take the `head()` of it, and we can use the function `inspect()` to see the individual rules:

```
rules |> head() |> inspect(linebreak = FALSE)

##      lhs
## [1] {}                             =>
```

```
## [2] {EDUCATION=Grad Study}             =>
## [3] {OCCUPATION=Clerical/Service Worker} =>
## [4] {INCOME=[30.000-40.000)}           =>
## [5] {UNDER18=Two}                      =>
## [6] {INCOME=[50.000-75.000)}           =>
##     rhs                  support    confidence
## [1] {LANGUAGE=English} 0.8666741 0.8666741
## [2] {LANGUAGE=English} 0.1000778 0.9316770
## [3] {LANGUAGE=English} 0.1046369 0.8860640
## [4] {LANGUAGE=English} 0.1111976 0.9009009
## [5] {LANGUAGE=English} 0.1073057 0.8405923
## [6] {LANGUAGE=English} 0.1329923 0.9143731
##     coverage  lift      count
## [1] 1.0000000 1.0000000 7794
## [2] 0.1074169 1.0750027  900
## [3] 0.1180918 1.0223728  941
## [4] 0.1234293 1.0394921 1000
## [5] 0.1276548 0.9699059  965
## [6] 0.1454465 1.0550368 1196
```

The `linebreak` = FALSE here splits the rules over several lines. I find it confusing too that to break the lines you have to set `linebreak` to FALSE, but that is how it is.

Each rule has a right-hand side, `rhs`, and a left-hand side, `lhs`. For a rule $X \Rightarrow Y$, X is the `rhs` and Y the `lhs`. The quality of a rule is measured by the following three columns:

- support: The fraction of the data where both X and Y hold true. Think of it as $\Pr(X, Y)$.

- confidence: The fraction of times where X is true that Y is also true. Think of it as $\Pr(Y \mid X)$.

- lift: How much better than random is the rule, in the sense that how much better is it compared to X and Y being independent. Think $\Pr(X, Y) / \Pr(X) \Pr(Y)$.

Good rules should have high enough support to be interesting—if a rule only affects a tiny number of data points out of the whole data, it probably isn't that important—so you want both support and confidence to be high. It should also tell you more than what you would expect by random chance, which is captured by lift.

You can use the sort() function to rearrange the data according to the quality measures:

```
rules |>
sort(by = "lift") |> head() |>
inspect(linebreak = FALSE)
```

```
##      lhs
## [1] {MARITAL.STATUS=Married, OCCUPATION=Professional/Managerial,
         LANGUAGE=
## [2] {MARITAL.STATUS=Married, OCCUPATION=Professional/Managerial}
## [3] {DUAL.INCOMES=No, HOUSEHOLDER=Own}
## [4] {AREA=10+ years, DUAL.INCOMES=Yes, HOME.TYPE=House}
## [5] {DUAL.INCOMES=Yes, HOUSEHOLDER=Own, HOME.TYPE=House,
         LANGUAGE=English}
## [6] {DUAL.INCOMES=Yes, HOUSEHOLDER=Own, HOME.TYPE=House}
##   rhs     support
## [1] => {DUAL.INCOMES=Yes}    0.1091960
## [2] => {DUAL.INCOMES=Yes}    0.1176471
## [3] => {MARITAL.STATUS=Married} 0.1016346
## [4] => {MARITAL.STATUS=Married} 0.1003002
## [5] => {MARITAL.STATUS=Married} 0.1098632
## [6] => {MARITAL.STATUS=Married} 0.1209830
##      confidence  coverage   lift     count
## [1] 0.8069022   0.1353275 3.281986   982
## [2] 0.8033409   0.1464472 3.267501 1058
## [3] 0.9713071   0.1046369 2.619965   914
## [4] 0.9605964   0.1044145 2.591075   902
## [5] 0.9601555   0.1144223 2.589886   988
## [6] 0.9594356   0.1260981 2.587944 1088
```

You can combine this with the subset() function to filter the rules:

```
rules |>
    subset(support > 0.5) |>
    sort(by = "lift") |>
    head() |>
    inspect(linebreak = FALSE)
```

```
##    lhs
## [1] {ETHNIC.CLASS=White}     =>
## [2] {AREA=10+ years}     =>
## [3] {UNDER18=None}     =>
## [4] {}   =>
## [5] {DUAL.INCOMES=Not Married} =>
##    rhs                 support   confidence
## [1] {LANGUAGE=English} 0.6110308 0.9456204
## [2] {LANGUAGE=English} 0.5098410 0.8847935
## [3] {LANGUAGE=English} 0.5609919 0.8813767
## [4] {LANGUAGE=English} 0.8666741 0.8666741
## [5] {LANGUAGE=English} 0.5207384 0.8611622
##      coverage   lift   count
## [1] 0.6461692 1.0910911 5495
## [2] 0.5762260 1.0209069 4585
## [3] 0.6364951 1.0169644 5045
## [4] 1.0000000 1.0000000 7794
## [5] 0.6046925 0.9936402 4683
```

Exercises

Dealing with Missing Data in the HouseVotes84 Data

In the PCA analysis, we translated missing data into 0.5. This was to move things along but probably not an appropriate decision. People who do not cast a vote are not necessarily undecided and therefore equally likely to vote yea or nay; there can be conflicts of interests or other reasons. So we should instead translate each column into three binary columns.

You can use the transmute() function from dplyr to add new columns and remove old ones—it is a bit of typing since you have to do it 16 times, but it will get the job done.

If you feel more like trying to code your way out of this transformation, you should look at the mutate_at() function from dplyr. You can combine it with column name matches and multiple functions to build the three binary vectors (for the ifelse() calls, you have to remember that comparing with NA always gives you NA, so you need always to check for that first). After you have created the new columns, you can remove the old ones using select() combined with match().

Try to do the transformation and then the PCA again. Does anything change?

k-means

Rescaling for k-means clustering

Use the scale() function to rescale the iris data set, then redo the *k*-means clustering analysis.

Varying k

Analyze the iris data with kmeans() with *k* ranging from one to ten. Plot the clusters for each *k*, coloring the data points according to the clustering.

## CHAPTER 8

Project 1: Hitting the Bottle

To see a data analysis in action, I will use an analysis that my student, Dan Søndergaard, did the first year I held my data science class. I liked his analysis so much that I wanted to include it in the book. I am redoing his analysis in the following with his permission.

The data contains physicochemical features measured from Portuguese Vinho Verde wines, and the goal was to try to predict wine quality from these measurements.[1]

Importing Data

If we go to the data folder, we can see that the data is split into three files: the measurements from red wine, white wine, and a description of the data (the file `winequality.names`). To avoid showing large URLs, I will not list the code for reading the files, but it is in the form

```
read.table(URL, header=TRUE, sep=';')
```

That there is a header that describes the columns, and that fields are separated by semicolons we get from looking at the files.

We load the red and white wine data into separate data frames called `red` and `white`.

We can combine the two data frames using

```
wines <- bind_rows(tibble(type = "red", red),
                   tibble(type = "white", white))
```

---

[1] https://archive.ics.uci.edu/ml/datasets/Wine+Quality

© Thomas Mailund 2022
T. Mailund, *Beginning Data Science in R 4*, https://doi.org/10.1007/978-1-4842-8155-0_8

The `tibble` and `bind_rows` functions are from the `dplyr` package. We can now have a look at the summary with

```
summary(wines)
```

You will see that there are 11 measurements for each wine, and each wine has an associated quality score based on sensory data. At least three wine experts judged and scored the wine on a scale between zero and ten. No wine achieved a score below three or above nine. There are no missing values. There is not really any measurement that we want to translate into categorical data. The quality scores are given as discrete values, but they are ordered categories, and we might as well consider them as numerical values for now.

Exploring the Data

With the data loaded, we first want to do some exploratory analysis to get a feeling for it.

Distribution of Quality Scores

The first thing Dan did was to look at the distribution of quality scores for both types of wine (see Figure 8-1):

```
ggplot(wines) +
  geom_bar(aes(x = factor(quality), fill = type),
           position = 'dodge') +
  xlab('Quality') + ylab('Frequency')
```

There are very few wines with extremely low or high scores. The quality scores also seem normal distributed, if we ignore that they are discrete. This might make the analysis easier.

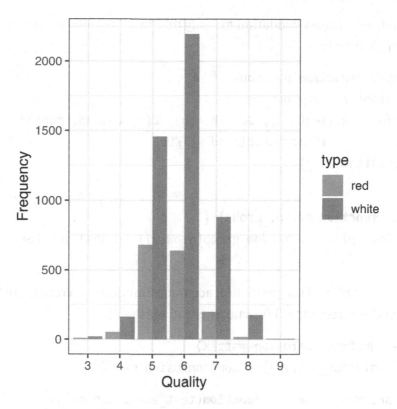

Figure 8-1. *Distribution of wine qualities*

Is This Wine Red or White?

The data set has two types of wine: red and white. As Dan noticed, these types are typically described by very different words by wine experts, but several experiments[2] have shown that even the best wine experts cannot distinguish red from white if the color is obscured or the experts are blindfolded. It is, therefore, interesting to see if the physicochemical features available in the data can help decide whether a wine is red or white.

Dan used the Naive Bayes method to explore this, so we need the e1071 package:

```
library(e1071)
```

---

[2] http://io9.com/wine-tasting-is-bullshit-heres-why-496098276

He used a fivefold cross-validation to study this, but I will just use the `partition()` function from Chapter 6:

```
random_group <- function(n, probs) {
  probs <- probs / sum(probs)
  g <- findInterval(seq(0, 1, length = n), c(0, cumsum(probs)),
                    rightmost.closed = TRUE)
  names(probs)[sample(g)]
}

partition <- function(df, n, probs) {
  replicate(n, split(df, random_group(nrow(df), probs)), FALSE)
}
```

and I will use a variation of the prediction accuracy function we wrote there for cars but using wines and the `accuracy()` function instead of `rmse()`:

```
accuracy <- function(confusion_matrix)
  sum(diag(confusion_matrix)) / sum(confusion_matrix)

prediction_accuracy_wines <- function(test_and_training) {
  test_and_training |>
    map_dbl(
      \(tt) {
        # Fit the model using training data
        model <- naiveBayes(type ~ ., data = tt$training)
        # Then make predictions on the test data
        predictions <- predict(model, newdata = tt$test)
        # Get accurracy of predictions
        accuracy(table(tt$test$type, predictions))
      }
    )
}
```

The formula `type ~ .` specifies that we wish to build a model for predicting type using all the remaining variables in the data, specified as “.”.

We get the following accuracy if we split the data randomly into training and test data 50/50:

```
random_wines <- wines |>
    partition(4, c(training = 0.5, test = 0.5))
random_wines |> prediction_accuracy_wines()
```

```
## [1] 0.9781471 0.9735303 0.9772238 0.9735303
```

This is a pretty good accuracy, so this raises the question of why experts cannot tell red and white wines apart.

Dan looked into this by determining the most significant features that divide red and white wines by building a decision tree:

```
library('party')
```

In `party`, we have a function, `ctree`, for building a decision tree. We still want to predict `type` based on all the other variables, so we want the formula `type ~ .`, but the `ctree` function won't like the `wines` data out of the box. It doesn't want to predict a variable that contains strings, so we must translate the strings in `type` into a factor first. No problem, we know how to do that; we can use `mutate` and the expression

```
wines |> mutate(type = as.factor(type))
```

and then fit the tree with this command:

```
tree <- ctree(type ~ ., data = wines |> mutate(type = as.factor(type)),
              control = ctree_control(minsplit = 4420))
```

(The `control` option is just for adjusting how the function should build the decision tree; you can experiment with it if you wish.)

The plot of the tree is too large for me to show here in the book with the size limit for figures, but try to plot it yourself.

He limited the number of splits made to get only the most important features. From the tree, we see that the total amount of sulfur dioxide, a chemical compound often added to wines to prevent oxidation and bacterial activity, which may ruin the wine, is chosen as the root split.

Sulfur dioxide is also naturally present in wines in moderate amounts. In the EU, the quantity of sulfur dioxide is restricted to 160 ppm for red wines and 210 ppm for white wines, so by law, we actually expect a significant difference of sulfur dioxide in the two types of wine. So Dan looked into that:

```
wines |>
  group_by(type) |>
  summarise(total.mean = mean(total.sulfur.dioxide),
            total.sd = sd(total.sulfur.dioxide),
            free.mean = mean(free.sulfur.dioxide),
            free.sd = sd(free.sulfur.dioxide),
            .groups = "drop")
```

```
## # A tibble: 2 × 5
##   type  total.mean total.sd free.mean free.sd
##   <chr>      <dbl>    <dbl>     <dbl>   <dbl>
## 1 red         46.5     32.9      15.9    10.5
## 2 white      138.      42.5      35.3    17.0
```

The average amount of total sulfur dioxide is indeed lower in red wines, and thus it makes sense that this feature is picked as a significant feature in the tree. If the amount of total sulfur dioxide in a wine is less than or equal to 67 ppm, we can say that it is a red wine with high certainty, which also fits with the summary statistics earlier.

Another significant feature suggested by the tree is the volatile acidity, also known as the vinegar taint. In finished (bottled) wine, a high volatile acidity is often caused by malicious bacterial activity, which can be limited by the use of sulfur dioxide as described earlier. Therefore, we expect a strong relationship between these features (see Figure 8-2):

```
qplot(total.sulfur.dioxide, volatile.acidity, data = wines,
      color = type,
      xlab = 'Total sulfur dioxide',
      ylab = 'Volatile acidity (VA)')
```

The plot shows the amount of volatile acidity as a function of the amount of sulfur dioxide. It also shows that, especially for red wines, the volatile acidity is low for wines with a high amount of sulfur dioxide. The pattern for white wine is not as clear.

However, Dan observed, as you can clearly see in the plot, a clear difference between red and white wines when considering the `total.sulfur.dioxide` and `volatile.acidity` features together.

So why can humans not taste the difference between red and white wines? It turns out that[3] sulfur dioxide cannot be detected by humans in free concentrations of less than 50 ppm. Although the difference in total sulfur dioxide is very significant between the two types of wine, the free amount is on average below the detection threshold, and thus humans cannot use it to distinguish between red and white.

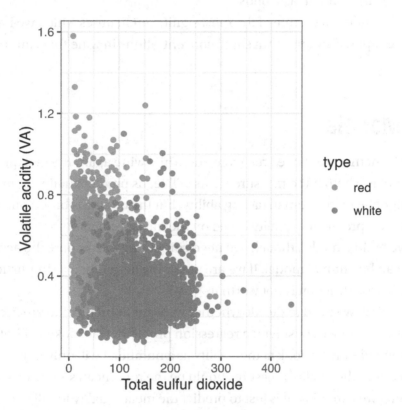

Figure 8-2. *Sulfur dioxide vs. volatile acidity*

```
wines |>
  group_by(type) |>
  summarise(mean = mean(volatile.acidity),
            sd = sd(volatile.acidity),
            .groups = "drop")
```

---

[3] http://en.wikipedia.org/wiki/Sulfur_dioxide#In_winemaking

```
## # A tibble: 2 × 3
##   type   mean    sd
##   <chr> <dbl> <dbl>
## 1 red    0.528 0.179
## 2 white 0.278 0.101
```

Similarly, acetic acid (which causes volatile acidity) has a detection threshold of 0.7 g/L, and again we see that the average amount is below this threshold and thus is undetectable by the human taste buds.

So Dan concluded that some of the most significant features which we have found to tell the types apart only appear in small concentrations in wine that cannot be tasted by humans.

Fitting Models

Regardless of whether we can tell red wine and white wine apart, the real question we want to explore is whether the measurements will let us predict quality. Some of the measures might be below human tasting ability, but the quality is based on human tasters, so can we predict the quality based on the measurements?

Before we build a model, though, we need something to compare the accuracy against that can be our null model. If we are not doing better than a simplistic model, then the model construction is not worth it.

Of course, first, we need to decide whether we want to predict the precise quality as categories or whether we consider it a regression problem. Dan looked at both options, but since we should mostly look at the quality as a number, I will only look at the latter.

For regression, the quality measure should be the root mean square error, and the simplest model we can think of is just to predict the mean quality for all wines:

```
rmse <- function(x,t) sqrt(mean(sum((t - x)^2)))

wines |>
  # predict the mean for all the wines, regardless of
  # parameters
  mutate(null_prediction = mean(quality)) |>
  # Summerise the predictions with a root mean square error
  summarise(rmse = rmse(null_prediction, quality)) |>
  # We have a data frame with a single number now, just
```

```
  # get that number
  as.numeric()
```

```
## [1] 70.38242
```

This is what we have to beat to have any model worth considering.

We do want to compare models with training and test data sets, though, so we do not use the mean for the entire data. So we need a function for comparing the results with split data.

To compare different models using rmse() as the quality measure, we need to modify our prediction accuracy function. We can give it as parameter the function used to create a model that works with predictions. It could look like this:

```
prediction_accuracy_wines <- function(test_and_training,
                                      model_function) {
  test_and_training |>
  map_dbl(
    \(tt) {
      # Fit the model using training data
      model <- model_function(quality ~ ., data = tt$training)
      # Then make predictions on the test data
      predictions <- predict(model, newdata = tt$test)
      # Get accuracy of predictions as a root mean square error
      rmse(predictions, tt$test$quality)
    }
  )
}
```

Here, we are hardwiring the formula to include all variables except for quality which is potentially leading to overfitting, but we are not worried about that right now.

To get this to work, we need a model function() that returns an object that works with predict(). To get this to work, we need to use generic functions, something we will not cover until later, but it mostly involves creating a so-called "class" and defining what predict() will do on objects of that class:

```
null_model <- function(formula, data) {
  # Here we just remember the mean of the input by putting it in a list
  # and by wrapping it in a `structure` with class "null_model" we can
```

```
  # define we want this model to make predictions
  structure(list(mean = mean(data$quality)),
          class = "null_model")
}

# The name predict.null_model says that if you call predict()
# on something with class "null_model", it is this function
# that R will call. Since "model" is the list we made above
# we can get the prediction by looking up "mean" in the object.
predict.null_model <- function(model, newdata) {
  rep(model$mean, each = nrow(newdata))
}
```

This null_model() function creates an object of class null_model and defines what the predict() function should do on objects of this class. We can use it to test how well the null model will perform on data:

```
test_and_training <- wines |>
    partition(4, c(training = 0.5, test = 0.5))
test_and_training |> prediction_accuracy_wines(null_model)
```

```
## [1] 48.85466 49.31583 49.74809 49.85921
```

Don't be too confused about these numbers being much better than the one we get if we use the entire data set. That is simply because the rmse() function will always give a larger value if there is more data, and we are giving it only half the data that we did when we looked at the entire data set.

We can instead compare it with a simple linear model:

```
test_and_training |> prediction_accuracy_wines(lm)
```

```
## [1] 41.13091 42.22651 41.71150 42.17854
```

Dan also tried different models for testing the prediction accuracy, but I have left that as an exercise. You can use this data set to explore the various methods we have seen in the last two chapters.

Exercises
Exploring Other Formulas

The `prediction_accuracy_wines()` function is hardwired to use the formula `quality ~ .` that uses all explanatory variables. Using all variables can lead to overfitting, so it is possible that using fewer variables can give better results on the test data. Add a parameter to the function for the formula and explore using different formulas.

Exploring Different Models

Try using different models than the null model and the linear model. Any model that can do regression and defines a `predict()` function should be applicable. Try it out.

Analyzing Your Own Data Set

Find a data set you are interested in investigating and go for it. To learn how to interpret data, you must use your intuition on what is worth exploring, and the only way to build that intuition is to analyze data.

## CHAPTER 9

Deeper into R Programming

In this chapter, we leave data analysis and return to programming and software development, topics that are the focus of the remaining chapters of the book. In the first chapter, you took a tutorial introduction to R programming, but we left out a lot of details. This chapter will cover many of those details, while the next two chapters will cover more advanced aspects of R programming: functional programming and object-oriented programming.

Expressions

We begin the chapter by going back to expressions. Everything we do in R involves evaluating expressions. Most expressions we evaluate to do a computation and get the result, but some expressions have side effects—like assignments—and those we usually evaluate because of the side effects.

Arithmetic Expressions

We saw the arithmetic expressions already, so we will just give a very short reminder here. The arithmetic expressions are operators that involve numbers and consist of the unary operators + and -:

```
+ x
- x
```

where + doesn't really do anything, while - changes the sign of its operand. Then there are the infix operators for addition, subtraction, multiplication, and division:

© Thomas Mailund 2022
T. Mailund, *Beginning Data Science in R 4*, https://doi.org/10.1007/978-1-4842-8155-0_9

```
x + y
x - y
x * y
x / y
```

Division will return a floating-point number even if both its operands are integers, so if you want to do integer division, you need the special operator for that:

```
x %/% y
```

If you want the remainder of integer division, you need this infix operator instead:

```
x %% y
```

Finally, there are operators for exponentiation. To compute x^y, you can use either of these two operators:

```
x ^ y
x ** y
```

In all these examples, x and y can be numbers or variables referring to numbers (actually, vectors of numbers since R always works on vectors), or they can be other expressions evaluating to numbers. If you compose expressions from infix operators, you have the same precedence rules you know from arithmetic. Exponentiation goes before multiplication that goes before addition, for example. This means that you will need to use parentheses if you need to evaluate the expressions in another order.

Since the rules are the ones you are used to, this is not likely to cause you troubles, except if you combine these expressions with the operator :. This isn't really an arithmetic operator, but it is an infix operator for generating sequences, and it has a higher precedence than multiplication but lower than exponentiation. This means that 1:2**2 will evaluate the 2**2 expression first to get 1:4 and then construct the sequence:

```
1:2**2
```

```
## [1] 1 2 3 4
```

while the expression 1:2*2 will evaluate the : expression first to create a vector containing 1 and 2 and then multiply this vector with 2:

```
1:2*2
```

```
## [1] 2 4
```

Since the unary - operator has higher precedence than :, it also means that -1:2 will give you the sequence from -1 to 2 and not the sequence containing -1 and -2. For that, you need parentheses:

```
-1:2
```

```
## [1] -1 0 1 2
```

```
-(1:2)
```

```
## [1] -1 -2
```

Functions are evaluated before the operators:

```
1:sqrt(4)
```

```
## [1] 1 2
```

Boolean Expressions

For boolean values—those that are either TRUE or FALSE—you also have logical operators. The operator ! negates a value:

```
!TRUE
```

```
## [1] FALSE
```

```
!FALSE
```

```
## [1] TRUE
```

and | and || are logical "or" operators, while & and && are logical "and" operators. The difference between | and || or & and && is how they deal with vectors. The one-character version will apply the operator element-wise and create a vector, while the two-character version will only look at the first value in vectors:

```
TRUE | FALSE
```

```
## [1] TRUE
```

```
FALSE | FALSE
```

```
## [1] FALSE
```

```
TRUE || FALSE
```

```
## [1] TRUE
```

```
FALSE || FALSE
```

```
## [1] FALSE
```

```
x <- c(TRUE, FALSE, TRUE, FALSE)
y <- c(TRUE, TRUE, FALSE, FALSE)
```

```
x | y
```

```
## [1] TRUE TRUE TRUE FALSE
```

```
x || y
```

```
## [1] TRUE
```

```
x & y
```

```
## [1] TRUE FALSE FALSE FALSE
```

```
x && y
```

```
## [1] TRUE
```

We typically use the two-character version in control structures like if—since these do not operate on vectors in any case—while we use the one-character version when we need to compute with boolean arithmetic, where we want our expressions to work as vectorized expressions.

Incidentally, all the arithmetic operators work like the | and & operators when operating on more than one value, that is, they operate element-wise on vectors. We saw that in Chapter 1 when we talked about vector expressions.

Basic Data Types

There are a few basic types in R: numeric, integer, complex, logical, and character.

Numeric

The numeric type is what you get any time you write a number into R. You can test if an object is numeric using the function is.numeric or by getting the object's class:

```
is.numeric(2)
```

```
## [1] TRUE
```

```
class(2)
```

```
## [1] "numeric"
```

Integer

The integer type is used for, well, integers. Surprisingly, the 2 earlier is not an integer in R. It is a numeric type which is the larger type that contains all floating-point numbers as well as integers. To get an integer, you have to make the value explicitly an integer, and you can do that using the function as.integer or writing L[1] after the literal:

```
is.integer(2)
```

```
## [1] FALSE
```

```
is.integer(2L)
```

```
## [1] TRUE
```

```
x <- as.integer(2)
is.integer(x)
```

```
## [1] TRUE
```

```
class(x)
```

```
## [1] "integer"
```

---

[1] L stands for long, and the reason is mostly historical. In old hardware, you worried about how many bits you should reserve for a number. Short integers had fewer bits than long integers, but could also represent fewer values.

If you translate a non-integer into an integer, you just get the integer part:

```
as.integer(3.2)
```

```
## [1] 3
```

```
as.integer(9.9)
```

```
## [1] 9
```

Complex

If you ever find that you need to work with complex numbers, R has those as well. You construct them by adding an imaginary number—a number followed by i—to any number or explicitly using the function as.complex. The imaginary number can be zero, 0i, which creates a complex number that has a zero imaginary part:

```
1 + 0i
```

```
## [1] 1+0i
```

```
is.complex(1 + 0i)
```

```
## [1] TRUE
```

```
class(1 + 0i)
```

```
## [1] "complex"
```

```
sqrt(as.complex(-1))
```

```
## [1] 0+1i
```

Logical

Logical values are what you get if you explicitly type in TRUE or FALSE, but it is also what you get if you make, for example, a comparison:

```
x <- 5 > 4
x
```

```
## [1] TRUE
```

```
class(x)
```

```
## [1] "logical"
```

```
is.logical(x)
```

```
## [1] TRUE
```

Character

Finally, characters are what you get when you type in a string such as "hello, world":

```
x <- "hello, world"
class(x)
```

```
## [1] "character"
```

```
is.character(x)
```

```
## [1] TRUE
```

Unlike in some languages, character here doesn't mean a single character but any text. So it is not like in C or Java where you have single character types, 'c', and multicharacter strings, "string"; in R, they are both just characters.

You can, similar to the other types, explicitly convert a value into a character (string) using as.character:

```
as.character(3.14)
```

```
## [1] "3.14"
```

I will not go further into string handling in R here. There are of course lots of functions for manipulating strings—and even though there are all those functions, I still find it a lot harder to manipulate strings in R than in scripting languages such as Python—but those are beyond the scope of this book.

Data Structures

From the basic types, you can construct other data structures, essentially by concatenating simpler types into more complex ones. The basic building blocks here are vectors—sequences of values all of the same type—and lists, sequences where the values can have different types.

Vectors

We have already seen vectors many times in this book, so you should be familiar with them. Whenever we have seen expressions involving single numbers, we have actually seen vectors containing a single value, so we have never seen anything that wasn't a vector. But we now consider more technical aspects of vectors.

What I have called vectors up till now is technically known as "atomic sequences." Those are any sequences of the basic types described earlier. You create these by concatenating basic values using the c function:

```
v <- c(1, 2, 3)
v
```

```
## [1] 1 2 3
```

or through some other operator or function, for example, the : operator or the rep function:

```
1:3
```

```
## [1] 1 2 3
```

```
rep("foo", 3)
```

```
## [1] "foo" "foo" "foo"
```

We can test if something is this kind of vector using the is.atomic function:

```
v <- 1:3
is.atomic(v)
```

```
## [1] TRUE
```

The reason I mention that "atomic sequences" is the technically correct term for what we have called vectors until now is that there is also something in R that is explicitly called a vector. In practice, there is no confusion because all the atomic sequences I have called vectors are also vectors.

```
v <- 1:3
is.vector(v)
```

```
## [1] TRUE
```

It is just that R only considers such a sequence a vector—in the sense that is.vector returns TRUE—if the object doesn't have any attributes (except for one, names, which it is allowed to have).

Attributes are meta-information associated with an object, and not something we will deal with much here, but you just have to know that is.vector will be FALSE if something that is a perfectly good vector gets an attribute:

```
v <- 1:3
is.vector(v)
```

```
## [1] TRUE
```

```
attr(v, "foo") <- "bar"
v
```

```
## [1] 1 2 3
## attr(,"foo")
## [1] "bar"
```

```
is.vector(v)
```

```
## [1] FALSE
```

So if you want to test if something is the kind of vector I am talking about here, use is.atomic instead.

When you concatenate (atomic) vectors, you always get another vector back. So when you combine several c() calls, you don't get any kind of tree structure if you do something like this:

```
c(1, 2, c(3, 4), c(5, 6, 7))
```

```
## [1] 1 2 3 4 5 6 7
```

The type might change; if you try to concatenate vectors of different types, R will try to translate the type into the most general type of the vectors:

```
c(1, 2, 3, "foo")

## [1] "1" "2" "3" "foo"
```

Matrix

If you want a matrix instead of a vector, what you really want is just a two-dimensional vector. You can set the dimensions of a vector using the dim function—it sets one of those attributes we talked about earlier—where you specify the number of rows and the number of columns you want the matrix to have:

```
v <- 1:6
attributes(v)

## NULL

dim(v) <- c(2, 3)
attributes(v)

## $dim
## [1] 2 3

dim(v)

## [1] 2 3

v

##      [,1] [,2] [,3]
## [1,]    1    3    5
## [2,]    2    4    6
```

When you do this, the values in the vector will go in the matrix column-wise, that is, the values in the vector will go down the first column first and then on to the next column and so forth.

You can use the convenience function matrix to create matrices, and there you can specify if you want the values to go by column or by row using the byrow parameter:

```
v <- 1:6
matrix(data = v, nrow = 2, ncol = 3, byrow = FALSE)

##      [,1] [,2] [,3]
## [1,]    1    3    5
## [2,]    2    4    6

matrix(data = v, nrow = 2, ncol = 3, byrow = TRUE)

##      [,1] [,2] [,3]
## [1,]    1    2    3
## [2,]    4    5    6
```

Once you have a matrix, there is a lot of support for doing linear algebra in R, but there are a few things you need to know. First, the * operator will not do matrix multiplication. You use * if you want to make element-wise multiplication; for matrix multiplication, you need the operator %*% instead:

```
(A <- matrix(1:4, nrow = 2))

##      [,1] [,2]
## [1,]    1    3
## [2,]    2    4

(B <- matrix(5:8, nrow = 2))
##      [,1] [,2]
## [1,]    5    7
## [2,]    6    8

A * B

##      [,1] [,2]
## [1,]    5   21
## [2,]   12   32

A %*% B

##      [,1] [,2]
## [1,]   23   31
## [2,]   34   46
```

If you want to transpose a matrix, you use the function t, and if you want to invert it, you use the function solve:

```
t(A)
```

```
##      [,1] [,2]
## [1,]    1    2
## [2,]    3    4
```

```
solve(A)
```

```
##      [,1] [,2]
## [1,]   -2  1.5
## [2,]    1 -0.5
```

```
solve(A) %*% A
```

```
##      [,1] [,2]
## [1,]    1    0
## [2,]    0    1
```

The solve function is really aimed at solving a set of linear equations, and it does that if it gets a vector argument as well, but you can check the documentation for the function to see how this is done.

You can also get higher-dimensional vectors, called arrays, by setting the dimension attribute with more than two dimensions as arguments, or you can use the function array.

Lists

Lists, like vectors, are sequences, but unlike vectors, the elements of a list can be any kind of objects, and they do not have to be the same type of objects. This means that you can construct more complex data structures out of lists.

For example, we can make a list of two vectors:

```
list(1:3, 5:8)
```

```
## [[1]]
## [1] 1 2 3
```

```
##
## [[2]]
## [1] 5 6 7 8
```

Notice how the vectors do not get concatenated like they would if we combined them with c(). The result of the preceding command is a list of two elements that happens to be both vectors.

They didn't have to have the same type either; we could make a list like this, which also consists of two vectors but vectors of different types:

```
list(1:3, c(TRUE, FALSE))
```

```
## [[1]]
## [1] 1 2 3
##
## [[2]]
## [1] TRUE FALSE
```

Since lists can contain other lists, you can build tree-like data structures quite naturally:

```
list(list(), list(list(), list()))
```

```
## [[1]]
## list()
##
## [[2]]
## [[2]][[1]]
## list()
##
## [[2]][[2]]
## list()
```

You can flatten a list into a vector using the function `unlist()`. This will force the elements in the list to be converted into the same type, of course, since that is required of vectors:

```
l <- list(1:4, 5:7)
l
```

```
## [[1]]
## [1] 1 2 3 4
##
## [[2]]
## [1] 5 6 7
```

```
unlist(l)
```

```
## [1] 1 2 3 4 5 6 7
```

Indexing

We saw basic indexing in the first chapter, but there is much more to indexing in R than that. Type "?[[" into the R prompt and prepare to be amazed.

We have already seen the basic indexing. If you want the nth element of a vector v, you use v[n]:

```
v <- c("foo", "bar", "baz", "qux", "qax")
v[2]
```

```
## [1] "bar"
```

But this you already knew. You also know that you can get a subsequence out of the vector using a range of indices:

```
v[2:3]
```

```
## [1] "bar" "baz"
```

which is really just a special case of using a vector of indices:

```
v[c(1,1,4,3,2)]
```

```
## [1] "foo" "foo" "qux" "baz" "bar"
```

Here, we are indexing with positive numbers, which makes sense since the elements in the vector have positive indices, but it is also possible to use negative numbers to index in R. If you do that, it is interpreted as specifying the complement of the values you want. So if you want all elements except the first element, you can use

```
v[-1]
```

```
## [1] "bar" "baz" "qux" "qax"
```

You can also use multiple negative indices to remove some values:

```
v[-(1:2)]
```

```
## [1] "baz" "qux" "qax"
```

You cannot combine positive and negative indices. I don't even know how that would even make sense, but in any case, you just can't.

Another way to index is to use a boolean vector. This vector should be the same length as the vector you index into, and it will pick out the elements where the boolean vector is true:

```
w <- 1:length(v)
w
```

```
## [1] 1 2 3 4 5
```

```
w %% 2 == 0
```

```
## [1] FALSE TRUE FALSE TRUE FALSE
```

```
v[w %% 2 == 0]
```

```
## [1] "bar" "qux"
```

If you want to assign to a vector, you just assign to elements you index; as long as the vector to the right of the assignment operator has the same length as the elements the indexing pulls out, you will be assigning to the vector:

```
v[w %% 2 == 0] <- "flob"
v
```

```
## [1] "foo" "flob" "baz" "flob" "qax"
```

If the vector has more than one dimension—remember that matrices and arrays are really just vectors with more dimensions—then you subset them by subsetting each dimension. If you leave out a dimension, you will get a whole range of values in that dimension, which is a simple way of getting rows and columns of a matrix:

```
m <- matrix(1:6, nrow = 2, byrow = TRUE)
m
```

```
##      [,1] [,2] [,3]
## [1,]    1    2    3
## [2,]    4    5    6
```

```
m[1,]
```

```
## [1] 1 2 3
```

```
m[,1]
```

```
## [1] 1 4
```

You can also index out a submatrix this way by providing ranges in one or more dimensions:

```
m[1:2,1:2]
```

```
##      [,1] [,2]
## [1,]    1    2
## [2,]    4    5
```

When you pull out a one-dimensional submatrix—as we did earlier with m[1,]—the result is a vector, not a matrix. Sometimes, that is what you want; sometimes, you don't really care if you get a matrix or a vector, but sometimes you want to do linear algebra, and then you definitely want that the submatrix you pull out is a matrix. You can tell R that it shouldn't reduce a one-dimensional matrix to a row by giving the indexing an option drop=FALSE:

```
m[1,,drop=FALSE]
```

```
##      [,1] [,2] [,3]
## [1,]    1    2    3
```

```
m[,1,drop=FALSE]
```

```
##      [,1]
## [1,]    1
## [2,]    4
```

If this looks weird to you (giving indexing an option), then what you need to know is that everything in R involves function calls. Indexing into a matrix is just another function call, and functions can take named arguments. That is all that is happening here.

When you subset a list using [ ], the result is always another list. If this surprises you, just remember that when you subset a vector, you also always get a vector back. You just don't think so much about it because the way we see single values are always as vectors of length one, so we are more used to that.

Anyway, you will always get a list out of subsetting a list with [ ]. Even if you are subsetting a single element, you are not getting that element; you are getting a list containing that one element:

```
L <- list(1,2,3)
L[1]
```

```
## [[1]]
## [1] 1
```

```
L[2:3]
```

```
## [[1]]
## [1] 2
##
## [[2]]
## [1] 3
```

If you want to get to the actual element in there, you need to use the [[ ]] operator instead:

```
L <- list(1,2,3)
L[[1]]
```

```
## [1] 1
```

Named Values

The elements in a vector or a list can have names. These are attributes that do not affect the values of the elements but can be used to refer to them.

You can set these names when you create the vector or list:

```
v <- c(a = 1, b = 2, c = 3, d = 4)
v
```

```
## a b c d
## 1 2 3 4
```

```
L <- list(a = 1:5, b = c(TRUE, FALSE))
L
```

```
## $a
## [1] 1 2 3 4 5
##
## $b
## [1] TRUE FALSE
```

or you can set the names using the names<- function. That weird name, by the way, means that you are dealing with the names() function combined with assignment. We will see how it works later.

```
names(v) <- LETTERS[1:4]
v
```

```
## A B C D
## 1 2 3 4
```

You can use names to index vectors and lists (where the [ ] and [[ ]] return either a list or the element of the list, as before):

```
v["A"]
```

```
## A
## 1
```

```
L["a"]
```

```
## $a
## [1] 1 2 3 4 5
L[["a"]]
## [1] 1 2 3 4 5
```

When you have named values, you can also use a third indexing operator, the $ operator. It essentially works like [[]] except that you don't have to put the name in quotes:

```
L$a
## [1] 1 2 3 4 5
```

There is never really any good time to introduce the [[]] operator for vectors but here goes: if you use the [[]] operator on a vector, it will only let you extract a single element, and if the vector has names, it will remove the name.

Factors

The factor type we saw in the first chapter is technically also a vector type, but it isn't a primitive type in the same sense as the previous types. It is stored as a vector of integers—the levels in the factor—and has associated attributes such as the levels. It is implemented using the class system we return to in two chapters' time, and we will not discuss it further here.

Formulas

Another data type is the formula. We saw these in Chapter 6, and we can create them using the ~ operator. Like factors, the result is an object defined using a class. We will see how we can use formulas to implement our own statistical models via model matrices in the last chapter of the book.

Control Structures

Control structures determine the flow of execution of a program. You can get far by just having one statement or expression after another, but eventually you will have to do one thing instead of another depending on the results of a calculation, and this is where control structures come in.

Like many other programming languages, you have two kinds of control structures in R: select (`if` statements) and loops (`for`, `while`, or `repeat` statements).

Selection Statements

If statements look like this:

```
if (boolean) {
    # do something
}
```

or like this

```
if (boolean) {
    # do one thing
} else {
    # do another thing
}
```

You can string them together like this:

```
if (x < 0) {
    # handle negative x
} else if (x > 0) {
    # handle positive x
} else {
    # handle if x is zero
}
```

In all the examples here I have put the statements you do if the condition is true or if it is false in curly brackets. Strictly speaking, this isn't necessary if we are talking about a single statement. This would work just fine:

```
if (x > 0) "positive" else if (x < 0) "negative" else "zero"
```

but it would fail if you put newlines in between the statements; the R parser would be confused about that, and there you do need curly brackets. This would be a syntax error:

```
if (x > 0)
    print("positive")
else if (x < 0)
    print("negative")
else
    print("zero")
```

while this would be okay:

```
if (x > 0) {
    print("positive")
} else if (x < 0) {
    print("negative")
} else {
    print("zero")
}
```

I recommend always using curly brackets until you get more familiar with R, since they work fine when you only have a single statement so you are not doing anything wrong in that case, and they are the only thing that works when you have more than one statement or when you have newlines in the if statement.

Loops

The most common looping construction you will use is probably the for loop. You use the for loop to iterate through the elements of a sequence, and the construction works like this:

```
for (i in 1:4) {
    print(i)
}

## [1] 1
## [1] 2
## [1] 3
## [1] 4
```

Keep in mind, though, that it is the elements in the sequence you are iterating through, so the variables you assign to the iteration variable are the elements in the sequence and not the index into the sequence:

```
for (i in c("foo", "bar", "baz")) {
    print(i) # i is foo, then bar, then baz, not index 1, 2, and then 3
}
## [1] "foo"
## [1] "bar"
## [1] "baz"
```

If you want to loop through the indices into the sequence, you can use the seq_along function:

```
x <- c("foo", "bar", "baz")
for (i in seq_along(x)) {
    print(i)
    print(x[i])
}

## [1] 1
## [1] "foo"
## [1] 2
## [1] "bar"
## [1] 3
## [1] "baz"
```

You will sometimes see code that uses this construction:

```
for (i in 1:length(x)) {
    # do stuff
}
```

Don't do that. It won't work if the sequence x is empty:

```
x <- c()
1:length(x)

## [1] 1 0
```

If you want to jump to the next iteration of a loop, you can use the next keyword. For example, the following will only print every second element of x:

```r
for (i in seq_along(x)) {
    if (i %% 2 == 0) {
        next
    }
    print(x[i])
}
```

If you want to terminate the loop completely, you can use break:

```r
for (i in 1:100) {
    if (i %% 2 == 0) {
        next
    }
    if (i > 5) {
        break
    }
    print(i)
}
## [1] 1
## [1] 3
## [1] 5
```

The two other loop constructs you won't use as often. They are the while and repeat loops.

The while loop iterates as long as a boolean expression is true and looks like this:

```r
i <- 1
while (i < 5) {
    print(i)
    i <- i + 1
}
## [1] 1
## [1] 2
## [1] 3
## [1] 4
```

The repeat loop simply goes on forever, at least until you break out of the loop:

```r
i <- 1
repeat {
    print(i)
    i <- i + 1
    if (i > 5) break
}
```

```
## [1] 1
## [1] 2
## [1] 3
## [1] 4
## [1] 5
```

A word of warning about looping...

If you read more about R, you will soon run into the statement that loops are slow in R. It isn't really as bad as some make it out to be, but it is somewhat justified. Because R is an extremely dynamic language—functions and variables can change at any time during program execution, if you want them to—it is hard for the interpreter to optimize code before it runs it, unlike in some other languages (but not that different from other dynamic languages such as Python). There haven't been many attempts at optimizing loops either, though, because there are typically better solutions in R than to use an explicit loop statement.

R is a so-called functional language (among other things), and in functional languages, you typically don't use loops. The way looping constructs work, you need to change the value of a looping variable or a boolean expression while you execute the code, and changing variables is considered "impure" in function languages (so, obviously, R is not a pure functional language, since it allows this). Instead, recursive functions are used for looping. Most functional languages don't even have looping constructs—and pure functional languages certainly do not. R is a bit more pragmatic, but you are typically better off with using alternatives to loops.

We will get more into that in the next chapter.

Functions

You define functions this way:

```
name <- function(arguments) expression
```

where name can be any variable name, arguments is a list of formal arguments to the function, and expression is what the function will do when you call it. It says expression because you might as well think about the body of a function as an expression, but typically it is a sequence of statements enclosed by curly brackets:

```
name <- function(arguments) { statements }
```

It is just that such a sequence of statements is also an expression; the result of executing a series of statements is the value of the last statement.

The following function will print a statement and return 5 because the statements in the function body are first a print statement and then just the value 5 that will be the return value of the function:

```
f <- function() {
    print("hello, world")
    5
}
f()

## [1] "hello, world"
## [1] 5
```

We usually don't write functions without arguments—like I just did earlier—but have one or more formal arguments. The arguments, in their simplest form, are just variable names. They are assigned values when you call the function, and these can then be used inside the function's body:[2]

```
plus <- function(x, y) {
    print(paste(x, "+", y, "is", x + y))
    x + y
}
```

_____

[2] I am actually lying here because the arguments to a function are not assigned values but expressions that haven't been evaluated yet. See lazy evaluation later.

```
div <- function(x, y) {
    print(paste(x, "/", y, "is", x / y))
    x / y
}

plus(2, 2)

## [1] "2 + 2 is 4"

## [1] 4

div(6, 2)

## [1] "6 / 2 is 3"

## [1] 3
```

Named Arguments

The order of arguments matters when you call a function because it determines which argument gets set to which value:

```
div(6,2)

## [1] "6 / 2 is 3"

## [1] 3

div(2,6)

## [1] "2 / 6 is 0.333333333333333"

## [1] 0.3333333
```

If a function has many arguments, though, it can be hard always to remember the order they should appear in, so there is an alternative way to specify which variable is given which values: named arguments. It means that when you call a function, you can make explicit which parameter each argument should be set to:

```
div(x = 6, y = 2)

## [1] "6 / 2 is 3"

## [1] 3
```

```
div(y = 2, x = 6)
```

```
## [1] "6 / 2 is 3"
```

```
## [1] 3
```

This makes explicit which parameter gets assigned which value, and you can think of it as an assignment operator. You shouldn't, though, because although you can use = as an assignment operator, you cannot use <- for specifying named variables. It looks like you can, but it doesn't do what you want it to do (unless you want something really weird):

```
div(x <- 6, y <- 2)
```

```
## [1] "6 / 2 is 3"
```

```
## [1] 3
```

```
div(y <- 2, x <- 6)
```

```
## [1] "2 / 6 is 0.333333333333333"
```

```
## [1] 0.3333333
```

The assignment operator <- returns a value, and that is passed along to the function as positional arguments. So in the second function call earlier, you are assigning 2 to y and 6 to x in the scope outside the function, but the values you pass to the function are positional, so inside the function you have given 2 to x and 6 to y.

Don't confuse the two assignment operators: the code most likely will run, but it is unlikely to do what you want it to do!

Default Parameters

When you define a function, you can provide default values to parameters like this:

```
pow <- function(x, y = 2) x^y
pow(2)
```

```
## [1] 4
```

```
pow(3)
```

```
## [1] 9
pow(2, 3)
## [1] 8
pow(3, 3)
## [1] 27
```

Default parameter values will be used whenever you do not provide the parameter at the function call.

Return Values

The return value of a function is the last expression in the statements executed in the function body. If the function is a sequence of statements, this is just the last statement in the sequence, but by using control structures, you can have different statements as the last statement:

```
safer_div <- function(x, y) {
    if (y == 0) {
        NA
    } else {
        x / y
    }
}
safer_div(4, 2)
## [1] 2
safer_div(4, 0)
## [1] NA
```

It is also possible to return explicitly from a function—similarly to breaking from a loop—using the return() statement:

```
safer_div <- function(x, y) {
    if (y == 0) {
        return(NA)
    }
```

```
    x / y
}
safer_div(4, 2)

## [1] 2

safer_div(4, 0)

## [1] NA
```

Notice that the `return()` statement has the return value in parentheses. Many programming languages would allow you to write

```
safer_div <- function(x, y) {
    if (y == 0) {
        return NA
    }
    x / y
}
```

but this would be an error in R.

Lazy Evaluation

Several places I have written about providing values to the function parameters when we call a function. In many programming languages, this is exactly how function calls work—the expressions provided for each parameter are evaluated, and the results are assigned to the function parameters so the function can use them in the function body—but in R it is actually the expressions that are assigned to the function parameters. And the expressions are not evaluated until they are needed, something called lazy evaluation.

There are some benefits to this way of handling function parameters and some weird consequences as well.

The first benefit is that it makes default parameters more flexible. We can write a function like this:

```
f <- function(x, y = x^2) y + x
```

where y has a default value that depends on the other parameter x. At the time where the function is declared, the value of x is not known, but y is not evaluated there, so it doesn't matter. Whenever we call the function, x is known inside the body of the function, and that is where we need the value of y, so that is where the expression will be evaluated:

```
f(2)
```

```
## [1] 6
```

Since y isn't evaluated before it is used, it does also mean that if you assign a different value to x before you use y, then y evaluates to a value that depends on the new value of x—not the value of x at the time the function was called!

```
g <- function(x, y = x^2) { x <- 0; y + x }
g(2)
```

```
## [1] 0
```

If, on the other hand, y is evaluated before we assign to x, then it will evaluate to the value that depends on x at the time we evaluate it and remain that value. It is evaluated the first time it is needed, and the result is then remembered for any later time we refer to y:

```
h <- function(x, y = x^2) { y; x <- 0; y + x }
h(2)
```

```
## [1] 4
```

So lazy evaluation lets you specify default parameters that depend on other parameters in a context where those parameters are unknown, but it comes at the prize of the value of the parameter depending on the context at the first time it gets evaluated.

If it was just to be able to specify variables this way, we could, of course, have a solution that doesn't involve the weirdness that we pay for it. This is what most programming languages have done, after all, but there are other benefits of lazy evaluation: you only evaluate an expression if you actually need it.

Scoping

Scope in R is lexical. This means that if a variable is used in a function but not defined in the function or part of the function's parameters, then R will start searching outward in the code from where the function was created. This essentially means searching outward and upward from the point in the code where the function is specified, since a function is created when the code is executed where the function is defined.

Consider the following code:

```
x <- "x"
f <- function(y) {
    g <- function() c(x, y)
    g()
}
f("y")
## [1] "x" "y"
```

Here, we have a global variable x and a function f that takes a parameter argument y. Inside f, we define the function g that neither defines nor takes as formal arguments variables x and y but does return them. We evaluate g as the last statement in f, so that becomes the result of calling f at the last line.

Inside g, we have not defined x or y, so to find their values, R will search outward from where g is created. It will find y as the argument of the function f and so get it from there and continue outward to find x at the global level.

The variables that g refers to are the variables and not the values at the time that g is created, so if we update the variables after we create g, we also change the value that g will return:

```
x <- "x"
f <- function(y) {
  g <- function() c(x, y)
  y <- "z"
  g()
}
f("y")

## [1] "x" "z"
```

317

This isn't just the lazy evaluation madness—it is not that g hasn't evaluated y yet, and it, therefore, can be changed. It does look up the value of y when it needs it:

```
x <- "x"
f <- function(y) {
  g <- function() c(x, y)
  g()
  y <- "z"
  g()
}
f("y")
```

```
## [1] "x" "z"
```

If we return the function g from f rather than the result of evaluating it, we see another feature of R's scoping—something called closures. R remembers the values of variables inside a function that we have returned from and that is no longer an active part of our computation. In the following example, we returned the function g at which point there is no longer an active f function. So there is not really an active instance of the parameter y any longer. Yet g refers to a y, so the parameter we gave to f is actually remembered:

```
x <- "x"
f <- function(y) {
  g <- function() c(x, y)
  g
}
g<- f("y")
g()
```

```
## [1] "x" "y"
```

We can see how this works if we invoke f twice, with different values for parameter y:

```
x <- "x"
f <- function(y) {
  g <- function() c(x, y)
  g
}
```

```
g<- f("y")
h <- f("z")
g()
```

```
## [1] "x" "y"
```

```
h()
```

```
## [1] "x" "z"
```

This creates two different functions. Inside f, they are both called g, but they are two different functions because they are created in two different calls to f, and they remember two different y parameters because the two instances of f were invoked with different values for y.

When looking outward from the point where a function is defined, it is looking for the values of variables at the time a function is invoked, not the values at the time where the function is created. Variables do not necessarily have to be defined at the time the function is created; they just need to be defined when the function is eventually invoked.

Consider this code:

```
f <- function() {
  g <- function() c(y, z)
  y <- "y"
  g
}
h <- f()
h()
```

```
## Error in h(): object 'z' not found
```

```
z <- "z"
h()
```

```
## [1] "y" "z"
```

Where the function g is defined—inside function f—it refers to variables y and z that are not defined yet. This doesn't matter because we only create the function g; we do not invoke it. We then set the variable y inside the context of the invocation of f and return g. Outside of the function call, we name the return value of f() h. If we call h at this point, it will remember that y was defined inside f—and it will remember its value at the point in time where we returned from f. There still isn't a value set for z, so calling h results in an error. Since z isn't defined in the enclosing scopes of where the inner function refers to it, it must be defined at the outermost global scope, but it isn't. If we do set it there, the error goes away because now R can find the variable by searching outward from where the function was created.

I shouldn't really be telling you this because the feature I am about to tell you about is dangerous. I will show you a way of making functions have even more side effects than they otherwise have, and functions really shouldn't have side effects at all. Anyway, this is a feature of the language, and if you are very careful with how you use it, it can be very useful when you just feel the need to make functions have side effects.

This is the problem: What if you want to assign to a variable in a scope outside the function where you want the assignment to be made? You cannot just assign to the variable because if you assign to a variable that isn't found in the current scope, then you create that variable in the current scope:

```r
f <- function() {
    x <- NULL
    set <- function(val) { x <- val }
    get <- function() x
    list(set = set, get = get)
}

x <- f()
x$get() # get x -- we haven't set it, so it is NULL

## NULL


x$set(5) # set x to five
x$get()  # now get the new value -- it doesn't work yet, though

## NULL
```

In this code—that I urge you to read carefully because there are a few neat ideas in it—we have created a getter and a setter function. The getter tells us what the variable x is, and the setter is supposed to update it. It doesn't quite work yet, though. When setting x in the body of the set function, we create a local variable inside that function—it doesn't assign to the x one level up.

There is a separate assignment operator, <<-, you can use for that. It will not create a new local variable but instead search outward to find an existing variable and assign to that. If it gets all the way to the outermost global scope, though, it will create the variable there if it doesn't already exist.

If we use that assignment operator in the preceding example, we get the behavior we were aiming for:

```
f <- function() {
    x <- NULL
    set <- function(val) { x <<- val }
    get <- function() x
    list(set = set, get = get)
}

x <- f()
x$get() # We get x, which is still NULL

## NULL

x$set(5) # We set x, and this time it works
x$get()  # as you can see here

## [1] 5
```

If we hadn't set the variable x inside the body of f in this example, both the getter and setter would be referring to a global variable, in case you are wondering. The first call to get would cause an error if there was no global variable. While this example shows you have to create an object where functions have side effects, it is quite a bit better to let functions modify variables that are hidden away in a closure like this than it is to work with global variables.

Function Names Are Different from Variable Names

One final note on scopes—which I am not sure should be considered a feature or a bug—is that if R sees something that looks like a function call, it is going to go searching for a function, even if searching outward from a function creation would get to a nonfunction first:

```
n <- function(x) x
f <- function(n) n(n)
f(5)

## [1] 5
```

Under the scoping rule that says that you should search outward, the n inside the f function should refer to the parameter to f. But it is clear that the first n is a function call and the second is its argument, so when we call f, it sees that the parameter isn't a function so it searches further outward and finds the function n. It calls that function with its argument. So the two n's inside f actually refer to different things.

Of course, if we call f with something that is actually a function, then it recognizes that n is a function, and it calls that function with itself as the argument:

```
f(function(x) 15)

## [1] 15
```

Interesting, right?

Recursive Functions

The final topic we will cover in this chapter is recursive functions. Some people find this a difficult topic, but in a functional programming language, it is one of the most basic building blocks, so it is really worth spending some time wrapping your head around, even though you are much less likely to need recursions in R than you are in most pure functional languages.

At the most basic level, though, it is just that we can define a function's computations in terms of calls to the same function—we allow a function to call itself, just with new parameters.

Consider the factorial operator $n! = n \times (n - 1) \times \cdots \times 3 \times 2 \times 1$. We can rewrite the factorial of n in terms of n and a smaller factorial, the factorial of $n - 1$, and get $n! = n \times (n - 1)!$. This is a classical case of where recursion is useful: we define the value for some n in terms of the calculations on some smaller value. As a function, we would write factorial(n) equals n * factorial(n-1).

There are two aspects to a recursive function, though. Solving a problem for size n involves breaking down the problem into something you can do right away and combining that with calls of the function with a smaller size, here $n - 1$. This part we call the "step" of the recursion. We cannot keep reducing the problem into smaller and smaller bits forever—that would be an infinite recursion which is as bad as an infinite loop in that we never get anywhere—at some point, we need to have reduced the problem to a size small enough that we can handle it directly. That is called the basis of the recursion.

For factorial, we have a natural basis in 1 since $1! = 1$. So we can write a recursive implementation of the factorial function like this:

```
factorial <- function(n) {
    if (n == 1) {
        1
    } else {
        n * factorial(n - 1)
    }
}
```

It is actually a general algorithmic strategy, called divide and conquer, to break down a problem into subproblems that you can handle recursively and then combine the results some way.

Consider sorting a sequence of numbers. We could sort a sequence using this strategy by first noticing that we have a simple basis—it is easy to sort an empty sequence or a sequence with a single element since we don't have to do anything there. For the step, we can break the sequence into two equally sized pieces and sort them recursively. Now we have two sorted sequences, and if we merge these two, we have combined them into a single sorted sequence.

Let's get started.

We need to be able to merge two sequences so we can solve that problem first. This is something we should be able to do with a recursive function because if either sequence is empty, we have a base case where we can just return the other sequence. If there are elements in both sequences, we can pick the sequence whose first element is smallest, pick that out as the first element we need in our final result, and just concatenate the merging of the remaining numbers:

```
merge <- function(x, y) {
  if (length(x) == 0) return(y)
  if (length(y) == 0) return(x)

  if (x[1] < y[1]) {
    c(x[1], merge(x[-1], y))
  } else {
    c(y[1], merge(x, y[-1]))
  }
}
```

A quick disclaimer here: Normally, this algorithm would run in linear time, but because of the way we call recursively, we are actually copying vectors whenever we are removing the first element, making it a quadratic time algorithm. Implementing a linear time merge function is left as an exercise.

Using this function, we can implement a sorting function. This algorithm is called merge sort, so that is what we call the function:

```
merge_sort <- function(x) {
  if (length(x) < 2) return(x)

  n <- length(x)
  m <- n %/% 2

  merge(merge_sort(x[1:m]), merge_sort(x[(m+1):n]))
}
```

So here, using two simple recursive functions, we solved a real algorithmic problem in a few lines of code. This is typically the way to go in a functional programming language like R. Of course, when things are easier done using loops, you shouldn't stick to the pure functional recursions. Use what is easiest in any situation you are in, unless you find that it is too slow. Only then do you start getting clever.

Exercises

Fibonacci Numbers

The Fibonacci numbers are defined as follows. The first two Fibonacci numbers are 1, F_1 = F_2 = 1. For larger Fibonacci numbers, they are defined as $F_i = F_{i-1} + F_{i-2}$.

Implement a recursive function that computes the n'th Fibonacci number.

The recursive function for Fibonacci numbers is usually quite inefficient because you are recomputing the same numbers several times in the recursive calls. So implement another version that computes the n'th Fibonacci number iteratively (i.e., start from the bottom and compute the numbers up to n without calling recursively).

Outer Product

The outer product of two vectors, v and w, is a matrix defined as

$$v \otimes w = vw^T = \begin{bmatrix} v_1 \\ v_2 \\ v_3 \end{bmatrix} \begin{bmatrix} w_1 & w_2 & w_3 & w_4 \end{bmatrix} = \begin{bmatrix} v_1 w_1 & v_1 w_2 & v_1 w_3 & v_1 w_4 \\ v_2 w_1 & v_2 w_2 & v_2 w_3 & v_2 w_4 \\ v_3 w_1 & v_3 w_2 & v_3 w_3 & v_3 w_4 \end{bmatrix}$$

Write a function that computes the outer product of two vectors.

There actually is a built-in function, `outer`, that you are overwriting here. You can get to it using the name `base::outer` even after you have overwritten it. You can use it to check that your own function is doing the right thing.

Linear Time Merge

The merge function we used earlier copies vectors in its recursive calls. This makes it slower than it has to be. Implement a linear time merge function.

Before you start, though, you should be aware of something. If you plan to append to a vector by writing something like

```
v <- c(v, element)
```

then you will end up with a quadratic time algorithm again. This is because when you do this, you are actually creating a new vector where you first copy all the elements in the old v vector into the first elements and then add the `element` at the end. If you do this n times, you have spent on average order n^2 per operation. It is because people do something like this in loops, more than the R interpreter, that has given R its reputation for slow loops. You should never append to vectors unless there is no way to avoid it.

In the case of the merge function, we already know how long the result should be, so you can preallocate a result vector and copy single elements into it. You can create a vector of length n like this:

```
n <- 5
v <- vector(length = n)
```

Should you ever need it, you can make a list of length n like this:

```
vector("list", length = n)
```

Binary Search

Binary search is a classical algorithm for finding out if an element is contained in a sorted sequence. It is a simple recursive function. The basic case handles a sequence of one element. There, you can directly compare the element you are searching for with the element in the sequence to determine if they are the same. If you have more than one element in the sequence, pick the middle one. If it is the element you are searching for, you are done and can return that the element is contained in the sequence. If it is smaller than the element you are searching for, then you know that if the element is in the list, then it has to be in the last half of the sequence, and you can search there. If it is larger than the element you are searching for, then you know that if it is in the sequence, it must be in the first half of the sequence, and you search recursively there.

If you implement this exactly as described, you have to call recursively with a subsequence. This involves copying that subsequence for the function call which makes the implementation much less efficient than it needs to be. Try to implement binary search without this.

More Sorting

In the merge sort we implemented earlier, we solve the sorting problem by splitting a sequence in two, sorting each subsequence, and then merging them. If implemented correctly, this algorithm will run in time $O(n \log n)$ which is optimal for sorting algorithms if we assume that the only operations we can do on the elements we sort are comparing them.

If the elements we have are all integers between 1 and n and we have m of them, we can sort them in time $O(n + m)$ using bucket sort instead. This algorithm first creates a vector of counts for each number between 1 and n. This takes time $O(n)$. It then runs through the m elements in our sequence, updating the counter for number i each time it sees i. This runs in time $O(m)$. Finally, it runs through these numbers from 1 up to n, outputting each number, the number of times indicated by the counters, in time $O(n + m)$.

Implement bucket sort.

Another algorithm that works by recursion, and that runs in expected time $O(n \log n)$, is quick sort. Its worst-case complexity is actual $O(n^2)$, but on average it runs in time $O(n \log n)$ and with a smaller overhead than merge sort (if you implement it correctly).

It works as follows: the base case—a single element—is the same as merge sort. When you have more than one element, you pick one of the elements in the sequence at random; call it the pivot. Now split the sequence into those elements that are smaller than the pivot, those that are equal to the pivot, and those that are larger. Sort the sequences of smaller and larger elements recursively. Then output all the sorted smaller elements, then the elements equal to the pivot, and then the sorted larger elements.

Implement quick sort.

Selecting the k Smallest Element

If you have n elements, and you want the k smallest, an easy solution is to sort the elements and then pick number k. This works well and in most cases is easily fast enough, but it is actually possible to do it faster. See, we don't actually need to sort the elements completely, we just need to have the k smallest element moved to position k in the sequence.

The quick sort algorithm from the previous exercise can be modified to solve this problem. Whenever we split a sequence into those smaller than, equal to, and larger than the pivot, we sort the smaller and larger elements recursively. If we are only interested in finding the element that would eventually end up at position k in the sorted lists, we don't need to sort the sequence that doesn't overlap this index. If we have $m < k$ elements smaller than the pivot, we can just put them at the front of the sequence without sorting them. We need them there to make sure that the k'th smallest element ends up at the right index, but we don't need them sorted. Similarly, if $k < m$, we don't need to sort the larger elements. If we sorted them, they would all end up at indices larger than k, and we don't really care about those. Of course, if there are $m < k$ elements smaller than the pivot and l equal to the pivot, with $m + l \geq k$, then the k smallest element is equal to the pivot, and we can return that.

Implement this algorithm.

CHAPTER 10

Working with Vectors and Lists

In this chapter, we explore working with vectors and lists a little further. We will not cover anything that is conceptually more complex that we did in the previous chapter. It is just a few more technical details we will dig into.

Working with Vectors and Vectorizing Functions

We start out by returning to expressions. In the previous chapter, we saw expressions on single (scalar) values, but we also saw that R doesn't really have scalar values; all the primitive data we have is actually vectors of data. What this means is that the expressions we use in R are actually operating on vectors, not single values.

When you write

```
(x <- 2 / 3)
```

```
## [1] 0.6666667
```

```
(y <- x ** 2)
```

```
## [1] 0.4444444
```

the expressions you write are, of course, working on single values—the vectors x and y have length 1—but it is really just a special case of working on vectors:

```
(x <- 1:4 / 3)
```

```
## [1] 0.3333333 0.6666667 1.0000000 1.3333333
```

```
(y <- x ** 2)
```

```
## [1] 0.1111111 0.4444444 1.0000000 1.7777778
```

© Thomas Mailund 2022
T. Mailund, *Beginning Data Science in R 4*, https://doi.org/10.1007/978-1-4842-8155-0_10

R works on vectors using two rules: operations are done element-wise, and vectors are repeated as needed.

When you write an expression such as x + y, you are really saying that you want to create a new vector that consists of the element-wise sum of the elements in vectors x and y. So for x and y like this

```
x <- 1:5
y <- 6:10
```

writing

```
(z <- x + y)
## [1]  7  9 11 13 15
```

amounts to writing

```
z <- vector(length = length(x))
for (i in seq_along(x)) {
    z[i] <- x[i] + y[i]
}
z

## [1]  7  9 11 13 15
```

This is the case for all arithmetic expressions or for logical expressions involving | or & (but not || or &&; these do not operate on vectors element-wise). It is also the case for most functions you can call, such as sqrt or sin:

```
sqrt((1:5)**2)

## [1] 1 2 3 4 5

sin(sqrt((1:5)**2))

## [1]  0.8414710  0.9092974  0.1411200 -0.7568025
## [5] -0.9589243
```

When you have an expression that involves vectors of different lengths, you cannot directly evaluate expressions element-wise. When this is the case, R will try to repeat the shorter vector(s) to create vectors of the same length. For this to work, the shorter vector(s) should have a length divisible in the length of the longest vector, that is, you should be able to repeat the shorter vector(s) an integer number of times to get the length of the longest vector. If this is possible, R repeats vectors as necessary to make all vectors the same length as the longest and then do operations element-wise:

```
x <- 1:12
y <- 1:2
x + y
```

```
## [1]  2  4  4  6  6  8  8 10 10 12 12 14
```

```
z <- 1:3
x + z
```

```
## [1]  2  4  6  5  7  9  8 10 12 11 13 15
```

If the shorter vector(s) cannot be repeated an integer number of times to match up, R will still repeat as many times as needed to match the longest vector, but you will get a warning since most times something like this happens, it is caused by buggy code:

```
z <- 1:5
x + z
```

```
## Warning in x + z: longer object length is not a
## multiple of shorter object length
```

```
## [1]  2  4  6  8 10  7  9 11 13 15 12 14
```

In the expression we saw a while back

```
(x <- 1:4 / 3)
```

```
## [1] 0.3333333 0.6666667 1.0000000 1.3333333
```

```
(y <- x ** 2)
```

```
## [1] 0.1111111 0.4444444 1.0000000 1.7777778
```

different vectors are repeated. When we divide 1:4 by 3, we need to repeat the (length one) vector 3 four times to be able to divide the 1:4 vector with the 3 vector. When we compute x ** 2, we must repeat 2 four times as well.

Whenever you consider writing a loop over vectors to do some calculations for each element, you should always consider using such vectorized expressions instead. It is typically much less error-prone, and since it involves implicit looping handled by the R runtime system, it is almost guaranteed to be faster than an explicit loop.

ifelse

Control structures are not vectorized. For example, if statements are not. If you want to compute a vector y from vector x such that y[i] == 5 if x[i] is even and y[i] == 15 if x[i] is odd, for example, you cannot write this as a vector expression:

```
x <- 1:10
if (x %% 2 == 0) 5 else 15
```

```
## Warning in if (x%%2 == 0) 5 else 15: the condition
## has length > 1 and only the first element will be
## used
```

```
## [1] 15
```

Instead, you can use the function ifelse that works like a vectorized selection; if the condition in its first element is true, it returns the value in its second argument, otherwise the value in its third argument, and it does this as vector operations:

```
x <- 1:10
ifelse(x %% 2 == 0, 5, 15)
```

```
##  [1] 15  5 15  5 15  5 15  5 15  5
```

Vectorizing Functions

When you write your own functions, you can write them such that they can also be used to work on vectors, that is, you can write them such that they can take vectors as input and return vectors as output. If you write them this way, then they can be used in vectorized expressions the same way as built-in functions such as sqrt and sin.

The easiest way to make your function work on vectors is to write the body of the function using expressions that work on vectors:

```
f <- function(x, y) sqrt(x ** y)
f(1:6, 2)
```

```
## [1] 1 2 3 4 5 6
```

```
f(1:6, 1:2)
```

```
## [1] 1.000000 2.000000 1.732051 4.000000 2.236068
## [6] 6.000000
```

If you write a function where you cannot write the body this way, but where you still want to be able to use it in vector expressions, you can typically get there using the Vectorize function.

As an example, say we have a table mapping keys to some values. We can imagine that we want to map names in a class to the roles the participants in the class have. In R, we would use a list to implement that kind of tables, and we can easily write a function that uses such a table to map names to roles:

```
role_table <- list("Thomas" = "Instructor",
                   "Henrik" = "Student",
                   "Kristian" = "Student",
                   "Randi" = "Student",
                   "Heidi" = "Student",
                   "Manfred" = "Student")
map_to_role <- function(name) role_table[[name]]
```

This works the way it is supposed to when we call it with a single name:

```
map_to_role("Thomas")
```

```
## [1] "Instructor"
```

```
map_to_role("Henrik")
```

```
## [1] "Student"
```

but it fails when we call the function with a vector because we cannot index the list with a vector in this way:

```
x <- c("Thomas", "Henrik", "Randi")
map_to_role(x)
```

```
## Error in role_table[[name]]: recursive indexing failed at level 2
```

So we have a function that maps a single value to a single value but doesn't work for a vector. The easy way to make such a function work on vectors is to use the Vectorize function. This function will wrap your function so it can work on vectors, and what it will do on those vectors is what you would expect: it will calculate its value for each of the elements in the vector, and the result will be the vector of all the results:

```
map_to_role <- Vectorize(map_to_role)
map_to_role(x)
```

```
##        Thomas     Henrik      Randi
##  "Instructor"  "Student"  "Student"
```

In this particular example with a table, the reason it fails is that we are using the` [[` index operator. Had we used the` [` operator, we would be fine (except that the result would be a list rather than a vector):

```
role_table[c("Thomas", "Henrik", "Randi")]
```

```
## $Thomas
## [1] "Instructor"
##
## $Henrik
## [1] "Student"
##
## $Randi
## [1] "Student"
```

So we could also have handled vector input directly by indexing differently and then flattening the list:

```
map_to_role_2 <- function(names) unlist(role_table[names])

x <- c("Thomas", "Henrik", "Randi")
map_to_role_2(x)
```

```
##        Thomas       Henrik       Randi
## "Instructor"    "Student"    "Student"
```

It's not always that easy to rewrite a function to work on vector input, though, and when we cannot readily do that, then the `Vectorize` function can be very helpful.

As a side note, the issue with using `` `[[` `` with a vector of values isn't just that it doesn't work. It actually does work, but it does something else than what we are trying to do here. If you give `` `[[` `` a vector of indices, it is used to do what is called recursive indexing. It is a shortcut for looking up in the list using the first variable and pulling out the vector or list found there. It then takes that sequence and looks up using the next index and so on. Take as an example the following code:

```
x <- list("first" = list("second" = "foo"), "third" = "bar")
x[[c("first", "second")]]
```

```
## [1] "foo"
```

Here, we have a list of two elements, the first of which is a list with a single element. We can look up the index "first" in the first list and get the list stored at that index. This list we can then index with the "second" index to get "foo" out.

The result is analogous to this:

```
x[["first"]][["second"]]
```

```
## [1] "foo"
```

This can be a useful feature—although to be honest I have rarely found much use for it in my own programming—but it is not the effect we wanted in our mapping to roles example.

The `apply` Family

Vectorizing a function makes it possible for us to use it implicitly on vectors. We simply give it a vector as input, and we get a vector back as output. Notice though that it isn't really a vectorized function just because it takes a vector as input—many functions take vectors as input and return a single value as output, for example, `sum` and `mean`—but we use those differently than vectorized functions. If you want that kind of function, you do have to handle explicitly how it deals with a sequence as input.

Vectorized functions can be used on vectors of data exactly the same way as on single values with exactly the same syntax. It is an implicit way of operating on vectors. But we can also make it more explicit when calling a function on all the elements in a vector which give us a bit more control of exactly how it is called. This, in turn, lets us work with those functions that do not just map from vectors to vectors but also from vectors to single values.

There are many ways of doing this—because it is a common thing to do in R—and we will see some general functions for working on sequences and calling functions on them in various ways later. In most of the code you will read, though, the functions that do this are named something with `apply` in their name, and those functions are what we will look at here.

Let's start with `apply`. This is a function for operating on vectors, matrices (two-dimensional vectors), or arrays (higher-order dimensional vectors).

apply

This function is easiest explained on a matrix, I think, so let's make one:

```
m <- matrix(1:6, nrow = 2, byrow = TRUE)
m
```

```
##      [,1] [,2] [,3]
## [1,]    1    2    3
## [2,]    4    5    6
```

The `apply` function takes (at least) three parameters. The first is the vector/matrix/array, the second which dimension(s) we should marginalize along, and the third the function we should apply.

What is meant by marginalization here is that you fix an index in some subset of the dimensions and pull out all values with that index. If we are marginalizing over rows, we will extract all the rows, so for each row, we will have a vector with an element per column, which is what we will pass the function.

We can illustrate this using the paste function that just creates a string of its input by concatenating it.[1]

If we marginalize on rows, it will be called on each of the two rows and produce two strings:

---

[1] So this is a case of a function that takes a vector as input but outputs a single value; it is not a vectorized function as those we talked about earlier.

```
apply(m, 1, \(x) paste(x, collapse = ":"))
```

```
## [1] "1:2:3" "4:5:6"
```

If we marginalize on columns, it will be called on each of the three columns and produce three strings:

```
apply(m, 2, \(x) paste(x, collapse = ":"))
```

```
## [1] "1:4" "2:5" "3:6"
```

If we marginalize on both rows and columns, it will be called on each single element instead:

```
apply(m, c(1, 2), \(x) paste(x, collapse = ":"))
```

```
##      [,1] [,2] [,3]
## [1,] "1"  "2"  "3"
## [2,] "4"  "5"  "6"
```

The output here is two-dimensional. That is of course because we are marginalizing over two dimensions, so we get an output that corresponds to the margins.

We can get higher-dimensional output in other ways. If the function we apply produces vectors (or higher-dimensional vectors) as output, then the output of apply will also be higher-dimensional. Consider a function that takes a vector as input and duplicates it by concatenating it with itself. If we apply it to rows or columns, we get a vector for each row/column, so the output has to be two-dimensional:

```
apply(m, 1, \(x) c(x,x))
```

```
##      [,1] [,2]
## [1,]    1    4
## [2,]    2    5
## [3,]    3    6
## [4,]    1    4
## [5,]    2    5
## [6,]    3    6
```

337

```
apply(m, 2, \(x) c(x,x))
```

```
##      [,1] [,2] [,3]
## [1,]    1    2    3
## [2,]    4    5    6
## [3,]    1    2    3
## [4,]    4    5    6
```

What `apply` does here is that it creates a matrix as its result, where the results of applying the function are collected as columns from left to right. The result of calling the function on the two rows is a matrix with two columns, the first column containing the result of applying the function to the first row and the second column the result of applying it to the second row. Likewise, for columns, the result is a vector with three columns, one for each column in the input matrix.

If we marginalize over more than one dimension—and get multidimensional output through that—and at the same time produce more than one value, the two effects combine and we get even higher-dimensional output:

```
apply(m, c(1,2), \(x) c(x,x))
```

```
## , , 1
##
##      [,1] [,2]
## [1,]    1    4
## [2,]    1    4
##
## , , 2
##
##      [,1] [,2]
## [1,]    2    5
## [2,]    2    5
##
## , , 3
##
##      [,1] [,2]
## [1,]    3    6
## [2,]    3    6
```

I admit that this output looks rather confusing. What happens, though, is the same as we saw when we marginalized on rows or columns. We get output for each margin we call the function on—in this case, each of the six cells in the input—and it gets collected "column-wise," except that this is at higher dimensions, so it gets collected at the highest dimension (which is the columns for two-dimensional matrices). So to get to the result of the six values the function was called with, we need to index these the same way they were indexed in the input matrix—that is what the margins were—but we need to do it in the highest dimensions. So we can get the six concatenations of input values this way:

```
x <- apply(m, c(1,2), \(x) c(x,x))
k <- dim(x)[3]
n <- dim(x)[2]
for (i in 1:n) {
   for (j in 1:k) {
      print(x[,i,j])
   }
}

## [1] 1 1
## [1] 2 2
## [1] 3 3
## [1] 4 4
## [1] 5 5
## [1] 6 6
```

So what happens if the function to apply takes arguments besides those we get from the matrix?

```
sumpow <- \(x, n) sum(x) ** n
apply(m, 1, sumpow)

## Error in FUN(newX[, i], ...): argument "n" is missing, with no default
```

Nothing Good, It Would Seem

The apply function expects to give you its values, but it doesn't a priori knows how to provide additional arguments. You have to give those additional arguments to apply if you want it to pass them onto your function. You can give these arguments as additional parameters to `apply`; they will be passed on to the function in the order you give them to `apply`:

```
apply(m, 1, sumpow, 2)
```

```
## [1]   36 225
```

It helps readability a lot, though, to explicitly name such parameters:

```
apply(m, 1, sumpow, n = 2)
```

```
## [1]   36 225
```

lapply

The `lapply` function is used for mapping over a list. Given a list as input, it will apply the function to each element in the list and output a list of the same length as the input containing the results of applying the function:

```
(l <- list(1, 2, 3))
```

```
## [[1]]
## [1] 1
##
## [[2]]
## [1] 2
##
## [[3]]
## [1] 3
```

```
lapply(l, \(x) x**2)
```

```
## [[1]]
## [1] 1
##
## [[2]]
## [1] 4
##
## [[3]]
## [1] 9
```

If the elements in the input list have names, these are preserved in the output vector:

```
l <- list(a=1, b=2, c=3)
lapply(l, \(x) x**2)

## $a
## [1] 1
##
## $b
## [1] 4
##
## $c
## [1] 9
```

If the input you provide is a vector instead of a list, it will just convert it into a list, and you will always get a list as output:

```
lapply(1:3, \(x) x**2)

## [[1]]
## [1] 1
##
## [[2]]
## [1] 4
##
## [[3]]
## [1] 9
```

Of course, if the elements of the list are more complex than a single number, you will still just apply the function to the elements:

```
lapply(list(a=1:3, b=4:6), \(x) x**2)

## $a
## [1] 1 4 9
##
## $b
## [1] 16 25 36
```

`sapply` **and** `vapply`

The `sapply` function does the same as `lapply` but tries to simplify the output. Essentially, it attempts to convert the list returned from `lapply` into a vector of some sort. It uses some heuristics for this and guesses as to what you want as output, simplifies when it can, but gives you a list when it cannot figure it out:

```
sapply(1:3, \(x) x**2)
```

```
## [1] 1 4 9
```

The guessing is great for interactive work but can be unsafe when writing programs. It isn't a problem that it guesses and can produce different types of output when you can see what it creates, but that is not safe deep in the guts of a program.

The function `vapply` essentially does the same as `sapply` but without the guessing. You have to tell it what you want as output, and if it cannot produce that, it will give you an error rather than produce an output that your program may or may not know what to do with.

The difference in the interface between the two functions is just that `vapply` expects a third parameter that should be a value of the type the output should be:

```
vapply(1:3, \(x) x**2, 1)
```

```
## [1] 1 4 9
```

Advanced Functions

We now get to some special cases for functions. I call the section "advanced functions," but it is not because they really are that advanced, they just require a little bit more than the basic functions we have already seen.

Special Names

But first a word on names. Functions can have the same kind of names that variables have—after all, when we name a function, we are really just naming a variable that happens to hold a function—but we cannot have all kinds of names to the right of the assignment operator. For example, `if` is a function in R, but you cannot write `if` to the left of an assignment.

Functions with special names, that is, names that you couldn't normally put before an assignment, can be referred to by putting them in backticks, so the function `if` we can refer to as `` `if` ``.

Any function can be referred to by its name in backticks, and furthermore you can use backticks to refer to a function in a context where you usually couldn't use its name. This works for calling functions where you can use, for example, infix operators as normal function calls:

```
2 + 2
## [1] 4
`+`(2, 2)
## [1] 4
```

or when assigning to a variable name for a function:

```
`%or die%` <- function(test, msg) if (!test) stop(msg)

x <- 5
(x != 0) %or die% "x should not be zero"

x <- 0
(x != 0) %or die% "x should not be zero"

## Error in (x != 0) %or die% "x should not be zero": x should not be zero
```

Infix Operators

If the last example looks weird to you, it may just be because you don't know about R's infix operators. In R, any variable that starts and ends with % is considered an infix operator, so calling x %foo% y amounts to calling `` `%foo%` ``(x,y). Several built-in infix operators do not have this type of name, + and * are two, but this naming convention makes it possible to create your own infix operators. We have seen this come to good use in the dplyr package for the %>% pipe operator.

Replacement Functions

Replacement functions are functions that pretend to be modifying variables. We saw one early where we assigned names to a vector:

```
v <- 1:4
names(v) <- c("a", "b", "c", "d")
v

## a b c d
## 1 2 3 4
```

What happens here is that R recognizes that you are assigning to a function call and goes looking for a function named `names<-`. It calls this function with the vector v and the vector of names, and the result of the function call gets assigned back to the variable v.

So what I just wrote means that

```
names(v) <- c("a", "b", "c", "d")
```

is short for

```
v <- `names<-`(v, c("a", "b", "c", "d"))
```

Replacement functions are generally used to modify various attributes of an object, and you can write your own just by using the convention that their names must end with `<-`:

```
`foo<-` <- function(x, value) {
  x$foo <- value
  x
}

`bar<-` <- function(x, value) {
  x$bar <- value
  x
}

x <- list(foo = 1, bar = 2)

x$foo
```

344

```
## [1] 1
foo(x) <- 3
x$foo

## [1] 3

x$bar

## [1] 2

bar(x) <- 3
x$bar

## [1] 3
```

Keep in mind that it is just shorthand for calling a function and then reassigning the result to a variable. It is not actually modifying any data. This means that if you have two variables referring to the same object, only the one you call the replacement function on will be affected. The replacement function returns a copy that is assigned the first variable, and the other variable still refers to the old object:

```
y <- x
foo(x) <- 5
x

## $foo
## [1] 5
##
## $bar
## [1] 3

y

## $foo
## [1] 3
##
## $bar
## [1] 3
```

Because replacement functions are just syntactic sugar on a function call and then a reassignment, you cannot give a replacement function, as its first argument, some expression that cannot be assigned to.

There are a few more rules regarding replacement functions. First, the parameter for the value you are assigning has to be called value. You cannot give it another name:

```
`foo<-` <- function(x, val) {
  x$foo <- val
  x
}

x <- list(foo = 1, bar = 2)
foo(x) <- 3

## Error in `foo<-`(`*tmp*`, value = 3): unused argument (value = 3)
```

The way R rewrites the expression assumes that you called the value parameter value, so do that.

You don't have to call the first parameter x, though:

```
`foo<-` <- function(y, value) {
  y$foo <- value
  y
}

x <- list(foo = 1, bar = 2)
foo(x) <- 3
x$foo

## [1] 3
```

You should also have the value parameter as the last parameter if you have more than two parameters. And you are allowed to, as long as the object you are modifying is the first and the value parameter the last:

```
`modify<-` <- function(x, variable, value) {
  x[variable] <- value
  x
}
```

```
x <- list(foo = 1, bar = 2)

modify(x, "foo") <- 3
modify(x, "bar") <- 4
x$foo

## [1] 3

x$bar

## [1] 4
```

How Mutable Is Data Anyway?

We just saw that a replacement function creates a new copy, so if we use it to modify an object, we are not actually changing it at all. Other variables that refer to the same object will see the old value and not the updated one. So we can reasonably ask: What does it take actually to modify an object?

The short, and almost always correct, answer is that you cannot modify objects ever.[2] Whenever you "modify" an object, you are creating a new copy and assigning that new copy back to the variable you used to refer to the old value.

This is also the case for assigning to an index in a vector or list. You will be creating a copy, and while it looks like you are modifying it, if you look at the old object through another reference, you will find that it hasn't changed:

```
x <- 1:4
f <- function(x) {
    x[2] <- 5
    x
}
x
```

---

[2] It is possible to do depending on what you consider an object. You can modify a closure by assigning to local variables inside a function scope. This is because namespaces are objects that can be changed. One of the object-oriented systems in R, RC, also allows for mutable objects, but we won't look at RC in this book. In general, you are better off thinking that every object is immutable, and any modification you are doing is actually creating a new object because, generally, that is what is happening.

```
## [1] 1 2 3 4
```

```
f(x)
```

```
## [1] 1 5 3 4
```

```
x
```

```
## [1] 1 2 3 4
```

Unless you have changed the `` `[` `` function (which I urge you not to do), it is a so-called primitive function. This means that it is written in C, and from C you actually can modify an object. This is important for efficiency reasons. If there is only one reference to a vector, then assigning to it will not make a new copy, and you will modify the vector in place as a constant time operation. If you have two references to the vector, then when you assign to it the first time, a copy is created that you can then modify in place. This approach to have immutable objects and still have some efficiency is called copy on write.

To write correct programs, always keep in mind that you are not modifying objects but creating copies—other references to the value you "modify" will still see the old value. To write efficient programs, also keep in mind that for primitive functions you can do efficient updates (updates in constant time instead of time proportional to the size of the object you are modifying) as long as you only have one reference to that object.

Exercises

between

Write a vectorized function that takes a vector x and two numbers, lower and upper, and replaces all elements in x smaller than lower or greater than upper with NA.

rmq

A range minimum query, rmq, extracts from a list the indices that have minimal values. Can you write a vectorized function that gives you the indices where the minimal value occurs? Hint: You can use min(x) to find the minimal value, you can compare it with x to get a logical vector, and you can get the values in vector where a logical vector is TRUE by indexing. Also, seq_along(x) gives you a vector of the indices in x.

CHAPTER 11

Functional Programming

In this chapter, we explore the programming paradigm called functional programming and how it relates to R. There are many definitions of what it means for a language to be a functional programming language, and there have been many language wars over whether any given feature is "pure" or not. I won't go into such discussions, but some features, I think everyone would agree, are needed. You should be able to define higher-order functions, you should be able to create closures, and you probably want anonymous functions as well.

Let's tackle anonymous functions right away, as these are pretty simple in R.

Anonymous Functions

In R, it is pretty easy to create anonymous functions: just don't assign the function definition to a variable name.

Instead of doing this:

```
square <- function(x) x^2
```

you simply do this:

```
function(x) x^2
```

If you want an even shorter expression, we have seen those as well:

```
\(x) x^2
```

In other languages where function definitions have a different syntax than variable assignment, you will have a different syntax for anonymous functions, but in R it is really as simple as this.

Why would you want an anonymous function?

© Thomas Mailund 2022
T. Mailund, *Beginning Data Science in R 4*, https://doi.org/10.1007/978-1-4842-8155-0_11

There are two common cases:

- We want to use a one-off function and don't need to give it a name.

- We want to create a closure.

Both cases are typically used when a function is passed as an argument to another function or when returned from a function. The first case is something we would use together with functions like `apply`. If we want to compute the sum of squares over the rows of a matrix, we can create a named function and apply it:

```
m <- matrix(1:6, nrow=3)
sum_of_squares <- function(x) sum(x^2)
apply(m, 1, sum_of_squares)
```

```
## [1] 17 29 45
```

but if this is the only time we need this sum of squares function, there isn't really any need to assign it a variable; we can just use the function definition direction:

```
apply(m, 1, \(x) sum(x^2))
```

```
## [1] 17 29 45
```

Of course, in this example, we could do even better by just exploiting that ^ is vectorized and write

```
apply(m^2, 1, sum)
```

```
## [1] 17 29 45
```

Using anonymous functions to create closures is what we do when we write a function that returns a function (and more about that later). We could name the function

```
f <- function(x) {
    g <- function(y) x + y
    g
}
```

but there really isn't much point if we just want to return it:

```
f <- function(x) function(y) x + y
```

That is all there is to anonymous functions, really. You can define a function without giving it a name, and that is about it.

Higher-Order Functions

In the terminology often used with functional programming, higher-order functions refer to functions that either take other functions as arguments, return functions, or both. The `apply` function that we have used multiple times is thus a higher-order function, as it takes a function as one of its inputs.

Functions Taking Functions As Arguments

We generally use function arguments to influence how a function is executed. If the arguments did not affect what the function did, we would have no need for them. Sometimes, the arguments are simple, like a boolean flag that determines part of what the function should do or a number it uses in its computation. But there are times where such static data doesn't suffice or at least makes it harder to implement the functionality we need. Sometimes, we want an argument that has some dynamic behavior, for example, an argument that can determine for each of a multiple of data pointers whether to skip the data or do some computation. Such dynamic behavior fits functions very well, and in these cases, a function argument is what you need.

In general, if some subcomputation of a function should be parameterized dynamically, then you do this by taking a function as one of its parameters. Say we want to write a function that works like `apply` but only apply an input function on elements that satisfy a predicate. We can implement such a function by taking the vector and two functions as input:

```
apply_if <- function(x, p, f) {
    result <- vector(length = length(x))
    n <- 0
    for (i in seq_along(x)) {
        if (p(x[i])) {
            n <- n + 1
            result[n] <- f(x[i])
        }
    }
    head(result, n)
}

apply_if(1:8, \(x) x %% 2 == 0, \(x) x^2)

## [1] 4 16 36 64
```

351

This isn't the most elegant way to solve this particular problem—we get back to the example in the exercises—but it illustrates the use of functions as parameters.

Functions Returning Functions (and Closures)

We create closures when we create a function inside another function and return it. Because this inner function can refer to the parameters and local variables inside the surrounding function, even after we have returned from it, we can use such inner functions to specialize generic functions. It can work as a template mechanism for describing a family of functions.

We can, for instance, write a generic power function and specialize it for squaring or cubing numbers:

```
power <- function(n) function(x) x^n
square <- power(2) # square fixes n to 2, so it compute squares
cube <- power(3)   # cube fixes n to 3, so it compute cubes
square(1:5)
```

```
## [1] 1 4 9 16 25
```

```
cube(1:5)
```

```
## [1] 1 8 27 64 125
```

This works because the functions returned by power(2) and power(3) live in a context—the enclosure—where n is known to be 2 and 3, respectively. We have fixed that part of the function we return.

Generally, functions can see the variables you provide as arguments and the variables you create inside them, but also variables in any enclosing function. So, when we write

```
power <- function(n) {
    # inside power we can see n because we get it as an argument
    f <- function (x) {
        # inside f we can see x because it is an argument,
        # but *also* n, since when we are inside f, we are also
```

```
    # inside power.
    x^2
  }
  return(f)
}
```

we create a function that will know its argument n. When we call the function, power(2), that n gets the value 2, and while running the function, we create the function f. We are not running it yet, we are not calling f inside power, but we have created the function. This is similar to how, when we have defined power, we haven't executed its body yet; that doesn't happen until we call power(2). When we call power(2), we create f, and we create it in an enclosure where it knows the n value in the call to power(2). When we return f from power(2), the function we return still remembers the values in its enclosure—and we call it a closure because of that. When we did

```
square <- power(2)
```

we get the inner function f out of the call and assign it to square. Now, we are no longer running power; we just remember n through the function we returned, the function that we now call square. When we call square(1:5), we call the function f inside the call to power(2). This function gets x as the argument, 1:5, and remembers n from when it was created.

This might appear a little complicated the first time you see it, and there are times where you have to study a function carefully to see how closures and variables interact, but it is a common programming trick because it is extremely useful. With closures like this, we can bundle up a little bit of data with a function, without having to provide it as arguments each time we call the function. This makes it easier to write reusable functions, because they do not need to know much about the function arguments we provide them, and it makes it easier to use other functions, because it can keep their interface simple.

Let's take another example. Say I have some algorithm in mind where I regularly need to find the index of the first element in a sequence that satisfies some property. Immediately, you should think of the property as some predicate function like in the apply_if function earlier, and that is how I will write it:

```
first_index <- function(x, p) {
  for (i in seq_along(x)) { # Run through x
    if (p(x[i])) { # return the first index that satisfy p
```

353

```
        return (i)
    }
  }
}
```

The function (implicitly) returns NULL if we do not find any value that satisfies p, and the loop terminates. When it does, the result of the function is the result of running the loop, and the result of a for loop is NULL. We could handle misses differently, but I want to keep the example simple.

Let's check it out. We can use it to look for the first even or odd number, for example:

```
x <- 1:10
first_index(x, \(x) x %% 2 == 0)
```

```
## [1] 2
```

```
first_index(x, \(x) x %% 2 != 0)
```

```
## [1] 1
```

This is simple enough, because the predicate doesn't have to know anything beyond what even or odd means. But what if I wanted to find the first element in some given range, say between four and seven? I could modify the function, of course:

```
first_index <- function(x, range) {
  for (i in seq_along(x)) { # Run through x
    if ((range[1] <= x[i]) && (x[i] <= range[2])) {
      # return the first index where x[i] is in the range
      return (i)
    }
  }
}
first_index(x, c(4, 7))
```

```
## [1] 4
```

It gets the job done, but this function is a lot less general than the first one we wrote. This one can only perform range queries and nothing else. It is a lot less general than when we used a predicate function.

We can try to provide generic data to the function, similar to how the `apply` family of functions permit us to provide data:

```r
first_index <- function(x, p, p_data) {
  for (i in seq_along(x)) { # Run through x
    if (p(x[i], p_data)) {
      # return the first index where x[i] is in the range
      return (i)
    }
  }
}
```

and then write a range predicate based on that:

```r
range_pred <- function(x, range) {
  (range[1] <= x) && (x <= range[2])
}
first_index(x, range_pred, c(4, 7))
```

```
## [1] 4
```

It works, but now we need to pass range data along as well as the predicate, we must always be careful that the extra predicate data matches what the predicate expects, and we have to provide predicate data even for predicate functions that do not need them (although there are ways around that problem in R).

The simplest search function we wrote was superior to the more complicated ones:

```r
first_index <- function(x, p) {
  for (i in seq_along(x)) { # Run through x
    if (p(x[i])) { # return the first index that satisfy p
      return (i)
    }
  }
}
```

It does one thing, and one thing only. The interface is trivial—give it a sequence and a predicate—and it is hard to get confused about its functionality.

If we want to do range queries, we can still use it. We can hardwire a range in the predicate if we want:

```
first_index(x, \(x) (4 <= x) && (x <= 7))
```

```
## [1] 4
```

and when we can't hardwire the range, we can bundle the range up with the predicate:

```
in_range <- function(from, to) { # A function for creating a predicate
  \(x) (from <= x) && (x <= to) # the predicate function it returns
}
```

```
p <- in_range(4, 7)
first_index(x, p)
```

```
## [1] 4
```

Because closures are as flexible as they are, we can even use the simple search function for more complex computations. What about finding the first repeated value?

```
repeated <- function() { # we don't need initial data for this
  # We will remember previously seen values here
  seen <- c()

  # The predicate goes here, it will check if we have seen a
  # given value before and update the `seen`
  # values if we haven't
  function(x) {
    if (x %in% seen) {
      TRUE # We have a repeat!
    } else {
      seen <<- c(seen, x) # append `x` to `seen`
      FALSE # this was the first time we saw x
    }
  }
}
```

This function is, I admit, a little complicated to wrap your head around, but it isn't too bad. The repated function creates an environment that has the variable seen. It is initially empty. Then it returns the closure. The closure will check if its input is in seen (x %in% seen does this) and, if it is, return true. Otherwise, it updates seen. Here, we need the <<- assignment to update the outer seen; if we used seen <- c(...), we would create a local variable that would be lost as soon as we returned. The <<- gives us a way to store data in the enclosing function. So, we store the new value x in the enclosing seen and return false. The next time we see x, if we ever do, we will find it in seen and report that.

```
x <- c(1:4, 1:5) # We see 1 a second time at index 5
first_index(x, repeated())
```

```
## [1] 5
```

The important point here is that we didn't have to change our search function, first_index, to make it do something complicated. It is still simple, it can be used in many different contexts, and when we need it to do something complicated, we can make it do so by writing the appropriate closure.

Filter, Map, and Reduce

Three patterns are used again and again in functional programming: filtering, mapping, and reducing. In R, all three are implemented in different functions, but you can write all your programs using the Filter, Map, and Reduce functions.

The Filter function takes a predicate and a vector or list and returns all the elements that satisfy the predicate:

```
is_even <- \(x) x %% 2 == 0
Filter(is_even, 1:8)
```

```
## [1] 2 4 6 8
```

```
Filter(is_even, as.list(1:8))
```

```
## [[1]]
## [1] 2
##
```

```
## [[2]]
## [1] 4
##
## [[3]]
## [1] 6
##
## [[4]]
## [1] 8
```

The Map function works like lapply: it applies a function to every element of a vector or list and returns a list of the result. Use unlist to convert it into a vector if that is what you want:

```
square <- \(x) x^2
Map(square, 1:4)
```

```
## [[1]]
## [1] 1
##
## [[2]]
## [1] 4
##
## [[3]]
## [1] 9
##
## [[4]]
## [1] 16
```

```
unlist(Map(square, 1:4))
```

```
## [1]  1  4  9 16
```

You can do slightly more with Map, though, since Map can be applied to more than one sequence. If you give Map more arguments, then these are applied to the function calls as well:

```
plus <- \(x, y) x + y
unlist(Map(plus, 0:3, 3:0))
```

```
## [1] 3 3 3 3
```

These constructions should be very familiar to you by now, so we will leave it at that.

The Reduce function might look less familiar. We can describe what it does in terms of adding or multiplying numbers, and it is in a way a generalization of this. When we write an expression like

```
a + b + c
```

or

```
a * b * c
```

we can think of this as a series of function calls:

```
`+`(`+`(a, b), c)
```

or

```
`*`(`*`(a, b), c)
```

The Reduce function generalizes this:

```
Reduce(f, c(a, b, c))
```

It is evaluated as

```
f(f(a, b), c)
```

which we can see by constructing a function that captures how it is called:

```
add_parenthesis <- \(a, b) paste("(", a, ", ", b, ")", sep = "")
Reduce(add_parenthesis, 1:4)
```

```
## [1] "(((1, 2), 3), 4)"
```

Using Reduce, we could thus easily write our own sum function:

```
mysum <- \(x) Reduce(`+`, x)
sum(1:4)
```

```
## [1] 10
```

```
mysum(1:4)
```

```
## [1] 10
```

There are a few additional parameters to the Reduce function—to give it an additional initial value instead of just the leftmost elements in the first function call or to make it apply the function from right to left instead of left to right—but you can check its documentation for details.

Functional Programming with `purrr`

The three functions from the previous section are the basic building blocks for functional programming. They are somewhat limited in what data they operate on, however, basically preferring the list type for everything they do. This is less limiting than it sounds, as most data structures in R are build around lists, and if you want to use R as a programming language, operating on lists is the way to go. However, if you want to do data science, which I assume you do if you are reading this book, then lists are rather primitive, and you will find yourself converting to and from them all the time.

The package `purrr` provides the same basic functions, and more, for a more convenient functional programming toolkit for data analysis. In this package, you get the same functions we covered earlier, but in different flavors depending on what data you need to do calculations on, and some functionality for converting between data types. If you are interested in functional programming in R, this package is well worth your time to familiarize yourself with.

A full exploration of `purrr` is beyond the scope of an introductory book such as this, but I will give you an idea about how it works by going through some of the variants of Filter, Map, and Reduce.

Start by loading purrr:

```
library(purrr)
```

For Filter, we have two variants, keep that keeps all the elements that satisfy a predicate, just as Filter does, and discard that removes the elements that satisfy the predicate:

```
1:10 |> keep(\(x) x %% 2 == 0) # get the even numbers

## [1]  2  4  6  8 10

1:10 |> discard(\(x) x %% 2 == 0) # remove the even numbers

## [1] 1 3 5 7 9
```

In purrr, there is an alternative syntax for anonymous functions based on formulas:

```
1:10 |> keep(~ .x %% 2 == 0) # get the even numbers
```

```
## [1]  2  4  6  8 10
```

```
1:10 |> discard(~ .x %% 2 == 0) # remove the even numbers
```

```
## [1] 1 3 5 7 9
```

The ~ prefix makes the argument a formula, but purrr will interpret it as a function. Then, it will interpret .x (notice the dot) as the first argument to the function. The second will be .y. You can also use .1, .2, .3, and so on. The alternative syntax is there for historical reasons; purrr was developed before R got the short function syntax \(x) ..., and it is cumbersome to write code such as keep(function(x) x %% 2 == 0). With \(x) ... expressions, there is less need for the alternative syntax, and I will not use it here, but you are likely to run into it if you read R code in the future.

The closest equivalent to Map in purrr is map. It basically works the same way:

```
1:4 |> map(\(x) x^2) # square the numbers
```

```
## [[1]]
## [1] 1
##
## [[2]]
## [1] 4
##
## [[3]]
## [1] 9
##
## [[4]]
## [1] 16
```

The map function will return a list, just as Map does, but you can convert the output to other data formats using a family of functions whose name starts with map_ and ends with a type specifier:

```
# get a vector of integers (The values we compute must be integers)
1L:4L |> map_int(\(x) x + 2L)
```

```
## [1] 3 4 5 6
```

```
# get a vector of nummerics; any number will work here
1:4 |> map_dbl(\(x) x^2)
```

```
## [1]  1  4  9 16
```

```
# get a vector of logical (boolean) values (we must compute booleans)
1:4 |> map_lgl(\(x) x %% 2 == 0)
```

```
## [1] FALSE  TRUE FALSE  TRUE
```

You can also map over data frames with map_df. It will map over the data frames and then merge them, so you get one data frame as output, with one row for each of the rows you produce with your mapping function:

```
dfs <- list(
    tibble(x = 1:2, y = 1:2, z = 1:2),
    tibble(x = 3:4, y = 3:4),
    tibble(x = 4:5, z = 4:5)
)
```

```
# mapping the identifier to see what map_df does with that
dfs |> map_df(\(df) df)
```

```
## # A tibble: 6 × 3
##       x     y     z
##   <int> <int> <int>
## 1     1     1     1
## 2     2     2     2
## 3     3     3    NA
## 4     4     4    NA
## 5     4    NA     4
## 6     5    NA     5
```

```
# modifying the data frames
mut_df <- \(df) df |> mutate(w = 2 * x ) # add column w
dfs |> map_df(mut_df) # now add w for all and merge them
```

```
## # A tibble: 6 × 4
##       x     y     z     w
##   <int> <int> <int> <dbl>
```

```
## 1     1     1     1     2
## 2     2     2     2     4
## 3     3     3    NA     6
## 4     4     4    NA     8
## 5     4    NA     4     8
## 6     5    NA     5    10
```

Like Map/map, you get the `purrr` Reduce by changing it to lowercase: reduce:

```
add_parenthesis <- \(a, b) paste("(", a, ", ", b, ")", sep = "") 1:4 |>
reduce(add_parenthesis)
```

```
## [1] "(((1, 2), 3), 4)"
```

You can change the order of applications, so you reduce from right to left instead of left to right, if you provide the additional argument `.dir = "backward"`:

```
1:4 |> reduce(add_parenthesis, .dir = "backward")
```

```
## [1] "(1, (2, (3, 4)))"
```

Functions As Both Input and Output

Functions can, of course, also both take functions as input and return functions as output.

This lets us modify functions and create new functions from existing functions.

First, let us consider two old friends, the factorial and the Fibonacci numbers. We have computed those recursively and using tables. What if we could build a generic function for caching results?

Here is an attempt:

```
cached <- function(f) {
    # ensures that we get f as it is when we call cached (see text)
    force(f)
    table <- list()
```

```r
function(n) {
    key <- as.character(n)
    if (key %in% names(table)) {
        print(paste("I have already computed the value for", n))
        table[[key]]
    } else {
        print(paste("Going to compute the value for", n))
        res <- f(n)
        print(paste("That turned out to be", res))
        table[key] <<- res # NB: <<- to update the closure table!
        print_table(table) # see function below
        res
    }
}
}

# pretty-printing the table
print_table <- function(tbl) {
    print("Current table:")
    for (key in names(tbl)) {
        print(paste(key, "=>", tbl[key]))
    }
}
```

I have added some output so it is easier to see what it does in the following.

It takes a function f and will give us another function back that works like f but
remembers functions it has already computed. First, it remembers what the input
function was by forcing it. This is necessary for the way we intend to use this cached
function. The plan is to replace the function in the global scope with a cached version so
the function out there will refer to the cached version. If we don't force f here, the lazy
evaluation means that when we eventually evaluate f, we are referring to the cached
version, and we will end up in an infinite recursion. You can try removing the force(f)
call and see what happens.

Next, we create a table—we are using a list which is the best choice for tables in R
in general. A list lets us use strings for indices, and doing that we don't need to have all
values between one and *n* stored to have an element with key *n* in the table.

The rest of the code builds a function that first looks in the table to see if the key is there. If so, we have already computed the value we want and can get it from the table. If not, we compute it, put it in the table, and return.

We can try it out on the `factorial` function:

```r
factorial <- function(n) {
  if (n == 1) {
    1
  } else {
    n * factorial(n - 1)
  }
}

factorial <- cached(factorial)
factorial(4)
```

```
## [1] "Going to compute the value for 4"
## [1] "Going to compute the value for 3"
## [1] "Going to compute the value for 2"
## [1] "Going to compute the value for 1"
## [1] "That turned out to be 1"
## [1] "Current table:"
## [1] "1 => 1"
## [1] "That turned out to be 2"
## [1] "Current table:"
## [1] "1 => 1"
## [1] "2 => 2"
## [1] "That turned out to be 6"
## [1] "Current table:"
## [1] "1 => 1"
## [1] "2 => 2"
## [1] "3 => 6"
## [1] "That turned out to be 24"
## [1] "Current table:"
## [1] "1 => 1"
## [1] "2 => 2"
## [1] "3 => 6"
```

```
## [1] "4 => 24"
```

```
## [1] 24
```

```
factorial(1)
```

```
## [1] "I have already computed the value for 1"
```

```
## [1] 1
```

```
factorial(2)
```

```
## [1] "I have already computed the value for 2"
```

```
## [1] 2
```

```
factorial(3)
```

```
## [1] "I have already computed the value for 3"
```

```
## [1] 6
```

```
factorial(4)
```

```
## [1] "I have already computed the value for 4"
```

```
## [1] 24
```

and on fibonacci

```
fibonacci <- function(n) {
   if (n == 1 || n == 2) {
      1
   } else {
      fibonacci(n-1) + fibonacci(n-2)
   }
}
```

```
fibonacci <- cached(fibonacci)
fibonacci(4)
```

```
## [1] "Going to compute the value for 4"
## [1] "Going to compute the value for 3"
## [1] "Going to compute the value for 2"
```

```
## [1] "That turned out to be 1"
## [1] "Current table:"
## [1] "2 => 1"
## [1] "Going to compute the value for 1"
## [1] "That turned out to be 1"
## [1] "Current table:"
## [1] "2 => 1"
## [1] "1 => 1"
## [1] "That turned out to be 2"
## [1] "Current table:"
## [1] "2 => 1"
## [1] "1 => 1"
## [1] "3 => 2"
## [1] "I have already computed the value for 2"
## [1] "That turned out to be 3"
## [1] "Current table:"
## [1] "2 => 1"
## [1] "1 => 1"
## [1] "3 => 2"
## [1] "4 => 3"

## [1] 3

fibonacci(1)

## [1] "I have already computed the value for 1"

## [1] 1

fibonacci(2)

## [1] "I have already computed the value for 2"

## [1] 1

fibonacci(3)

## [1] "I have already computed the value for 3"

## [1] 2
```

```
fibonacci(4)
```

```
## [1] "I have already computed the value for 4"
```

```
## [1] 3
```

Ellipsis Parameters...

Before we see any more examples of function operations, we need to know about a special function parameter, the ellipsis or "three-dot" parameter.

This is a magical parameter that lets you write a function that can take any number of named arguments and pass them on to other functions.

Without it, you would get an error if you provide a parameter to a function that it doesn't know about:

```
f <- function(a, b) NULL
f(a = 1, b = 2, c = 3)
```

```
## Error in f(a = 1, b = 2, c = 3): unused argument (c = 3)
```

With it, you can provide any named parameter you want:

```
g <- function(a, b, ...) NULL
g(a = 1, b = 2, c = 3)
```

```
## NULL
```

Of course, it isn't much of a feature to allow a function to take arguments that it doesn't know what to do with. But you can pass those arguments on to other functions that maybe do know what to do with them, and that is the purpose of the "..." parameter.

We can see this in effect with a very simple function that just passes the "..." parameter on to list. This works exactly like calling list directly with the same parameters, so nothing magical is going on here, but it shows how the named parameters are being passed along:

```
tolist <- function(...) list(...)
```

```
tolist()
```

```
## list()
```

```
tolist(a = 1)
```

```
## $a
## [1] 1
```

```
tolist(a = 1, b = 2)
```

```
## $a
## [1] 1
##
## $b
## [1] 2
```

This parameter has some uses in itself because it lets you write a function that calls other functions, and you can provide those functions parameters without explicitly passing them along. It is particularly important for generic functions (a topic we will cover in the next chapter) and for modifying functions in function operators.

Here, we will just have a quick second example, taken from Wickham's *Advanced R* programming book (that I cannot praise high enough), of modifying a function— wrapping a function to time how long it takes to run.

The following function wraps the function f into a function that times it and returns the time usage rather than the result of the function. It will work for any function since it just passes all parameters from the closure we create to the function we wrap (although the error profile will be different since the wrapping function will accept any named parameter, while the original function f might not allow that):

```
time_it <- function(f) {
    force(f)
    function(...) {
        system.time(f(...))
    }
}
```

We can try it out like this:

```
ti_mean <- time_it(mean)
ti_mean(runif(1e6))
```

```
##    user  system elapsed
## 0.024   0.000   0.023
```

Exercises

`apply_if`

Consider the function `apply_if` we implemented earlier. There, we use a loop. Implement it using `Filter` and `Map` instead.

For the specific instance we used in the example:

```
apply_if(v, function(x) x %% 2 == 0, function(x) x^2)
```

we only have vectorized functions. Rewrite this function call using a vectorized expression.

power

Earlier, we defined the generic power function and the instances `square` and `cube` this way:

```
power <- function(n) function(x) x^n
square <- power(2)
cube <- power(3)
```

If we instead defined

```
power <- function(x, n) x^n
```

how would you then define `square` and `cube`?

Row and Column Sums

Using `apply`, write functions `rowsum` and `colsum` that compute the row sums and column sums, respectively, of a matrix.

Factorial Again...

Write a vectorized factorial function. It should take a vector as input and compute the factorial of each element in the vector.

Try to make a version that remembers factorials it has already computed so you don't need to recompute them (without using the cached function from before, of course).

Function Composition

For two functions f and g, the function composition creates a new function $f \circ g$ such that $(f \circ g)(x) = f(g(x))$.

There isn't an operator for this in R, but we can make our own. To avoid clashing with the outer product operator, %o%, we can call it %.%.

Implement This Operator

Using this operator, we should, for example, be able to combine Map and unlist once and for all to get a function for the unlist(Map(...)) pattern:

```
uMap <- unlist %.% Map
```

So this function works exactly like first calling Map and then unlist:

```
plus <- function(x, y) x + y
unlist(Map(plus, 0:3, 3:0))
```

```
## [1] 3 3 3 3
```

```
uMap(plus, 0:3, 3:0)
```

```
## [1] 3 3 3 3
```

With it, you can build functions by stringing together other functions (not unlike how you can create pipelines in magrittr).

For example, you can compute the root mean square error function like this:

```
error <- function(truth) function(x) x - truth
square <- function(x) x^2

rmse <- function(truth)
    sqrt %.% mean %.% square %.% error(truth)
```

```
mu <- 0.4
x <- rnorm(10, mean = 0.4)
rmse(mu)(x)
```

```
## [1] 1.249175
```

Combining a sequence of functions like this requires that we read the operations from right to left, so I personally prefer the approach in magrittr, but you can see the similarity.

CHAPTER 12

Object-Oriented Programming

In this chapter, we look at R's flavor of object-oriented programming. Actually, R has three different systems for object-oriented programming: S3, S4, and RC. We will only look at S3, which is the simplest and (as far as I know) the most widely used.

Immutable Objects and Polymorphic Functions

Object orientation in S3 is quite different from what you might have seen in Java or Python or that class of languages. Naturally so, since data in R is immutable and the underlying model in OO in languages such as Java and Python is that you have objects with states that you can call methods on to change the state. You don't have a state as such in S3; you have immutable objects. Just like all other data in R.

What's the point then of having object orientation if we don't have object states? What we get from the S3 system is polymorphic functions, called "generic" functions in R. These are functions whose functionality depends on the class of an object—similar to methods in Java or Python where methods defined in a class can be changed in a subclass to refine behavior.

You can define a function foo to be polymorphic and then define specialized functions, say foo.A and foo.B. Then, calling foo(x) on an object x from class A will actually call foo.A(x) and for an object from class B will actually call foo.B(x). The names foo.A and foo.B were not chosen at random here, we shall see, since it is precisely how we name functions that determine which function gets called.

We do not have objects with states, we simply have a mechanism for having the functionality of a function depend on the class an object has—something often called "dynamic dispatch" or "polymorphic methods." Here, of course, since we don't have states, we can call it polymorphic functions.

© Thomas Mailund 2022
T. Mailund, *Beginning Data Science in R 4*, https://doi.org/10.1007/978-1-4842-8155-0_12

Data Structures

Before we get to making actual classes and objects, though, we should have a look at data structures. We discussed the various built-in data structures in R in earlier chapters. Those built-in data types are the basic building blocks of data in R, but we never discussed how we can build something more complex from them.

More important than any object-oriented system is the idea of keeping related data together so we can treat it as a whole. If we are working on several pieces of data that somehow belongs together, we don't want it scattered out in several different variables, perhaps in different scopes, where we have little chance of keeping it consistent. Even with immutable data, such a program in a consistent state would be a nightmare.

For the data that we analyze, we therefore typically keep it in a data frame. This is a simple idea for keeping data together. All the data we are working on is in the same data frame, and we can call functions with the data frame and know that they are getting all the data in a consistent state. At least as consistent as we can guarantee with data frames; we cannot promise that the data itself is not messed up somehow, but we can write functions under the assumption that data frames behave a certain way.

What about something like a fitted model? If we fit a model to some data, that fit is stored variables capturing the fit. We certainly would like to keep those together when we do work with the model because we would not like accidentally to use a mix of variables fitted to two different models. We might also want to keep other data together with the fitted model, for example, some information about what was actually fitted, if we want to check that in the R shell later, or the data it was fitted to.

The only option we really have for collecting heterogeneous data together as a single object is a list. And that is how you do it in R.

Example: Bayesian Linear Model Fitting

Project 2, described in the last chapter of the book, concerns Bayesian linear models. To represent such, we would wrap data for a model in a list. For fitting data, let us assume that you have a function like the one in the following (refer to the last chapter for details of the mathematics).

It takes the model specification in the form of a formula as its parameter model and the prior precision alpha and the "precision" of the data beta. It then computes the mean and the covariance matrix for the model fitted to the data. That part is left out here since you are supposed to implement that yourself as an exercise, but you can see the details in the last chapter. It then wraps up the fitted model together with some related data—the formula used to fit the model, the data used in the model fit (here assumed to be in the variable frame)—and put them in a list, which the function returns:

```r
blm <- function(model, alpha = 1, beta = 1, ...) {

    # Here goes the mathematics for computing the fit.
    frame <- model.frame(model, ...)
    phi <- model.matrix(frame)
    no_params <- ncol(phi)
    target <- model.response(frame)

    covar <- solve(diag(alpha, no_params) +
                        beta * t(phi) %*% phi)
    mean <- beta * covar %*% t(phi) %*% target
    list(formula = model,
         frame = frame,
         mean = mean,
         covar = covar)
}
```

We can see it in action by simulating some data and calling the function:

```r
# fake some data for our linear model
x <- rnorm(10)

a <- 1 ; b <- 1.3
w0 <- 0.2 ; w1 <- 3
y <- rnorm(10, mean = w0 + w1 * x, sd = sqrt(1/b))

# fit a model
model <- blm(y ~ x, alpha = a, beta = b)
model

## $formula
## y ~ x
##
## $frame
```

```
##                 y              x
## 1     -1.5185867 -0.5588255
## 2     -1.1664514 -0.6097720
## 3     -1.2509896  0.1808256
## 4     -1.9412380 -0.6595195
## 5     -0.5012965 -0.3030505
## 6      0.5057768  0.2613801
## 7      2.1098271  0.7792971
## 8      5.0790285  1.3976716
## 9     -3.2896676 -1.6030947
## 10    -0.7780154 -0.2601806
##
## $mean
##                     [,1]
## (Intercept) 0.04588209
## x           2.36006846
##
## $covar
##              (Intercept)          x
## (Intercept) 0.07319446 0.01382804
## x           0.01382804 0.10828240
```

Collecting the relevant data of a model fit like this together in a list, so we always know we are working on the values that belong together, makes further analysis of the fitted model much easier to program.

Classes

The output we got when we wrote

```
model
```

is what we get if we call the print function on a list. It just shows us everything that is contained within the list. The print function is an example of a polymorphic function, however, so when you call print(x) on an object x, the behavior depends on the class of the object x.

If you want to know what class an object has, you can use the class function:

```
class(model)
```

```
## [1] "list"
```

and if you want to change it, you can use the `class<-` replacement function:

```
class(model) <- "blm"
```

We can use any name for a class; here, I've used blm for Bayesian linear model.

By convention, we usually call the class and the function that creates elements of that class the same name, so since we are creating this type of objects with the blm function, convention demands that we call the class of the object blm as well. It is just a convention, though, so you can call the class anything.

We can always assign a class to an object in this way, but changing the class of an existing object is considered bad style. We keep the data that belongs together in a list to make sure that it is kept consistent, but the functionality we want to provide for a class is as much a part of the class as the data, so we also need to make sure that the functions that operate on objects of a given class always get data that is consistent with that class. We cannot do that if we go around changing the class of objects willy-nilly.

The function that creates the object should assign the class, and then we should leave the class of the object alone. We can set the class with the `class<-` function and then return it from the blm function:

```
blm <- function(model, alpha = 1, beta = 1, ...) {

  # stuff happens here...

  object <- list(formula = model,
                 frame = frame,
                 mean = mean,
                 covar = covar)
  class(object) <- "blm"
  object
}
```

The class is represented by an attribute of the object, however, and there is a function that sets these for us, structure, and using that we can create the object and set the class at the same time, which is a little better:

```
blm <- function(model, alpha = 1, beta = 1, ...) {

    # stuff happens here...

    structure(list(formula = model,
                   frame = frame,
                   mean = mean,
                   covar = covar),
              class = "blm")
}
```

Now that we have given the model object a class, let's try printing it again:

```
model
```

```
## $formula
## y ~ x
##
## $frame
##             y          x
## 1   -1.5185867 -0.5588255
## 2   -1.1664514 -0.6097720
## 3   -1.2509896  0.1808256
## 4   -1.9412380 -0.6595195
## 5   -0.5012965 -0.3030505
## 6    0.5057768  0.2613801
## 7    2.1098271  0.7792971
## 8    5.0790285  1.3976716
## 9   -3.2896676 -1.6030947
## 10  -0.7780154 -0.2601806
##
## $mean
##                     [,1]
## (Intercept) 0.04588209
## x           2.36006846
```

```
##
## $covar
##            (Intercept)        x
## (Intercept) 0.07319446 0.01382804
## x           0.01382804 0.10828240
##
## attr(,"class")
## [1] "blm"
```

The only difference from before is that it has added information about the `"class"` attribute toward the end. It still just prints everything that is contained within the object. This is because we haven't told it to treat any object of class blm any differently yet.

Polymorphic Functions

The `print` function is a polymorphic function. This means that what happens when it is called depends on the class of its first parameter. When we call `print`, R will get the class of the object, let's say it is blm as in our case, and see if it can find a function named `print.blm`. If it can, then it will call this function with the parameters you called `print` with. If it cannot, it will instead try to find the function `print.default` and call that.

We haven't defined a print function for the class blm, so we saw the output of the default print function instead.

Let us try to define a blm-specific print function:

```
print.blm <- function(x, ...) {
    print(x$formula)
}
```

Here, we just tell it to print the formula we used for specifying the model rather than the full collection of data we put in the list.

If we print the model now, this is what happens:

```
model
```

```
## y ~ x
```

That is how easy it is to provide your own class-specific `print` function. And that is how easy it is to define your own class-specific polymorphic function in general. You just take the function name and append `.classname` to it, and if you define a function with that name, then that function will be called when you call a polymorphic function on an object with that class.

One thing you do have to be careful about, though, is the interface to the function. By that I mean the parameters the function takes (and their order). Each polymorphic function takes some arguments; you can see which by checking the function documentation:

`?print`

When you define your specialized function, you can add more parameters to your function, but you should define it such that you at least take the same parameters as the generic function does. R will not complain if you do not define it that way, but it is bound to lead to problems later on when someone calls the function with assumptions about which parameters it takes based on the generic interface and then runs into your implementation of a specialized function that behaves a different way. Don't do that. Implement your function so it takes the same parameters as the generic function. This includes using the same names for parameters. Someone might provide named parameters to the generic function, and that will only work if you call the parameters the same names as the generic function. That is why we used x as the input parameter for the `print.blm` function earlier.

Defining Your Own Polymorphic Functions

To define a class-specific version of a polymorphic function, you just need to write a function with the right name. There is a little bit more to do if you want to define your very own polymorphic function. Then you also need to write the generic function—the function you will use when you have objects of different types, and that is responsible for dispatching the function call to class-specific functions.

You do this using the `UseMethod` function. The generic function typically just does this and looks something like this:

```
foo <- function(x, ...) UseMethod("foo")
```

You specify a function with the parameters the generic function should accept and then just call UseMethod with the name of the function to dispatch to. Then it does it magic and finds out which class-specific function to call and forwards the parameters to there.

When you write the generic function, it is also good style to define the default function as well:

```
foo.default <- function(x, ...) print("default foo")
```

With that, we can call the function with all types of objects. If you don't want that to be possible, a safe default function would be one that throws an error:

```
foo("a string")
```

```
## [1] "default foo"
```

```
foo(12)
```

```
## [1] "default foo"
```

And of course, with the generic function in place, we can define class-specific functions just like before:

```
foo.blm <- function(x, ...) print("blm foo")
foo(model)
```

```
## [1] "blm foo"
```

You can add more parameters to more specialized functions when the generic function takes ... as an argument; the generic will just ignore the extra parameters, but the concrete function that is called might be able to do something about it:

```
foo.blm <- function(x, upper = FALSE, ...) {
  if (upper) {
    print("BLM FOO")
  } else {
    print("blm foo")
  }
}
```

```
foo("a string")
```

```
## [1] "default foo"
```

```
foo(model)
```

```
## [1] "blm foo"
```

```
foo("a string", upper = TRUE)
```

```
## [1] "default foo"
```

```
foo(model, upper = TRUE)
```

```
## [1] "BLM FOO"
```

Class Hierarchies

Polymorphic functions are one aspect of object-oriented programming; another is inheritance. This is the mechanism used to build more specialized classes out of more general classes.

The best way to think about this is as levels of specialization. You have some general class of objects, say "furniture," and within that class are more specific categories, say "chairs," and within that class even more specific types of objects, say "kitchen chairs." A kitchen chair is also a chair, and a chair is also furniture. If there is something you can do to all furniture, then you can definitely also do it to chairs. For example, you can set furniture on fire; you can set a chair on fire. It is not the case, however, that everything you can do to chairs you can do to all furniture. You can throw a chair at unwelcome guests, but you cannot throw a piano at them (unless you are the Hulk).

The way specialization like this works is that there are some operations you can do for the general classes. Those operations can be done on all instances of those classes, including those that are really objects of more specialized classes.

The operations might not do exactly the same thing—like we can specialize print, an operation we can call on all objects, to do something special for blm objects—but there is some meaningful way of doing the operation. Quite often, the way a class is specialized is exactly by doing an operation that can be done by all objects from the general class, but just in a more specialized way.

The specialized classes, however, can potentially do more, so they might have more operations that are meaningful to do to them. That is fine, as long as we can treat all objects of a specialized class as we can treat objects of the more general class.

This kind of specialization is partly interface and partly implementation.

Specialization As Interface

The interface is the set of which functions we can call on objects of a given class. It is a kind of protocol for how we interact with objects of the class. If we imagine some general class of "fitted models," we might specify that for all models, we should be able to get the fitted parameters, and we should be able to make predictions for new values. In R, such functions exist, `coef` and `predict`, and any model is expected to implement them.

This means that I can write code that interacts with a fitted model through these general model functions, and as long as I stick to the interface they provide, I could be working on any kind of model. If, at some point, I find out that I want to replace a linear regression model with a decision tree regression model, I can just plug in a different fitted model and communicate with it through the same polymorphic functions. The actual functions that will be called when I call the generic functions `coef` and `predict` will, of course, be different, but the interface is the same.

R will not enforce such interfaces for you. Classes in R are not typed in the same way as they are in, for example, Java, where it would be a type error to declare something as an object satisfying a certain interface if it does in fact not implement the necessary functions. R doesn't care. Not until you call a function that isn't there; then you might be in trouble, of course. But it is up to you to implement an interface to fit the kind of class or protocol you think your class should match.

If you implement the functions that a certain interface expects (and these functions actually do something resembling what the interface expects the functions to do and are not just named the same things),[1] then you have a specialization of that interface. You can do the same operations as every other class that implements the interface, but, of course, your operations are uniquely fitted to your actual class.

---

[1] To draw means something very different when you are a gunslinger compared to when you are an artist, after all.

You might implement more functions, making your class capable of more than the more general class of objects, but that is just fine. And other classes might implement those operations as well, so now you have more than one class with the more specialized operations—a new category that is more general and can be specialized further.

You have a hierarchy of classes defined by which functions they provide implementations of.

Specialization in Implementations

Providing general interfaces and then specializing them to specific kinds of objects—in the case of R by providing implementations of polymorphic functions—is the essential feature of the concept of class hierarchies in object-oriented programming. It is what lets you treat objects of different kinds as a more general class.

There is another aspect of class hierarchies, though, that has to do with code reuse. You already get a lot of this just by providing interfaces to work with objects, of course, since you can write code that works on a general interface and then reuse it on all objects that implement this interface. But there is another type of reuse you get when you build a hierarchy of classes where you go from abstract, general classes to more specialized and concrete classes. When you are specializing a class, you are taking functionality that exists for the more abstract class and defining a new class that implements the same interface except for a few differences here and there.

When you refine a class in this way, you don't want to implement new versions of all the polymorphic functions in its interface. Many of them will do exactly the same as the implementation for their more general class.

Let's say we want to have a class of objects where you can call functions foo and bar. We can call that class A and define it as follows:

```
foo <- function(object, ...) UseMethod("foo")
foo.default <- function(object, ...) stop("foo not implemented")

bar <- function(object, ...) UseMethod("bar")
bar.default <- function(object, ...) stop("bar not implemented")

A <- function(f, b) structure(list(foo = f, bar = b), class = "A")
foo.A <- function(object, ...) paste("A::foo ->", object$foo)
bar.A <- function(object, ...) paste("A::bar ->", object$bar)
```

```
a <- A("qux", "qax")
foo(a)
```

```
## [1] "A::foo -> qux"
```

```
bar(a)
```

```
## [1] "A::bar -> qax"
```

For a refinement of that, we might want to change how bar works and add another function baz:

```
baz <- function(object, ...) UseMethod("baz")
baz.default <- function(object, ...) stop("baz not implemented")
```

```
B <- function(f, b, bb) {
  a <- A(f, b)
  a$baz <- bb
  class(a) <- "B"
  a
}
```

```
bar.B <- function(object, ...) paste("B::bar ->", object$bar)
baz.B <- function(object, ...) paste("B::baz ->", object$baz)
```

The function foo we want to leave just the way it is, but if we define the class B as before, calling foo on a B object gives us an error because it will be calling the foo. default function:

```
b <- B("qux", "qax", "quux")
foo(b)
```

```
## Error in foo.default(b): foo not implemented
```

This is because we haven't told R that we consider the class B a specialization of class A. We wrote the constructor function—the function where we make the object, the function B—such that all B objects contain the data that is also found in an A object. We never told R that we intended B objects also to be A objects.

We could, of course, make sure that foo called on a B object behaves the same way as if called on an A object by defining foo.B such that it calls foo.A. This wouldn't be too much work for a single function, but if there are many polymorphic functions that work on A objects, we would have to implement B versions for all of them—a tedious and error-prone work.

If only there were a way of telling R that the class B is really an extension of the class A. And there is. The class attribute of an object doesn't have to be a string. It can be a vector of strings. If, for B objects, we say that the class is B first and A second, like this:

```r
B <- function(f, b, bb) {
  a <- A(f, b)
  a$baz <- bb
  class(a) <- c("B", "A")
  a
}
```

then calling foo on a B object—where foo.B is not defined—will call foo.A as its second choice and before defaulting to foo.default:

```r
b <- B("qux", "qax", "quux")
foo(b)
```

```
## [1] "A::foo -> qux"
```

```r
bar(b)
```

```
## [1] "B::bar -> qax"
```

```r
baz(b)
```

```
## [1] "B::baz -> quux"
```

The way the class attribute is used with polymorphic functions is that R will look for functions with the class names appended in the order of the class attributes. The first it finds will be the one that is called, and if it finds no specialized version, it will go for the .default version. When we set the class of B objects to be the vector c("B", "A"), we are saying that R should call .B functions first, if it can find one, but otherwise call the .A function.

This is a very flexible system that lets you implement multiple inheritances from classes that are otherwise not related, but you do so at your own peril. The semantics of how this works—functions are searched for in the order of the class names in the vector—the actual code that will be run can be hard to work out if these vectors get too complicated.

Another quick word of caution is this: if you give an object a list of classes, you should include the classes all the way up the class hierarchy. If we define a new class, C, intended as a specialization of B, we cannot just say that it is an object of class c("C", "B") if we also want it to behave like an A object:

```
C <- function(f, b, bb) {
  b <- B(f, b, bb)
  class(b) <- c("C", "B")
  b
}

c <- C("foo", "bar", "baz")
foo(c)
```

```
## Error in foo.default(c): foo not implemented
```

When we call foo(c) here, R will try the functions, in turn: foo.C, foo.B, and foo.default. The only one that is defined is the last, and that throws an error if called.

So what we have defined here is an object that can behave like B but only in cases where B differs from A's behavior! Our intention is that B is a special type of A, so every object that is a B object we should also be able to treat as an A object. Well, with C objects, that doesn't work.

We don't have a real class hierarchy here like we would find in languages like Python, C++, or Java. We just have a mechanism for calling polymorphic functions, and the semantic here is just to look for them by appending the names of the classes found in the class attribute vector. Your intentions might very well be that you have a class hierarchy with A being the most general class, B a specialization of that, and C the most specialized class, but that is not what you are telling R—because you cannot. You are telling R how it should look for dynamic functions, and with the preceding code, you told it to look for .C functions first, then .B functions, and you didn't tell it any more, so the next step it will take is to look for .default functions, not .A functions. It doesn't know that this is where you want it to look.

If you add this to the class attribute, it will work, though:

```
C <- function(f, b, bb) {
  b <- B(f, b, bb)
  class(b) <- c("C", "B", "A")
  b
}

c <- C("foo", "bar", "baz")
foo(c)

## [1] "A::foo -> foo"

bar(c)

## [1] "B::bar -> bar"

baz(c)

## [1] "B::baz -> baz"
```

You are slightly better off getting the class attribute from the object you create in the constructor, though. If, at some point, you changed the class attribute of the object returned from the B() constructor, you don't want to have to change the class vector in all classes that are extending the class:

```
C <- function(f, b, bb) {
  b <- B(f, b, bb)
  class(b) <- c("C", class(b))
  b
}
```

Exercises

Shapes

Let us imagine that we need to handle some geometric shapes for a program. These could be circles, squares, triangles, etc. Properties we need to know about the shapes are their circumference and area. These properties can be calculated from properties of the shapes, but the calculations are different for each shape.

388

So for our shapes, we want (at least) an interface that gives us two functions: `circumference` and `area`. The default functions, where we have no additional information about an object aside from the fact that it is a shape, are meaningless and so should raise an error (check the `stop` function for this), but each specialized shape should implement these two functions.

Implement this protocol/interface and the two functions for at least circles and rectangles—by all means, more shapes if you want to.

Polynomials

Write a class that lets you represent polynomial objects. An n-degree polynomial is on the form $c_0 + c_1x + c_2x^2 + \cdots + c_nx^n$ and can be represented by the $n + 1$ coefficients $(c_0, c_1, \ldots, c_n)$. Write the interface such that you can evaluate polynomials in any point x, that is, with a function `evaluate_polynomial(poly, x)` that gives you the value of the polynomial at the point x.

The function `uniroot` (built into R) lets you find the roots of a general function. Use it to write a function that finds the roots of your polynomials. This function works by numerically finding the points where the polynomial is zero. For lines and quadratic polynomials, though, there are analytical solutions. Write special cases for such polynomials such that calling the root finding function on the special cases exploits that solutions are known there.

CHAPTER 13

Building an R Package

Now we know how to write functions and create classes in R, but neither functions nor classes are the unit we use for collecting and distributing R code. That unit is the package. It is packages you load and import into your namespace when you write

```
library(something)
```

and it is packages you download when you write

```
install.packages("something")
```

The topic for this chapter is how to make your own packages. In the space available, I can only give a very broad overview of the structure of R packages, but it should be enough to get you started. If you want to read more, I warmly recommend Hadley Wickham's book *R Packages*.

Creating an R Package

I am going to assume that you use RStudio for this. If you don't, you can have a look at the package devtools. It provides functions for doing everything we can do through the GUI in RStudio.

To create a new package, go to the menu File and choose New Project…, and you should get a dialog that asks you whether your new project should be in a new directory, in an existing directory, or checked out of a version control repository. Pick the New Directory.

After that, you get the choice between an empty project, a package, or a Shiny application. Not surprisingly, you want to pick R Package here.

Now you get to a dialog window where you can set the details of the package. You can choose the Type of the package (where you can choose between a plain package and one that uses Rcpp to make C++ extensions), you can specify the Name of the package,

© Thomas Mailund 2022
T. Mailund, *Beginning Data Science in R 4*, https://doi.org/10.1007/978-1-4842-8155-0_13

and you can provide existing source files to include in the package. Further, you need to choose a location to put the new package and whether you want to use a `git` repository for the package.

Choose a plain package and click yes to creating a `git` repository (we will return to `git` later). You now just need to pick a name and a place to put the package. Where you put it is up to you, but there are some guidelines for package names.

Package Names

A package name can consist of letters, numbers, and ".", but must start with a letter and must not have "." as the last character. You cannot use other characters, such as underscores or dashes.

Whenever you build software that you intend for other people to be able to use, be careful with the name you give it. Give it a name that is easy to remember and easy to Google.

For experimenting with packages, you can just create one called `test`.

Create a new package and have a look at the result.

The Structure of an R Package

In the directory that RStudio built for you, you should have two directories, `R` and `man`; three text files, `.Rbuildignore`, `DESCRIPTION`, and `NAMESPACE`; and one project file (its name will be the name of your package followed by `.Rproj`).

The last of these files is used by RStudio, and all you need to know about it is that if you open this file in RStudio, you get an open version of the state of the project you had last time you worked on it.

Inside the R directory, you have an example file, `R/hello.R`, and inside the `man` directory, you have an example documentation[1] file, `man/hello.Rd`.

The text files and the two directories are part of what an R package looks like, and they must always be there with exactly those names. There are a few more directories that also have standard names,[2] but they are not required, and we don't have them here for now.

---

[1] man stands for manual, and the abbreviation man is a legacy from UNIX.

[2] For example, `vignettes/` for documentation vignettes, `data/` for data you want to include with your package, and `src/` for C/C++ extensions.

.Rbuildignore

The directory you have created contains the source code for the package, but it isn't the actual package. The package is something you need to build and install from this source code. We will get to how to do that shortly.

The `.Buildignore` file tells R what not to include when it builds a package. Files that are not mentioned here will automatically be included. This isn't a disaster as such, but it does lead to messy packages for others to use, and if you upload a package to CRAN,[3] the filters there do enforce a strict directory and file structure where you will not be allowed to include files or directories that do not follow that structure.

The automatically generated `.Buildignore` file looks like this:

```
^.*\.Rproj$
^\.Rproj\.user$
```

These are two regular expressions that prevent R from including the RStudio files in compiled packages.

The ^ character here matches the beginning of a file name, while $ matches the end. A non-escaped `.` matches any character, while an escaped `\.` matches an actual dot. The * specifies that the previous symbol can be repeated any number of times. So the first regular expression specifies any file name that ends in `.Rproj` and the second expression any file name that ends in `.Rproj.user`.

Description

This file contains meta-information about your package. If you called your package `test` and created it the same day I did (November 11, 2015), it should now look like this:

```
Package: test
Type: Package
Title: What the Package Does (Title Case)
Version: 0.1
Date: 2015-11-22
Author: Who wrote it
```

---

[3] CRAN is the official repository for R package and the place where the `install.packages` function finds them.

```
Maintainer: Who to complain to <yourfault@somewhere.net>
Description: More about what it does (maybe more than one line)
License: What license is it under?
LazyData: TRUE
```

You need to update it to describe your new package.

I give a short description of the metadata in the following, but you can also read more about it in Hadley Wickham's *R Packages* book.

Title

The title field is pretty self-explanatory. You need to give your package a title. Here, (Title Case) means that you need to use capital first letters in the words there like you would for the title of a book.

If you read the documentation for a package on CRAN, it will look like this: packagename: This is the Title. Don't include the package name in your title here; that is automatically added to the documentation page. You just want the title.

Version

This is just a number to track which version of your package people have installed. Whenever you make changes to your package and release them, this number should go up.

The version numbers are not only used to indicate that you have updated a version, but they are also necessary for specifying dependencies between packages sometimes. If a feature was introduced in version 1.2 but didn't exist in version 1.1, then other packages that use this feature need to know whether they have access to version 1.2 or higher. We will return to dependencies later.

There are some conventions for version numbers but nothing that is strictly enforced. The convention here is that a released version has the numbering scheme major.minor. patch, so the version 1.2.3 means that the major version number is 1, the minor 2, and that this is patched version 3. Patches are smaller changes, typically bug fixes and such, while minor revisions usually include some new functionality. The difference between what is considered minor and major is very subjective, but any time the interface changes—that is, you change the way a function is called such that the old types of calls are now incorrect—you definitely should increase the major version number.

If you have a development version of your package that you are distributing for those adventurous enough to work with a beta release, the convention is to add a development release number as well. Then the version number looks like `major.minor.patch.develop-number` where by convention the last number starts at 9000 and is increased with every new release.

You are just beginning developing your new package, so change the version number to `0.0.0.9000`.

Description

This field should describe the package. It is typically a one-paragraph short description. To make R parse the `DESCRIPTION` file correctly, you must indent the lines following `Description:` if the description spans over multiple lines.

Author and Maintainer

Delete these two fields. There is a better way to specify the same information that makes sure that it is provided in a more structured form. You should use a new field called `Authors@R:` instead. There is nothing wrong with the `Author` or `Maintainer` fields, and you can keep them without any trouble; the new field just lets you provide the same information more succinctly.

This field takes an R expression specifying one or more authors where the author information is provided by a call to the function `person`—which is how we make sure that it is structured appropriately. Check the documentation for the function (`?person`) for more details.

You are a single author, so you should use something like this:

```
Authors@R: person("First Name", "Last Name",
                email = "your.email@your.domain.com",
                role = c("aut", "cre"))
```

The roles here mean author and creator. The documentation for the `person` function lists other options.

If there is more than one person involved as an author or a maintainer or another sort of contributor, you can have a sequence of `persons` by concatenating them with the `c` function.

License

This specifies the software license the package is released under. It can really be anything, but if you want to put your package on CRAN, you have to pick one of the licenses that CRAN accepts.

You specify which of the recognized licenses you want to use by their abbreviation, so to specify that your package is released under the GPL version 2 license, you write

```
License: GPL-2
```

Type, Date, LazyData

The Type and LazyData fields are not essential. You can delete them if you want. Type is just saying that you have a package, but we sort of know that already. LazyData tells R that if you include data in your package, it should load it lazily. Again, this is not something that is of particular importance (unless you plan to include extremely large data sets with your package; if you do that, then Google for the documentation of LazyData).

The Date of course includes the date. This should be the last date you modified the package, that is, the last time you updated the version.

URL and BugReports

If you have a web page for the package and a URL for reporting bugs, these are the fields you want to use. They are not required for a package but are of course very helpful for a user to have.

Dependencies

If your package has dependencies, you have three fields you can specify them in: Depends, Imports, and Suggests.[4]

---

[4] There are a few more fields for, e.g., linking to external C/C++ code, but these three fields are the most important fields.

With Depends, you can specify both packages that need to be installed for your package to work and which version of R is required for your package to work. For packages, though, it is better to use Imports and Suggests than Depends, so use Depends only to specify which version of R you need.

You specify this like

```
Depends: R (>= 4.1)
```

This is saying that you need R to work (not surprisingly, but the syntax is the same for packages), and it has to be at least version 4.1.

The syntax for dependencies is a comma-separated list of package names (or R as before) with optional version number requirements in parentheses after the package name.

Imports and Suggests fields could look like this:

```
Imports:
  ggplot2,
  dplyr (>= 1.0.7),
  pracma
Suggests:
  testthat,
  knitr
```

specifying that we import three packages, ggplot2, dplyr, and pracma, and we use testthat and knitr in some functions if these packages are available. We require that dplyr has at least version 1.0.7, but do not put any demands on the versions of the other packages. (The required version for dplyr is completely arbitrary here; it just happens to be the version I have installed as I am writing this. Don't read anything more into it.)

The difference between Imports and Suggests is that requirements in Imports must be installed for your package to be installed (or they will be installed if you tell R to install with dependencies), while requirements in Suggests do not.

Using an Imported Package

Packages in the Imports or Suggests lists are not imported into your namespace the way they would be if you call library(package). This is to avoid contaminating your package namespace, and you shouldn't break that by calling library yourself. If you want to use functions from other packages, you must do so by explicitly accessing them through their package namespace or by explicitly importing them at a single function level.

The way to access a function from another package without importing the package namespace is using the :: notation. If you want to get to the filter function in dplyr without importing dplyr, you can get the function using the name dplyr::filter.

If you access names from a package that you have listed in your Imports field, then you know that it exists even if it isn't imported into your namespace, so you just need to use the long name.

An alternative way of importing functions is using Roxygen—which we will discuss later—where you can import the namespace of another package or just the name of a single function in another package for a single function at a time.

Using a Suggested Package

Accessing functions in a suggested package—the packages named in the Suggests field—is done using the :: notation, just as you would for imported packages. There is just one more complication: the package might not be installed on the computer where your package is installed. That is the difference between suggesting a dependency and requiring it by putting it in the Imports field.

The purpose of suggesting packages instead of importing them is that the functionality your package provides doesn't strictly depend on the other package, but you can do more or do things more efficiently if a suggested package is there.

So you need a way of checking if a package is installed before you use it, and that way is the function requireNamespace. It returns TRUE if the namespace (package) you ask for is installed and FALSE otherwise, so you can use it like this:

```
if (requireNamespace("package", quietly = TRUE)) {
    # use package functionality
} else {
    # do something that doesn't involve the package
    # or give up and throw an exception with stop()
}
```

The quietly option is to prevent it from printing warnings—you are handling the cases where the package is not installed, so there is no need for it to do it.

NAMESPACE

The NAMESPACE file provides information about which of the functions you implement in your package should be exported to the namespace of the user when they write library(test).

Each package has its own namespace. It is similar to how each function has a namespace in its body where we can define and access local variables. Functions you write in a package will look for other functions first in the package namespace and then in the global namespace.

Someone who wants to use your package can get access to your function by loading it into their namespace using

library(test)

or by explicitly asking for a function in your namespace

test::function_name()

but they can only get access to functions (and other objects) explicitly exported.[5] If a function is not explicitly exported, it is considered an implementation detail of the package that code outside the package should not be able to access.

The NAMESPACE file is where you specify what should be exported from the package.[6]

The auto-generated file looks like this:

exportPattern("^[[:alpha:]]+")

It is just exporting anything that has an alphanumeric name. This is definitely too much, but we ignore it for now. We are not going to edit this file manually since we can export functions (and all other objects) much easier using Roxygen as described in the following.

---

[5] Strictly speaking, this is not true. You can actually get to internal functions if you use the ::: operator instead of the :: operator, so if function_name is not exported but still implemented in package test, then you can access it with test:::function_name. But you shouldn't. You should keep your damned dirty paws away from internal functions! They can change at any time, with no warning from the package maintainer, and no one will feel sorry for you when they do and your own code breaks because of it.

[6] It is also used to import selected functions or packages, but using Roxygen @import and @importFrom are better solutions for that.

R/ and man/

The R/ directory is where you should put all your R code, and the man/ directory is where the package documentation goes. There is one example file in both directories just after RStudio has generated your new package. You can have a look at them and then delete them afterward.

All the R code you write for a package should go in files in the R/ directory to be loaded into the package. All documentation will go in man/, but we are not going to write the documentation there manually. Instead, we will use Roxygen to document functions, and then Roxygen will automatically make the files that go in man/.

Checking the Package

Before we look at Roxygen, and start adding functionality to our package, I want you to check that it is in a consistent state. There are a number of consistency requirements that packages should satisfy, mostly related to file names, naming conventions, and such, and it is best to frequently check if your package looks the way it should. To check this, go to the Build menu and pick Check Package. You can also do it on the command line using the devtools package:

```
install.packages("devtools)
```

and the function check():

```
devtools::check()
```

It will run a bunch of checks, more every time they update devtools it seems, but at the end, it will tell you if your package is okay. If you run it now, you should get zero errors, zero warnings, and zero notes.

As we start modifying the package, run check() from time to time. If something breaks the package's consistency, it is better to know early, so you know what you broke and can easily fix it. If you have made tons of changes, it can be harder to track down what changes were a problem.

Roxygen

Roxygen is a system for writing documentation for your packages, and if you are familiar with Javadoc, you will recognize its syntax. It does a few things more, however, including handling your namespace import and export, as we will see.

To use it, you first have to install it, so run

```
install.packages("roxygen2")
```

Now go into the Build menu and select Configure Build Tools.... There, pick Build Tools and check Generate documentation with Roxygen, and in the dialog that pops up, check Build & Reload. This makes sure that Roxygen is used to generate documentation and that the documentation is generated when you build the package. This will also make sure that Roxygen handles the import and export of namespaces.

Documenting Functions

We can see how Roxygen works through an example:

```
#' Add two numbers
#'
#' This function adds two numbers together.
#'
#' @param x A number
#' @param y Another number
#' @return The sum of x and y
#'
#' @export
add <- function(x, y) x + y
```

The documentation for this function, add, is provided in comments above the function, but comments starting with the characters #' instead of just #. This is what tells Roxygen that these comments are part of the documentation that it should process.

The first line becomes the title of the documentation for the function. It should be followed by an empty line (still in #' comments).

The text that follows is a description of the function. It is a bit silly with the documentation for this simple function, but normally you will have a few paragraphs describing what the function does and how it is supposed to be used. You can write as much documentation here as you think is necessary.

The lines that start with an @ tag—for example, @param and @return—contain information for Roxygen. They provide information that is used to make individual sections in the documentation.

The @param tags are used for describing parameters. That tag is followed by the name of a parameter and after that a short description of the parameter.

The @return tag provides a description of what the function returns.

After you have written some documentation in Roxygen comments, you can build it by going into the menu Build and choosing Document. Roxygen will not overwrite the existing NAMESPACE file, because it didn't generate the file itself, so delete it before you run Document. That way, Roxygen is free to write to it. You only need to do this once; after that, Roxygen recognizes that it is a file that it controls.

After you have built the documentation, take a look at the NAMESPACE file and the man/ directory. In the NAMESPACE file, you should see that the function has been exported:

```
# Generated by roxygen2: do not edit by hand

export(add)
```

and in the man/ directory, there should be a file, add.Rd, documenting the function.

Import and Export

In the NAMESPACE file, you should see that your documented function is explicitly exported. That is because we provided the @export tag with the documentation. It tells Roxygen to export it from the package namespace.

This is the easiest way to handle the namespace export, so, if for nothing else, you should use Roxygen for this rather than manually editing the NAMESPACE file.

Roxygen will also make sure that polymorphic functions and other kinds of objects are correctly exported if you use the @export tag—something that requires different kinds of commands in the NAMESPACE file. You don't have to worry about it as long as you use Roxygen.

Roxygen can also handle import of namespaces. Remember that the packages you list in your Imports field in the DESCRIPTION file are guaranteed to be installed on the computer where your package is installed but that the namespaces of these packages are not imported. You have to use the :: notation to access them.

Well, with Roxygen you can use the tag @importFrom package object to import object (typically a function) into your namespace in a function that you give that tag to. For normal functions, I don't really see the point of using this feature since it isn't shorter to write than just using the :: notation. For infix functions, though, it makes them easier to use since then you can actually use the infix function as an infix operator.

So in the following function, we can use the %>% operator from dplyr because we import it explicitly. You cannot really get to infix operators otherwise.

```
#' Example of using dplyr
#'
#' @param data A data frame containing a column named A
#' @param p    A predicate function
#' @return The data frame filtered to those rows where p is true on A
#'
#' @importFrom dplyr filter
#' @importFrom dplyr %>%
#' @export
filter_on_A <- function(data, p) {
    data %>% filter(p(A))
}
```

If you write a function that uses a lot of functionality from a package, you can also import the entire namespace of that package. That is similar to using library(package) and is done with the @import tag:

```
#' @import dplyr
#' @export
filter_on_A <- function(data, p) {
    data %>% filter(p(A))
}
```

Package Scope vs. Global Scope

A quick comment is in order about the namespace of a package when you load it with `library(package)`. I mentioned it earlier, but I just want to make it entirely clear. A package has its own namespace where its functions live. Functions that are called from other functions written inside a package are first looked for in the package namespace before they are looked for in the global namespace.

If you write a function that uses another function from your package and someone redefines the function in the global namespace after loading your package, it doesn't change what function is found inside your package.

It doesn't matter if a function is exported or local to a package for this to work. R will always look in a package namespace before looking in the global namespace.

Internal Functions

You might not want to export all functions you write in a package. If there are some functions, you consider implementation details of your package design, you shouldn't export them. If you do, people might start to use them, and you don't want that if it is functionality you might change later on when you refine your package.

Making functions local, though, is pretty easy. You just don't use the `@export` tag. Then they are not exported from the package namespace when the package is loaded, and then they cannot be accessed from outside the package.[7]

File Load Order

Usually, it shouldn't matter in how many files you write your package functionality. It is usually easiest to find the right file to edit if you have one file for each (major) function or class, but it is mostly a matter of taste.

It also shouldn't matter in which files various functions are put—whether internal or exported—since they will all be present in the package namespace. And if you stick to using functions (and S3 polymorphic functions), the order in which files are processed when building packages shouldn't matter.

---

[7] Except through the `:::` operator, of course, but people who use this to access the internals of your package know—or should know—that they are accessing implementation details that could change in the future, so it is their own fault if their code is broken sometime down the line.

It does matter for S4 classes and such, and in case you ever run into it being an issue, I will just quickly mention that package files are processed in alphabetical order. Alphabetical for the environment you are in, though, since alphabetical order actually depends on which language you are in, so you shouldn't rely on this.

Instead, you can use Roxygen. It can also make sure that one file is processed before another. You can use the @include field to make a dependency between a function and another file:

```
#' @import otherfile.R
```

I have never had the need for this myself, and you probably won't either, but now you know.

Adding Data to Your Package

It is not uncommon for packages to include some data, either data used by the package implementation or more commonly data used for example purposes.

Such data goes in the data/ directory. You don't have this directory in your freshly made package, but it is where data should go if you want to include data in your package.

You cannot use any old format for your data. It has to be in a file that R can read, typically .RData files. The easiest way to add data files, though, is using functionality from the devtools package. If you don't have it installed, then

```
install.packages("devtools")
```

and then you can use the use_data function to create a data file.

For example, I have a small test data set in my admixturegraph package that I made using the command

```
bears <- read.table("bears.txt")
devtools::use_data(bears)
```

This data won't be directly available once a package is loaded, but you can get it using the data function:

```
library(admixturegraph)
data(bears)
bears
```

You cannot add documentation for data directly in the data file, so you need to put it in an R file in the R/ directory. I usually have a file called data.R that I use for documenting my package data.

For the bears data, my documentation looks like this:

```
#' Statistics for populations of bears
#'
#' Computed $f_4(W,X;Y,Z)$ statistics for different
#' populations of bears.
#'
#' @format A data frame with 19 rows and 6 variables:
#' \describe{
#'   \item{W}{The W population}
#'   \item{X}{The X population}
#'   \item{Y}{The Y population}
#'   \item{Z}{The Z population}
#'   \item{D}{The $D$ ($f_4(W,X;Y,Z)$) statistics}
#'   \item{Z.value}{The blocked jacknife Z values}
#' }
#'
#' @source \url{http://onlinelibrary.wiley.com/doi/10.1111/
mec.13038/abstract}
#' @name bears
#' @docType data
#' @keywords data
```

NULL

The NULL after the documentation is needed because Roxygen wants an object after documentation comments, but it is the @name tag that tells it that this documentation is actually for the object bears. The @docType tells it that this is data that we are documenting.

The @source tag tells us where the data is from; if you have generated it yourself for your package, you don't need this tag.

The @format tag is the only complicated tag here. It describes the data, which is a data frame, and it uses markup that looks very different from Roxygen markup text.

The documentation used by R is actually closer to (La)TeX than the formatting we have been using, and the data description reflects this.

You have to put your description inside curly brackets marked up with \ `description{}`, and inside it, you have an item per data frame column. This has the format \item{column name}{column description}.

Building an R Package

In the frame to the upper right in RStudio, you should have a tab that says Build. Select it.

Inside the tab, there are three choices in the toolbar: Build & Reload, Check, and More. They all do just what it says on the tin: the first builds and (re)loads your package, the second checks it—this means running the consistency checks we saw earlier—and the third gives you various other options in a drop-down menu.

You use Build & Reload to recompile your package when you have made changes to it. It loads all your R code (and various other things) to build the package, and then it installs it and reloads it into your terminal so you can test the new functionality.

A package you have built and installed this way can also be used in other projects afterward.

When you have to send a package to someone, you can make a source package in the More drop-down menu. It creates an archive file (`.tar.gz`).

Exercises

In the last chapter, you wrote functions for working with shapes and polynomials. Now try to make a package for each with documentation and correct exporting of the functions. If you haven't implemented all the functionality for those exercises, this is your chance to do so.

CHAPTER 14

Testing and Package Checking

Without testing, there is little guarantee that your code will work at all. You probably test your code when you write it by calling your functions with a couple of chosen parameters, but to build robust software, you will need to approach testing more rigorously. And to prevent bugs from creeping into your code over time, you should test often. Ideally, you should check all your code anytime you have made any changes to it.

There are different ways of testing software—software testing is almost a science in itself—but the kind of testing we do when we want to make sure that the code we just wrote is working as intended is called unit testing. The testing we do when we want to ensure that changes to the code do not break anything is called regression testing.

Unit Testing

Unit testing is called that because it tests functional units—in R, that essentially means single functions or a few related functions. Whenever you write a new functional unit, you should write test code for that unit as well. The test code is used to check that the new code is actually working as intended, and if you write the tests such that they can be run automatically later on, you have also made regression tests for the unit at the same time. Whenever you make any changes to your code, you can run all your automated tests, and that will check each unit and make sure that everything works as it did before.

Most programmers do not like to write tests. It is exciting to write new functionality, but to probe new features for errors is a lot less interesting. However, you really do need the tests, and you will be happy that you have them in the long run. Don't delay writing tests until after you have written all your functions. That is leaving the worst for last, and that is not the way to motivate you to write the tests. Instead, you can write your unit tests while you write your functions; some even suggest writing them before you write your

© Thomas Mailund 2022
T. Mailund, *Beginning Data Science in R 4*, https://doi.org/10.1007/978-1-4842-8155-0_14

functions, something called test-driven programming. The idea here is that you write the tests that specify how your function should work, and you know that your function works as intended when it passes the tests you wrote for it.

I have never found test-driven programming that useful for myself. It doesn't match the way I work where I like to explore different interfaces and uses of a function while I am implementing it, but some prefer to work that way. I do, however, combine my testing with my programming in the sense that I write small scripts calling my functions and fitting them together while I experiment with the functions. I write that code in a way that makes it easy for me to take the experiments and then use them as automated tests for later.

Take, for example, the shape exercise we had earlier, where you had to write functions for computing the area and circumference of different shapes. I have written a version where I specify rectangles by width and height. A test of the two functions could then look like this:

```
area <- function(x) UseMethod("area")
circumference <- function(x) UseMethod("circumference")

rectangle <- function(width, height) {
    structure(list(width = width, height = height),
              class = c("rectangle", "shape"))
}

area.rectangle <- function(x) x$height * x$width
circumference.rectangle <- function(x) 2 * x$height + 2 * x$width

r <- rectangle(width = 2, height = 4)
area(r)
```

```
## [1] 8
```

```
circumference(r)
```

```
## [1] 12
```

The area is 2 × 4 and the circumference is 2 × 2 + 2 × 4, so this looks fine. But I am testing the code by calling the functions and looking at the printed output. I don't want to test the functions that way forever—I cannot automate my testing this way because I then have to sit and look at the output of my tests. But they are okay tests. I just need to automate them.

Automating Testing

All it takes to automate the test is to check the result of the functions in code rather than looking at it, so code that resembles the following code would be an automated test:

```
r <- rectangle(width = 2, height = 4)
if (area(r) != 2*4) {
    stop("Area not computed correctly!")
}
if (circumference(r) != 2*2 + 2*4) {
    stop("Circumference not computed correctly!")
}
```

It is a little more code, yes, but it is essentially the same test, and this is something I can run automatically later on. If it doesn't complain about an error, then the tests are passed, and all is good.

You can write your own test this way, put them in a directory called `tests/` (which is where R expects tests to live), and then run these tests whenever you want to check the status of your code, that is, whenever you have made modifications to it.

Scripts in the `tests/` directory will also be automatically run whenever you do a consistency check of the package (something we return to later).

That is what happens when you click Check in the Build tab on the right in RStudio or select Check Package in the Build menu, but it does a lot more than just run tests, so it is not the most efficient way of running the tests.

There are some frameworks for formalizing this type of testing in R. I use a framework called `testthat`. Using this framework, it is easy to run tests (without the full package check) and easy to write tests in a more structured manner—of course at the cost of having a bit more code to write for each test.

Install it now, if you do not have it already:

```
install.packages("testthat")
```

Using `testthat`

The `testthat` framework provides functions for writing unit tests and makes sure that each test is run in a clean environment (so you don't have functions defined in one test leak into another because of typos and such). It needs a few modifications to your `DESCRIPTION` file and your directory structure, but you can automatically make these adjustments by running

```
usethis::use_testthat()
```

This adds testthat to the Suggests packages and makes the directory tests/testthat and the file `tests/testthat.R`. You can have a look at the file, but it isn't that interesting. Its purpose is to make sure that the package testing—that runs all scripts in the `tests/` directory—will also run all the `testthat` tests.

The `testthat` tests should all go in the `tests/testthat` directory and in files whose names start with `test`. Otherwise, `testthat` cannot find them. The tests are organized in contexts and tests to make the output of running the tests more readable—if a test fails, you don't just want to know that some test failed somewhere, but you want some information about which test where, and that is provided by the contexts.

At the top of your test files, you set a context using the `context` function. It just gives a name to the following batch of tests. This context is printed during testing, so you can see how the tests are progressing, and if you keep to one context per file, you can see in which files tests are failing.

The next level of tests is wrapped in calls to the `test_that` function. This function takes a string as its first argument which should describe what is being tested and as its second argument a statement that will be the test. The statement is typically more than one single statement, and in that case, it is wrapped in {} brackets.

At the beginning of the test statements, you can create some objects or whatever you need for the tests, and after that, you can do the actual tests. Here, `testthat` also provides a whole suite of functions for testing if values are equal, almost equal, if an expression raises a working, triggers an error, and much more. All these functions start with `expect_`, and you can check the documentation for them in the `testthat` documentation.

The test for computing the area and circumference of rectangles earlier would look like this in a `testthat` test:

```
context("Testing area and circumference")

test_that("we compute the correct area and circumference", {
  r <- rectangle(width = 2, height = 4)

  expect_equal(area(r), 2*4)
  expect_equal(circumference(r), 2*2 + 2*4)
})
```

Try to add this test to your shapes packet from the last chapter's exercises and see how it works. Try modifying it to trigger an error and see how that works.

You should always worry a little bit when testing equality of numbers, especially if it is floating-point numbers. Computers do not treat floating-point numbers the way mathematics treat real numbers. Because floating-point numbers have to be represented in finite memory, the exact number you get will depend on how you compute it, even if mathematically two expressions should be identical.

For the tests we do with the preceding rectangle, this is unlikely to be a problem. There isn't really that many ways to compute the two quantities we test for, and we would expect to get exactly these numbers. But how about the quantities for circles?

```
circle <- function(radius) {
    structure(list(r = radius),
              class = c("circle", "shape"))
}
area.circle <- function(x) pi * x$r**2
circumference.circle <- function(x) 2 * pi * x$r

test_that("we compute the correct area and circumference", {
  radius <- 2
  circ <- circle(radius = radius)

  expect_equal(area(circ), pi * radius^2)
  expect_equal(circumference(circ), 2 * radius * pi)
})
```

Here, I use the built-in pi, but what if the implementation used something else? Here, we are definitely working with floating-point numbers, and we shouldn't ever test for exact equality. Well, the good news is that expect_equal doesn't. It actually tests for equality within some tolerance of floating-point uncertainty—that can be modified using an additional parameter to the function—so all is good. To check exact equality, you should instead use the function `expect_identical`, but it is usually `expect_equal` you want.

Writing Good Tests

The easiest way to get some tests written for your code is to take the experiments you make when developing the code and translate them into unit tests like this right away— or even put your checks in a unit test file, to begin with. By writing the tests at the same time as you write the functions—or at least immediately after—you don't build a backlog of untested functionality (and it can be very hard to force yourself to go and spend hours just writing tests later on). Also, it doesn't really take that much longer to take the informal testing you write to check your functions while you write them and put them into a `testthat` file and get a formal unit test.

If this is all you do, then at least you know that the functionality that was tested when you developed your code is still there in the future—or you will be warned if it breaks at some point because their tests will start to fail.

But if you are writing tests anyway, you might as well be a little more systematic about it. We always tend to check for the common cases—the cases we have in mind when we write the function—and forget about special cases. Special cases are frequently where bugs hide, however, so it is always a good idea to put them in your unit tests as well.

Special cases are situations such as empty vectors and lists or `NULL` as a list. If you implement a function that takes a vector as input, make sure that it also works if that vector is empty. If it is not a meaningful value for the function to take, and you cannot think of a reasonable value to return if the input is empty, then make sure the function throws an error rather than just do something that it wasn't designed to do.

For numbers, exceptional cases are often zero or negative numbers. If your functions can handle these cases, excellent (but make sure you test it!); if they cannot handle these special situations, throw an error.

For the shapes, it isn't meaningful to have nonpositive dimensions, so in my implementation, I raise an error if I get that, and a test for it, for rectangles, could look like this:

```
test_that("Dimensions are positive", {
  expect_error(rectangle(width = -1, height = 4))
  expect_error(rectangle(width = 2, height = -1))
  expect_error(rectangle(width = -1, height = -1))

  expect_error(rectangle(width = 0, height = 4))
  expect_error(rectangle(width = 2, height = 0))
  expect_error(rectangle(width = 0, height = 0))
})
```

When you are developing your code and corresponding unit tests, it is always a good idea to think a little bit about what the special cases could be and make sure that you have tests for how you choose to handle them.

Using Random Numbers in Tests

Another good approach to testing is to use random data. With tests we manually set up, we have a tendency to avoid pathological cases because we simply cannot think them up. Random data doesn't have this problem. Using random data in tests can, therefore, be more efficient, but, of course, it makes the tests nonreproducible, which makes debugging extremely hard.

You can, of course, set the random number generator seed. That makes the test deterministic and reproducible, but defeats the purpose of having random tests, to begin with.

I don't really have a good solution to this, but I sometimes use this trick: I pick a random seed and remember it, set the seed, and since I now know what the random seed was, I can set it again if the test fails and debug from there.

You can save the seed by putting it in the name of the test. Then if the test fails, you can get the seed from the error message:

```
seed <- as.integer(1000 * rnorm(1))
test_that(paste("The test works with seed", seed), {
    set.seed(seed)
    # test code that uses random numbers
})
```

Testing Random Results

Another issue that pops up when we are working with random numbers is what the expected value that a function returns should be. If the function is not deterministic but depends on random numbers, we don't necessarily have an expected output.

If all we can do to test the result in such cases is statistical, then that is what we must do. If a function is doing something useful, it probably isn't completely random, and that means that we can do some testing on it, even if that test can sometimes fail.

As a toy example, we can consider estimating the mean of a set of data by sampling from it. It is a silly example since it is probably much faster to just compute the mean in the first place in this example, but let's consider it for fun anyway.

If we sample n elements, the standard error of the mean should be $s / \sqrt{n}$ where s is the sample standard error. This means that the difference between the true mean and the sample mean should be distributed as $N\left(0, s / \sqrt{n}\right)$, and that is something we can test statistically, if not deterministically.

In the following code, I normalize the distance between the two means by dividing it with $s / \sqrt{n}$, which should make it distributed as $Z \sim N(0, 1)$. I then pick a threshold for significance that should only be reached one time in a thousand. I actually pick one that is only reached one in two thousand, but I am only testing the positive value for Z, so there is another implicit one in two thousand at the negative end of the distribution:

```
seed <- as.integer(1000 * rnorm(1))
test_that(paste("Sample mean is close to true, seed", seed), {
    set.seed(seed)

    data <- rnorm(10000)
    sample_size <- 100
    samples <- sample(data, size = sample_size, replace = TRUE)

    true_mean <- mean(data)
    sample_mean <- mean(samples)

    standard_error <- sd(samples) / sqrt(sample_size)
    Z <- (true_mean - sample_mean) / standard_error
    threshold <- qnorm(1 - 1/2000)

    expect_less_than(abs(Z), threshold)
})
```

This test is expected to fail one time in a thousand, but we cannot get absolute certainty when the results are actually random. If this test failed a single time, I wouldn't worry about it, but if I see it fail a couple of times, it becomes less likely that it is just a fluke, and then I would go and explore what is going on.

Checking a Package for Consistency

The package check you can do by clicking Check in the Build tab on the right in RStudio, or the Check Package in the Build menu, runs your unit tests but also a lot more.

It calls a script that runs a large number of consistency checks to make sure that your package is in tip-top shape. It verifies that all your functions are documented, that your code follows certain standards, that your files are in the right directories (and that there aren't files where there shouldn't be[1]), that all the necessary meta-information is provided, and many, many more things. You can check `http://r-pkgs.had.co.nz/check.html` for a longer list of the tests done when a package is being checked.

You should try and run a check for your packages. It will write a lot of output, and at the end, it will inform you how many errors, warnings, and notes it found.

In the output, every test that isn't declared to be OK is something you should look into. It might not be an error, but if the check raises any flags, you will not be allowed to put it on CRAN—at least not without a very good excuse.

Exercise

You have written two packages—for shapes and for polynomials—and your exercise now is to write unit tests for these and get them to a point where they can pass a package check.

---

[1] If there are, you should have a look at .Rbuildignore. If you have a file just the place you want it but the check is complaining, you can just add the file name to buildignore and it will stop complaining. If you have a README.Rmd file, for example, it will probably complain, but then you can add a line to .Rbuildignore that says ^README.Rmd$.

CHAPTER 15

Version Control

Version control, in its simplest form, is used for tracking changes to your software. It is also an efficient way of collaborating on software development since it often allows several developers to make changes to the software and merge it with changes from other developers. RStudio supports two version control systems, Subversion and git. Of these, git is the most widely used, and although these things are very subjective of course, I think that it is also the better system. It is certainly the system we will use here.

Version Control and Repositories

There are two main purposes of using a version control system when you develop software. One is simply to keep track of changes, such that you can later check when which modifications were made to your source code, and if you discover that they were in error, revert to earlier versions to try a different approach. It provides a log for your software development that allows you to go back in time and try again when you find that what you have done so far leads to some place you don't want to go.

The other job a version control system typically does is that it makes it easier for you to collaborate with others. Here, the idea is that you share some global repository of all code and code changes—the log that the version control system keeps of all changes—and each developer works on a copy when modifying the code and submits that code to the repository when they are done changing the code. In early version control systems, it was necessary to lock files when you wanted to modify them to prevent conflicts with other developers who might also be editing the same files. These days version control systems are more lenient when it comes to the concurrent editing of the same files, and they will typically just merge changes as long as there are no changes in overlapping lines (in which case you will have to resolve conflicts manually).

© Thomas Mailund 2022
T. Mailund, *Beginning Data Science in R 4*, https://doi.org/10.1007/978-1-4842-8155-0_15

With this type of version control, different developers can work concurrently on different parts of the code without worrying about conflicts. Should there be conflicts, these will be recognized when you attempt to push changes to the global repository, and you will be told to resolve the conflicts.

The version control system git allows even more concurrent and independent development than this, by not even having a single global repository as such—at least in theory. In practice, having a global repository for the official version of your software is a good idea, and people do have that. The system just doesn't enforce a single global repository, but, instead, is built around having many repositories that can communicate changes to each other.

Whenever you are working with git, you will have a local repository together with your source code. You can use this repository as the log system mentioned earlier or create branches for different features or releases as we will see later. You make changes to your source code like normally and can then commit it to your local repository without any conflict with other people's changes. However, you can't see their changes, and they can't see yours because you are working on different local repositories. To make changes to another repository, you have to push your changes there, and to get changes from another repository, you have to pull them from there.

This is where you typically use a global repository. You make changes to your local repository while developing a feature, but when you are done, you push those changes to the global repository. Or if you do not have permission to make changes to the global repository—perhaps because you cloned someone else's code and made changes to that—ask someone who does have permission to pull your changes into the repository, known as a "pull request."

Using Git in RStudio

This is all very theoretical, and if it is hard for me to write, it is probably also hard for you to understand. Instead, let us see git in practice.

RStudio has some rudimentary tools for interacting with git: it lets you create repositories, commit to them, and push changes to other repositories. It does not support the full range of what you can do with git—for that, you need other tools or to use the command-line version of git—but for day-to-day version control, it suffices for most tasks.

Installing Git

If you haven't installed git already on your computer, you can download it from `http://git-scm.com`. There should be versions for Windows, OS X, and Linux, although your platform might have better ways of installing it.

For example, on a Debian/Ubuntu system, you should be able to use

```
sudo apt-get install git-core
```

while on a Red Hat/Fedora system, you should be able to use

```
sudo yum install git-core
```

You have to Google around to check how best to install git on other systems.

Once installed, you want to tell git who you are. It needs this to be able to tag changes to your code with your name. It isn't frightfully important if you are the only one working on the code, but if more people are collaborating on the software development, it is necessary to identify who made which changes. You tell git who you are by running the following commands in a terminal:[1]

```
git config --global user.name "YOUR FULL NAME"
git config --global user.email "YOUR EMAIL ADDRESS"
```

You also might have to tell RStudio where the git command you installed can be found. You do that by going to the Tools menu and select Global Options.... In the window that pops up, you should find, on the icons on the left, a panel with Git/SVN, and in there you can tell RStudio where the git command can be found.

The git you have installed is a command-line tool. RStudio has some GUI to work with git, but you can't do everything from the GUI. There are a few GUI tools that allow you to do a lot more with git than RStudio, and I recommend getting one of those—I often find it easier using them than the command lines myself since I am getting old and forget the exact commands.

---

[1] Not the R terminal. You need to run this in an actual shell terminal for it to work. In RStudio, next to the R terminal, called Console, there should be a tab that says Terminal. That is where you want to go. If you do not have that tab, there will be another way to get it on your system, but how you do it depends on your platform. I can't help you there. If you don't know how to, it is time to fire up Google once again.

Some good choices are

- Sourcetree (www.sourcetreeapp.com): For Windows and OS X

- GitHub Desktop (https://desktop.github.com): For Linux, Windows, and OS X (for working with GitHub repositories)

- GitG (https://wiki.gnome.org/Apps/Gitg): For Linux

Sometimes, though, you do need to use the command-line version. There is a very nice interactive web tutorial for the command-line git program here: https://try. github.io/levels/1/challenges/1.

Making Changes to Files, Staging Files, and Committing Changes

If you checked that your project should use git when you created your package, you should have a Git tab on the top right of RStudio, next to the Build tab. Click it.

In the main part of this panel, there is a list of files. There are three columns, staged, status, and path; the latter is the names of modified files (or directories), see Figure 15-1.

If this is the first time you access this panel, the status will contain a yellow question mark for all files you have modified since you created the object (including files that RStudio made during the package creation). This status means that git doesn't know about these files yet. It can see that the files are there, but you have never told it what to do about them. We will do something about that now.

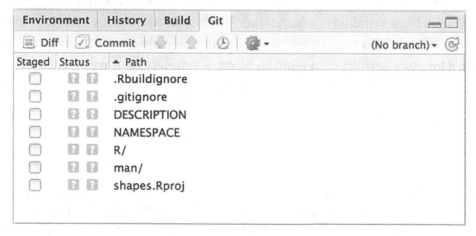

Figure 15-1. *Git panel showing modified files*

The staged column has tick buttons for all the files. If you tick one, the status for that file changes to a green "A." This means that you have staged the file to be added to the git repository. Do this for all of the files. When you do it for a directory, all the files in that directory will also be staged for adding. This is also what we want for now.

The process of committing changes to git involves staging changes to be committed before we actually commit them. What we just did was telling git that next time we commit changes, we want these files added. Generally, committing will only affect changes we have staged. This lets you commit only some of the changes you have made to your source code, which can be helpful at times. You might have made several changes to many files, but at some point, you only want to commit a particular bug fix and not changes for a new feature that you are not quite done with yet. Staging only the changes you want to commit allows for this.

Anyway, we have staged everything, and to commit the changes, you now have to click the Commit button in the toolbar. This opens a new window that shows you the changes you are about to commit and lets you write a commit message (on the upper right). This message is what goes into the change log. Give a short and meaningful description of your changes here. You will want it if you need to find the changes in your log at some later time. Then click Commit and close the window. The Git panel should now be empty of files. This is because there are no more changes since the last commit, and the panel only shows the files that have changed between your current version of your software and the version that is committed to git.

To do what we just did in the terminal instead, we would stage files using the git add command:

```
git add filename
```

and we would commit staged changes using the git commit command:

```
git commit -m "message"
```

Now try modifying a file. After you have done that, you should see the file displayed in the Git panel again, see Figure 15-2, this time with a status that is a blue "M." This, not surprisingly, stands for modified.

If you stage a file for commit here, the status is still "M," but RStudio indicates that it is now staged by moving the "M" to the left a little, see Figure 15-3. Not that you really need that feedback, you can also see that it is staged from the ticked staged button of course.

Committing modified files works exactly like committing added files.

In the terminal, you use `git add` for staging modified files as well. You don't have a separate command for staging adding new files and staging modified files. It is `git add` for both.

Adding Git to an Existing Project

If you didn't create your project with a git repository associated with it—and you have just learned about git now, so unless you have always just ticked the "git" button when creating projects, you probably have many projects without git associated—you can still set up git for an existing directory. You just have to do it on the command line.

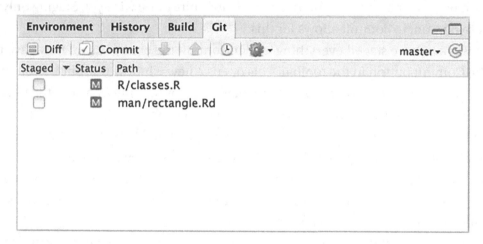

Figure 15-2. *Modified files in the Git panel*

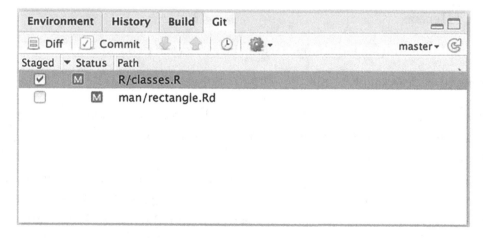

Figure 15-3. *Modified files with one file staged for commit*

Go to the directory where the project is and run the command:

```
git init
```

This sets up an empty repository. The files already in the directory can then be added just as we saw earlier.

Bare Repositories and Cloning Repositories

Most of the material in this section is not something you will ever have to worry about if you use a repository server such as GitHub. There, creating a repository and interacting with it is handled through a web interface, and you won't have to worry about the details, except for "cloning" a repository. We will create a so-called "bare" repository manually here and see how we can communicate changes in different local repositories through this.

The repositories we made when we created R projects or used `git init` in a directory are local repositories used for version control of the source code in the project directory. They are not really set up for collaboration between developers. While it is technically possible to merge changes in one such repository into another, it is a bit cumbersome and not something you want to deal with on an everyday basis.

To synchronize changes between different repositories, we want a bare repository. This is just a repository where we don't have the local source code included; it isn't really special, but it prevents you from making local changes to it, and you can only update it with changes from other repositories.

To create it, we need to use the command-line version of git. Create a directory where you want the repository, go in there, and type:

```
git --bare init
```

The repository now contains the various files that git needs to work with—your local repositories also include these; they are just hidden in a subdirectory called `.git/` when you have the local source code as well.

We are not going to do anything with this repository directly. We just need it to see how we work with other repositories connected to it.

Go to a directory where you want the working source code version of the repository and make a copy of the bare repository by writing

```
git clone /path/to/bare/repository
```

You will get a warning that you have cloned an empty repository. We already know that, so don't worry about it. We are going to add to it soon.

To see how we communicate between repositories, though, you need to make another working copy. You can either go to another directory and repeat the clone command or clone the repository but give it another name with

```
git clone /path/to/bare/repository name
```

We now have two clones of the bare repository and can see how we can push changes from a clone to the cloned repository and how we can pull updates in the cloned repository into the clone.

As I wrote earlier, going through a bare repository is not the only way to move changes from one repository to another, but it is the easiest way to work with git and the one you will be using if you use a server such as GitHub. If you do, and we see later how to, then GitHub will make the bare repository for you, and you just need to clone it to somewhere on your own computer to work with it.

Pushing Local Changes and Fetching and Pulling Remote Changes

Go into one of the clones you just made. It will look like an empty directory because we haven't made any changes to it yet. In fact, it does contain a hidden directory, `.git/`, where git keeps its magic, but we do not need to know about that.

Try to make some files, add them to git, and commit the changes:

```
touch foo bar
git add foo bar
git commit -m "added foo and bar"
```

If you now check the log

```
git log
```

you will see that you have made changes. If you look in the other clone of the bare repository, though, you don't yet see those changes.

There are two reasons for this: (1) we have only made changes to the cloned repository but never pushed them to the bare repository the two clones are connected to, and (2) even if we had done that, we haven't pulled the changes down into the other clone.

The first of these operations is done using git push. This will push the changes you have made in your local repository up to the repository you cloned it from:[2]

```
git push
```

You don't need to push changes up to the global (bare) repository after each commit; you probably don't want to do that, in fact. The idea with this workflow is that you make frequent commits to your local code to make the version control fine-grained, but you push these changes up when you have finished a feature—or at least gotten it to a stage where it is meaningful for others to work on your code. It isn't a major issue if you commit code that doesn't quite work to your local repository—although generally, you would want to avoid that—but it will not be popular if you push code that doesn't work onto others.

After pushing the changes in the first cloned repository, they are still not visible in the second repository. You need to pull them down.

There is a command

```
git fetch
```

that gets the changes made in the global repository and makes it possible for you to check them out before merging them with your own code. This can be useful because you can then check it out and make sure it isn't breaking anything for you before you merge it with your code. After running the fetch command, you can check out branches from the global repository, make changes there, and merge them into your own code using the branching mechanism described in the following. In most cases, however, we just want to merge the changes made to the global repository into our current code, and you don't really want to modify it before you do so. In that case, the command

```
git pull
```

---

[2] If we didn't have a bare repository we had cloned both repositories from, we could still have connected them to see changes made to them, but pushing changes would be much more complicated. With a bare repository that both are cloned from, pushing changes upward is as easy as git push.

will both fetch the latest changes and merge them into your local repository in a single operation. This is by far the most common operation for merging changes others have made and pushed to the global repository with your own.

Go to the repository clone without the changes and run the command. Check that you now have the changes there.

The general workflow for collaborating with others on a project is to make changes and commit them to your own repository. You use this repository to make changes you are not ready to share yet, and you are the only one who can see them. Then, when you are ready to share with your collaborators, you can push the changes to the shared repository, and when you need changes others have made, you can pull them.

If you try to push to the global repository, and someone else has pushed changes that you haven't pulled yet, you will get an error. Don't worry about that. Just pull the changes; after that, you can push your changes.

If you pull changes into your repository, and you have committed changes there that haven't been pushed yet, the operation becomes a merge operation, and this requires a commit message. There is a default message for this that you can just use.

You have your two repositories to experiment with, so try to make various variations of pushing and pulling changes into a repository where you have committed changes. The preceding explanation will hopefully make a lot more sense for you after you have experimented a bit on your own.

RStudio has some basic support for pushing and pulling. If you make a new RStudio project and choose to put it in an existing directory, you can try to make one that sits in your cloned repositories. If you do this, you will find that the Git panel now has two new buttons: push and pull.

Handling Conflicts

If it happens that someone has pushed changes to the global repository that overlap lines that you have been editing in your local repository, you will get a so-called conflict when you pull changes.

Git will inform you about this, whether you pull from RStudio or use the command line. It will tell you which files are involved, and if you open a file with a conflict, you will see that git has marked the conflict with text that looks like this:

```
<<<<<<< HEAD
your version of the code
```

```
=======
the remote version of the code
>>>>>>> 9a0e21ccd38f7598c05fe1e21e2b32091bb0839b
```

It shows you the version of the changes you have made and the version of the changes that are in the global repository. Because there are changes both places, git doesn't know how to merge the remote repository into your repository in the pull command.

You have to go into the file and edit it so it contains the version you want, which could be a merge of the two revisions. Get rid of the <<<</====/>>>> markup lines when you are done making the changes.

Once you have edited the file with conflicts, you need to stage it—running the git add filename on the command line or ticking the file in the staged column in the Git plane in RStudio—and commit it. This tells git that you have handled the conflict and will let you push your own changes if you want to do this.

Working with Branches

Branches are a feature of most version control systems, which allow you to work on different versions of your code at the same time. A typical example is having a branch for developing new features and another branch for the stable version of your software. When you are working on implementing new features, the code is in a state of flux, the implementation of the new feature might be buggy, and the interface to the feature could be changing between different designs. You don't want people using your package to use such a version of your software—at least not without being aware that the package they are using is unstable and that the interface they are using could be changed at a moment's notice. So you want the development code to be separate from the released code.

If you just made releases at certain times and then implemented new features between making releases, that wouldn't be much of an issue. People should be using the version you have released and not commits that fall between released versions. But the world is not that simple if you make a release with a bug in it—and let's face it, that is not impossible—and you want to fix that bug when it is discovered. You probably don't want to wait with fixing the bug until you are done with all the new features you are working. So you want to make changes to the code in the release. If there are more bugs, you will commit more bug fixes onto the release code. And all this while you are still making changes to your development code. Of course, those bug fixes you make to the released code, you also want to merge into the development code. After all, you don't want the next release to reintroduce bugs you have already fixed.

This is where branches come in. RStudio has very limited support for branches, and it doesn't help you create them.[3] For that, we need to use the command line.

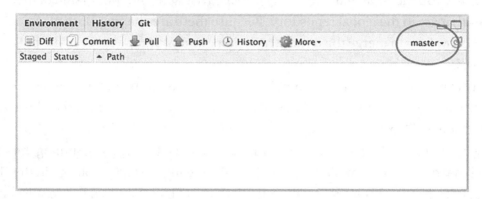

Figure 15-4. *Git panel when the code is on the* master *branch*

To create a branch, you use the command git branch name, so to create a development branch—called develop for lack of imagination—we use

```
git branch develop
```

This just creates the branch. We are not magically moved to the branch or anything. It just tells git that we have a new branch (and it branches off our current position in the list of commits done to the repository).

In RStudio, we can see which branch we are on in the Git panel, see Figure 15-4. In the project you have experimented on so far—and any project you have made where you created a git repository with git init or by ticking the git selection in the dialog window when you created the project—you will be on branch master. This is the default branch and the branch that is typically used for released versions.

If you click the branch drop-down in the Git panel, you get a list of the branches you have in your repository, see Figure 15-5. You will have a branch called origin/master. This is the master branch on the central repository and the one you merge with when pulling data. Ignore it; it is not important for us. If you ran the git branch develop command, you should also have a develop branch. If you select it, you move to that branch, see Figure 15-6.

---

[3] Some of the other GUIs for working with git have excellent support for working with branches. You should check them out.

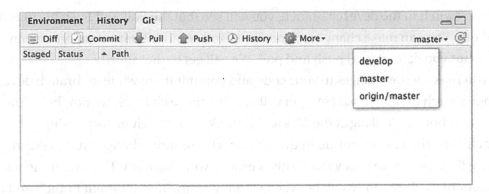

Figure 15-5. *Selecting a branch to switch to*

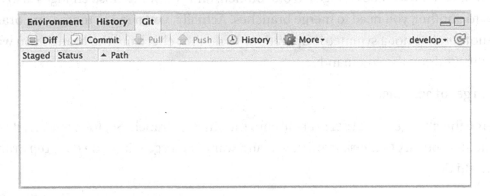

Figure 15-6. *After switching to branch* `develop`

You can also get a list of branches on the command line with

`git branch`

and you can switch to a branch using the command[4]

`git checkout branchname`

---

[4] You can also combine the creation and checkout of a branch using `git checkout -b branchname` if you want. That creates the branch first and then checks it out. To change between branches later on, though, you need the checkout command without option `-b`.

If you switch to the `develop` branch, you will see that the Pull and Push buttons are grayed out. You can make changes to your code and commit them when on a given branch, but you cannot (yet) push and pull. We will get to that shortly.

If you make some changes to your code and commit them while on branch `develop` and then switch to branch `master`, you will see that those changes are not there. You can see that both by looking at the files and by looking at the git history (using `git log` or clicking the History button in the Git panel). Similarly, changes you make in `master` will not show up in `develop`. This is exactly what we want. The two branches are independent, and we can switch between working on the development branch and the release version of our software by switching branches.

When you have made changes to one branch, and you want those changes also to be added to another, you need to merge branches. Actually, you need to merge one branch into another; it is not a symmetric operation. To do this, check out the branch you want to modify and run the command:

```
git merge otherbranch
```

to merge the changes in "otherbranch" into the current branch. So, for example, if you have made a bug fix to the `master` branch and want to merge it into the `develop` branch, you would do

```
git checkout develop
git merge master
```

If a merge causes conflicts, you resolve them the same was as if a pull causes conflicts, not surprisingly since a pull command is actually just a shortcut for fetching and merging.

Typical Workflows Involve Lots of Branches

Git is optimized for working with lots of branches (unlike some version control systems, where creating and merging branches can be rather slow operations). This is reflected in how many people use branches when working with git: you create many branches and work on a graph of different versions of your code and merge them together whenever you need to.

Having a development branch and a master branch is a typical core of the repository structure, but it is also very common to make a branch for each new feature you implement. Typically, you branch these off the `develop` branch when you start working on the feature and merge them back into `develop` when you are done. It is also common to have a separate branch for each bug fix—typically branched off `master` when you start implementing the fix and then branched back into `master` as well as into `develop` when you are done. See Atlassian's Git Tutorial (`www.atlassian.com/git/tutorials/comparing-workflows`) for different workflows that exploit having various branches.

If you create a lot of branches for each feature or bug fix, you don't want to keep them around after you are done with them—unlike the `develop` and `master` branches that you probably want to keep around forever. To delete a branch, you use the command:

```
git branch -d branchname
```

Pushing Branches to the Global Repository

You can work on as many branches as you like in your local repository, but they are not automatically found in the global repository. The `develop` branch we made earlier exists only in the local repository, and we cannot push changes made to it to the global repository—we can see this in RStudio since the push (and pull) buttons are grayed out.

If you want a branch to exist also on the global repository—so you can push to it, and so collaborators can check it out—you need to create a branch in that repository and set up a link between your local repository and the global repository.

You can do that for the `develop` branch by checking it out and running the command:

```
git push --set-upstream origin develop
```

This pushes the changes and also remembers for the future that branch is linked to the `develop` branch in `origin`. The name `origin` refers to the repository you cloned when you created this repository.[5]

Whether you want a branch you are working on also to be found in the global repository is a matter of taste. If you are working on a feature that you want to share when it is completed but not before, you probably don't want to push that branch to the global repository. For the `develop` and `master` branches, though, you definitely want those to be in the global repository.

---

[5] It is slightly more complex than this; you can have links to other repositories and pull from them or push to them (if they are bare repositories), and `origin` is just a default link to the one you cloned for. It is beyond the scope of these notes, however, to go into more details. If you always work with a single global repository that you push to and pull from, then you don't need to know any more about links to remote repositories.

GitHub

GitHub (https://github.com) is a server for hosting git repositories. You can think of it as a place to have your bare/global repository with some extra benefits. There are ways for automatically installing packages that are hosted on GitHub, there is web support for tracking bugs and feature requests, and there is support for sharing fixes and features in hosted projects through a web interface.

To use it, you first need to go to the home page and sign up. This is free, and you just need to pick a username and a password.

Once you have created an account on GitHub, you can create new repositories by clicking the big + in the upper-right corner of the home page, see Figure 15-7.

Clicking it, you get to a page where you can choose the name of the repository, create a short description, pick a license, and decide whether you want to add a README.md file to the repository. I recommend that you always have a README.md file—it works as the documentation for your package since it is displayed on the home page for the repository at GitHub. You probably want to set up a README.Rmd file to generate it, though, as we saw in Chapter 13. For now, though, you might as well just say yes to have one generated.

Once you have generated the repository, you go to a page with an overview of the code in the repository, see Figure 15-8.

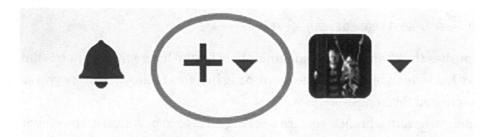

Figure 15-7. *Button to create a new repository at the GitHub home page. Found on the upper right of the home page*

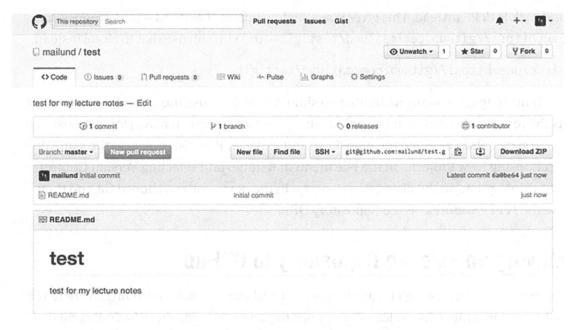

Figure 15-8. *New GitHub repository containing only a README.md file*

You can explore the web page and the features implemented there later—it is a good idea to know what it supports you doing—but for now we can just use the repository here as a remote global repository. To clone it, you need the address in the field next to the button that says SSH. In my test repository, it is `git@github.com:mailund/test.git`. This is an address you can use to clone the repository using the "ssh" protocol:

```
git clone git@github.com:mailund/test.git
```

This is a protocol that you will have access to on many machines, but it involves you having to deal with a public/private key protocol. Check the documentation for setting up the ssh key at GitHub for learning more about this (`https://help.github.com/articles/generating-ssh-keys/`). It is mostly automated by now, and you should be able to set it up just by making a push and answering yes to the question you get there.

It is not the easiest protocol to work with, though, if you are on a machine that has HTTPS—the protocol used by your web browser for secure communication. You will almost certainly have that on your own machine, but depending on how firewalls are set up, you might not have access to it on computer clusters and such, and then you need to use the ssh protocol. To use HTTPS instead of SSH, just click the SSH button drop-down and pick HTTPS instead. This gives you a slightly different address—in my repository, I get `https://github.com/mailund/test.git`—and you can use that to clone instead:

```
git clone https://github.com/mailund/test.git
```

If nothing goes wrong with this, you should be able to use the cloned repository just as the repositories we looked at earlier where we made our own bare/global repository.

You can also check out the repository and make an RStudio project at the same time by choosing New Project... in the File menu in RStudio and selecting Version Control (the third option) in the dialog that pops up. In the next window, choose Git and then use the HTTPS address as the Repository URL.

Moving an Existing Repository to GitHub

If you have already used git locally in a project and want to move it to GitHub, there is a little more you must do—at least if you want to move your repository including all the history stored in it and not just the current version of the source code in it.

First, you need to make a bare version of your repository. This is, as we saw a while ago, just a version of the repository without source code associated.

If your repository is called `repo`, we can make a bare version of it, called `repo.git`, by cloning it:

```
git clone --bare repo repo.git
```

To move this to GitHub, create an empty repository there and get the HTTPS address of it. Then go into the bare repository we just made and run the following command:

```
cd repo.git
git push --mirror <https address at github>
```

Now just delete the bare repository we used to move the code to GitHub and clone the version from GitHub. Now you have a version from there that you can work on:

```
rm -rf repo.git
git clone <https address at github>
```

Installing Packages from GitHub

A very nice extra benefit you get from having your R packages on GitHub—in addition to having version control—is that other people can install your package directly from there. The requirements for putting packages on CRAN are much stricter than for putting R packages on GitHub, and you are not allowed to upload new versions to CRAN very often, so for development versions of your R package, GitHub is an excellent alternative.

To install a package from GitHub, you need to have the devtools package installed:

```
install.packages("devtools")
```

after which you can install a package named "packagename" written by GitHub user "username" with the command

```
devtools::install_github("username/packagename")
```

Collaborating on GitHub

The repositories you make on GitHub are by default only editable by yourself. Anyone can clone them to get the source code, but only you can push changes to the repository. This is, of course, useful to prevent random people from messing with your code but prevents collaborations.

One way to collaborate with others is to give them write permissions to the repository. On the repository home page, you must select the Settings entry in the toolbar and then pick Collaborators in the menu on the left. After that, you get to a page where you can add collaborators identified by their user account on GitHub. Collaborators can push changes to the repository just as you can yourself. To avoid too much confusion, when different collaborators are updating the code, it is useful to have some discipline in how changes are merged into the master (and/or the develop) branch. One approach that is recommended and supported by GitHub is to make changes in separate branches and then use so-called pull requests to discuss changes before they are merged into the main branches.

Pull Requests

The workflow for making pull requests is to implement your new features or bug fixes or whatever you are implementing on separate branches from `develop` or `master`, and instead of merging them directly, you create what is called a pull request. You can start a pull request by switching to the branch on the repository home page and selecting the big green New pull request button, or if you just made changes, you should also have a green button saying Compare & pull request that lets you start a pull request.

Clicking the button takes you to a page where you can name the pull request and write a description of what the changes in the code you have made are doing. You also decide which branch you want to merge the pull into; above the title you give the pull request, you can select two branches, the one you want to merge into (base) and the branch you have your new changes on (compare). You should pick the one you branched out of when you made the new branch here. After that, you can create the pull request.

The only thing this is really doing is that it creates a web interface for having a discussion about the changes you have made. It is possible to see the changes on the web page and comment on them and to make comments to the branch in general. At the same time, anyone can check out the branch and make their own modifications to it. As long as the pull request is open, the discussion is going, and people can improve on the branch.

When you are done, you can merge the pull request (using the big green Merge pull request button you can find on the web page that contains the discussion about the pull request).

Forking Repositories Instead of Cloning

Making changes to separate branches and then making pull requests to merge in the changes still requires writing access to the repository. This is excellent for collaborating with a few friends, but not ideal for getting fixes from random strangers—or for making fixes to packages other people write, people who won't necessarily want to give you full write access to their software.

Not to worry, it is still possible to collaborate with people on GitHub without having write access to each other's repositories. The way that pull requests work, there is actually no need for branches to be merged to be part of the same base repository. You can merge branches from anywhere if you want to.

If you want to make changes to a repository that you do not have write access to, you can clone it and make changes to the repository you get as the clone, but you cannot push those changes back to the repository you cloned it from. And other users on GitHub can't see the local changes you made (they are on your personal computer, not on the GitHub server). What you want is a repository on GitHub that is a clone of the repository you want to modify and that is a bare repository, so you can push changes into it. You then want to clone that repository to your own computer. Changes you make to your own computer can be pushed to the bare repository you have on GitHub—because it is a bare repository and because you have writing access to it—and other users on GitHub can see the repository you have there.

Making such a repository on GitHub is called forking the repository. Technically, it isn't different from cloning—except that it is a bare repository you make—and the terminology is taking from open source software where forking a project means making your own version and developing it independent of previous versions.

Anyway, whenever you go to a repository home page on GitHub, you should see a button at the top right—to the right of the name and branch of the repository you are looking at—that says Fork. Clicking that will make a copy of the repository that you have writing access to. You cannot fork your own repositories—I'm not sure why you are not allowed to, but in most cases, you don't want to do that anyway, so it is not a major issue—but you can fork any repository at other users' accounts.

Once you have made the copy, you can clone it down to your computer and make changes to it, as you can with any other repositories. The only way this repository is different from a repository you made yourself is that when you make pull requests, GitHub knows that you forked it off another repository. So when you make a pull request, you can choose not only the base and compare branches but also the base fork and the head fork—the former being the repository you want to merge changes into, and the latter the repository where you made your changes. If someone forks your project and you make a pull request in the original repository, you won't see the base fork and head fork choices by default, but clicking the link compare across forks when you make pull requests will enable them there as well.

If you make a pull request with your changes to someone else's repository, the procedure is exactly the same as when you make a pull request on your own projects, except that you cannot merge the pull request after the discussion about the changes. Only someone with writing permission to the repository can do that.

The same goes if someone else wants to make changes to your code. They can start a pull request with their changes into your code, but only you can decide to merge the changes into the repository (or not) following the pull discussion.

This is a very flexible way of collaborating—even with strangers—on source code development and one of the great strengths of git and GitHub.

Exercises

Take any of the packages you have written earlier and create a repository on GitHub to host it. Push your code there.

## CHAPTER 16

Profiling and Optimizing

In this second chapter, we will briefly consider what to do when you find that your code is running too slow and, in particular, how to figure out why it is running too slow.

Before you start worrying about your code's performance, though, it is important to consider if it is worth speeding it up. It takes you time to improve performance, and it is only worth it if the improved performance saves you time when this extra programming is included. For an analysis you can run in a day, there is no point in spending one day making it faster, even if it gets much faster, because you still end up spending the same time, or more, to finally get the analysis done.

Any code you just need to run a few times during an analysis is usually not worth optimizing. We rarely need to run an analysis just once—optimistically, we might hope to, but, in reality, we usually have to run it again and again when data or ideas change—but we don't expect to run it hundreds or thousands of times. So even if it will take a few hours to rerun an analysis, your time is probably better spent working on something else while it runs. It is rarely worth it to spend a lot of time making it faster. The CPU's time is cheap compared to your own.

If you are developing a package, though, you often do have to consider performance to some extent. A package, if it is worth developing, will have more users, and the total time spent on running your code makes it worthwhile, up to a point, to make that code fast.

Profiling

Before you can make your code faster, you need to figure out why it is slow, to begin with. You might have a few ideas about where the code is slow, but it is actually surprisingly hard to guess at this. Quite often, I have found, it is nowhere near where I thought it would be, that most of the time is actually spend. On two separate occasions, I have tried working really hard on speeding up an algorithm only to find out later that the reason

© Thomas Mailund 2022
T. Mailund, *Beginning Data Science in R 4*, https://doi.org/10.1007/978-1-4842-8155-0_16

my program was slow was the code used for reading the program's input. The parser was slow. The algorithm was lightning fast in comparison. That was in C, where the abstractions are pretty low level and where it is usually pretty easy to glance from the code how much time it will take to run. In R, where the abstractions are very high level, it can be extremely hard to guess how much time a single line of code will take to run.

The point is, if you find that your code is slow, you shouldn't be guessing at where it is slow. You should measure the running time and get to know for sure. You need to profile your code to know which parts of it are taking up most of the running time. Otherwise, you might end up optimizing code that uses only a few percentages of the total running time and leaving the real time wasters alone.

In typical code, there are only a few real bottlenecks. If you can identify these and improve their performance, your work will be done. The rest will run fast enough. Figuring out where those bottlenecks are requires profiling.

We are going to use the package profvis for profiling. RStudio has built-in support for this package, and you should have a Profile item in your menu bar. That is a convenient interface to the package, but in case you want to use it from outside of RStudio, I will just use the package directly here.

A Graph-Flow Algorithm

For an example of some code, we could imagine we would wish to profile we consider a small graph algorithm. It is an algorithm for smoothing out weights put on nodes in a graph. It is part of a method used for propagating weights of evidence for nodes in a graph and is something we've been using to boost searching for disease-gene associations using gene-gene interaction networks. The idea is that if a gene is a neighbor to another gene in this interaction network, then it is more likely to have a similar association with a disease as the other gene. So genes with known association are given an initial weight, and other genes get a higher weight if they are connected to such genes than if they are not.

The details of what the algorithm is used for is not so important, though. All it does is to smooth out weights between nodes. Initially, all nodes, n, are assigned a weight $w(n)$. Then in one iteration of smoothing, this weight is updated as

$$w'(n) = \alpha w(n) + (1 - \alpha) \frac{1}{|N(n)|} \sum_{v \in N(n)} w(v)$$ where α is a number between zero and one and $N(n)$ denotes the neighbors of node n. If this is iterated enough times, the weights in a

graph become equal for all connected nodes in the graph, but if stopped earlier, it is just a slight smoothing, depending on the value of α.

To implement this, we need both a representation of graphs and the smoothing algorithm. We start with representing the graph. There are many ways to do this, but a simple format is a so-called incidence matrix. This is a matrix that has entry $M_{i,j} = 0$ if nodes i and j are not directly connected and $M_{i,j} = 1$ if they are. Since we want to work on a nondirected graph in this algorithm, we will have $M_{i,j} = M_{j,i}$.

We can implement this representation using a constructor function that looks like this:

```
graph <- function(n, edges) {
  m <- matrix(0, nrow = n, ncol = n)

  no_edges <- length(edges)
  if (no_edges >= 1) {
    for (i in seq(1, no_edges, by = 2)) {
      m[edges[i], edges[i+1]] <- m[edges[i+1], edges[i]] <- 1
    }
  }

  structure(m, class = "graph")
}
```

Here, I require that the number of nodes is given as an argument n and that edges are specified as a vector where each pair corresponds to an edge. This is not an optimal way of representing edges if graphs should be coded by hand, but since this algorithm is supposed to be used for very large graphs, I assume we can write code elsewhere for reading in a graph representation and creating such an edge vector.

There is not much to the function. It just creates the incidence matrix and then iterates through the edges to set it up. There is a special case to handle if the edge vector is empty. Then the seq() call will return a list going from one to zero. So we avoid this. We might also want to check that the length of the edge vector is a multiple of two, but I haven't bothered. I am going to assume that the code that generates the vector will take care of that.

Even though the graph representation is just a matrix, I give it a class in case I want to write generic functions for it later.

With this graph representation, the smoothing function can look like this:

```r
smooth_weights <- function(graph, node_weights, alpha) {
  if (length(node_weights) != nrow(graph))
    stop("Incorrect number of nodes")

  no_nodes <- length(node_weights)
  new_weights <- vector("numeric", no_nodes)

  for (i in 1:no_nodes) {
    neighbour_weights <- 0
    n <- 0
    for (j in 1:no_nodes) {
      if (i != j && graph[i, j] == 1) {
        neighbour_weights <- neighbour_weights + node_weights[j]
        n <- n + 1
      }
    }

    if (n > 0) {
      new_weights[i] <-
        alpha * node_weights[i] +
        (1 - alpha) * neighbour_weights / n
    } else {
      new_weights[i] <- node_weights[i]
    }

  }
  new_weights
}
```

It creates the new weights vector we should return and then iterate through the matrix in nested loops. If the incidence matrix says that there is a connection between i and j, and $i \neq j$—we don't want to add a node's own weight if there is a self-loop—we use it to calculate the mean. If there is something to update—which there will be if there are any neighbors to node i—we do the update.

The code is not particularly elegant, but it is a straightforward implementation of the idea.

To profile this code, we use the `profvis()` function from `profvis`. It takes an expression as its single argument, so to profile more than a single function call, we give it a code block, translating the sequence of statements into an expression.

I just generate a random graph with 1000 nodes and 300 edges and random weights. We are not testing the code here, only profiling it. However, if this was real code and not just an example, we should, of course, have unit tests—this is especially important if you start rewriting code to optimize it. Otherwise, you might end up getting faster but incorrect code for all your efforts.

```r
profvis::profvis({
  n <- 1000
  nodes <- 1:n
  edges <- sample(nodes, 600, replace = TRUE)
  weights <- rnorm(n)
  g <- graph(n, edges)
  smooth_weights(g, weights, 0.8)
})
```

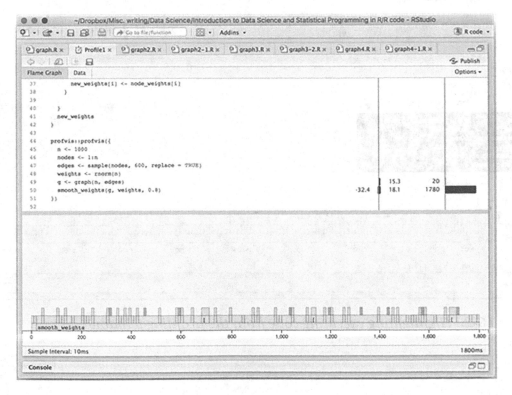

Figure 16-1. *Window showing profile results*

(If you are using RStudio, you get the same effect by highlighting the lines from n <-
1000 to smooth_weights(...) and selecting Profile Selected Line(s) in the Profile menu.)

Running this code will open a new tab showing the results; see Figure 16-1. The top
half of the tab shows your code with annotations showing memory usage first and time
usage second as horizontal bars. The bottom half of the window shows the time usage
plus a call stack.

We can see that the total execution took about 1800ms. The way to read the graph
is that, from left to right, you can see what was executed at any point in the run, with
functions called directly in the code block we gave profvis() at the bottom and code
they called directly above that and further function calls stacked even higher.

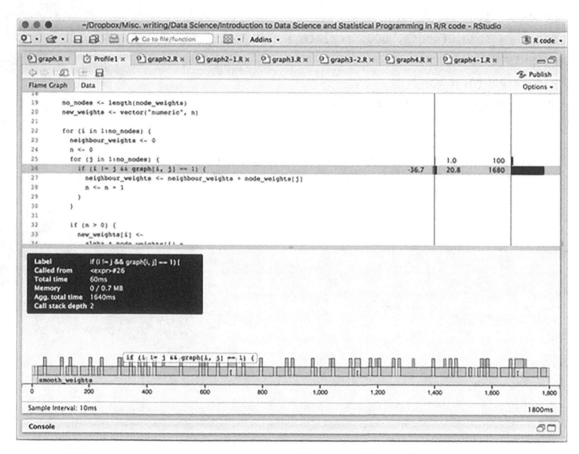

Figure 16-2. *Highlighting executing code from the profiling window*

We can also see that by far the most time was spent in the `smooth_weights()` function since that stretches almost all the way from the leftmost part of the graph and all the way to the rightmost. This is what we would expect, of course, but as I've mentioned, sometimes profile results can surprise you.

If you move your mouse pointer into the window, either in the code or in the bottom graph, it will highlight what you are pointing at; see Figure 16-2. You can use this to figure out where the time is being spent.

In this particular case, it looks like most of the time is spent in the inner loop, checking if an edge exists or not. Since this is the inner part of a double loop, this might not be so surprising. The reason that it is not all the body of the inner loop but the `if` statement is probably that we check the `if` expression in each iteration, but we do not execute its body unless it is true. And with 1000 nodes and 300 edges, it is only true with probability around $300/(1000*1000) = 3 \times 10\text{-}4$ (it can be less since some edges could be identical or self-loops).

Now if we had a performance problem with this code, this is where we should concentrate our optimization efforts. With 1000 nodes, we don't really have a problem. 1800ms is not a long time, after all. But the application I have in mind has around 30,000 nodes, so it might be worth optimizing a little bit.

If you need to optimize something, the first you should be thinking is: Is there a better algorithm or a better data structure? Algorithmic improvements are much more likely to give substantial performance improvements compared to just changing details of an implementation.

In this case, if the graphs we are working on are sparse, meaning they have few actual edges compared to all possible edges, then an incidence matrix is not a good representation. We could speed the code up by using vector expressions to replace the inner loop and hacks like that, but we are much better off considering another representation of the graph.

Here, of course, we should first figure out if the simulated data we have used is representative of the actual data we need to analyze. If the actual data is a dense graph and we do performance profiling on a sparse graph, we are not getting the right impression of where the time is being spent and where we can reasonably optimize. But the application I have in mind, I pigheadedly claim, is one that uses sparse graphs.

With sparse graphs, we should represent edges in a different format. Instead of a matrix, we will represent the edges as a list where for each node we have a vector of that node's neighbors.

We can implement that representation like this:

```
graph <- function(n, edges) {
  neighbours <- vector("list", length = n)

  for (i in seq_along(neighbours)) {
    neighbours[[i]] <- vector("integer", length = 0)
  }

  no_edges <- length(edges)

  if (no_edges >= 1) {
    for (i in seq(1, no_edges, by = 2)) {
      n1 <- edges[i]
      n2 <- edges[i+1]
      neighbours[[n1]] <- c(n2, neighbours[[n1]])
      neighbours[[n2]] <- c(n1, neighbours[[n2]])
    }
  }

  for (i in seq_along(neighbours)) {
    neighbours[[i]] <- unique(neighbours[[i]])
  }

  structure(neighbours, class = "graph")
}
```

We first generate the list of edge vectors, then we initialize them all as empty integer vectors. We then iterate through the input edges and update the edge vectors. The way we update the vectors is potentially computationally slow since we force a copy of the previous vector in each update, but we don't know the length of these vectors a priori, so this is the easy solution, and we can worry about it later if the profiling says it is a problem.

Now, if the edges we get as input contain the same pair of nodes twice, we will get the same edge represented twice. This means that the same neighbor to a node will be used twice when calculating the mean of the neighbor weights. If we want to allow such multi-edges in the application, that is fine, but we don't, so we explicitly make sure that the same neighbor is only represented once by calling the unique() function on all the vectors at the end.

With this graph representation, we can update the smoothing function to this:

```
smooth_weights <- function(graph, node_weights, alpha) {
  if (length(node_weights) != length(graph))
    stop("Incorrect number of nodes")

  no_nodes <- length(node_weights)
  new_weights <- vector("numeric", no_nodes)

  for (i in 1:no_nodes) {
    neighbour_weights <- 0
    n <- 0
    for (j in graph[[i]]) {
      if (i != j) {
        neighbour_weights <- neighbour_weights + node_weights[j]
        n <- n + 1
      }
    }

    if (n > 0) {
      new_weights[i] <-
        alpha * node_weights[i] +
        (1 - alpha) * neighbour_weights / n
    } else {
      new_weights[i] <- node_weights[i]
    }

  }
  new_weights
}
```

Very little changes. We just make sure that *j* only iterates through the nodes we know to be neighbors of node *i*.

The profiling code is the same as before, and if we run it, we get the results shown in Figure 16-3.

449

We see that we have gotten a substantial performance improvement. The execution time is now 20ms instead of 1800ms. We can also see that half the time is spent on constructing the graph and only half on smoothing it. In the construction, nearly all the time is spent in `unique()`, while in the smoothing function, the time is spent in actually computing the mean of the neighbors.

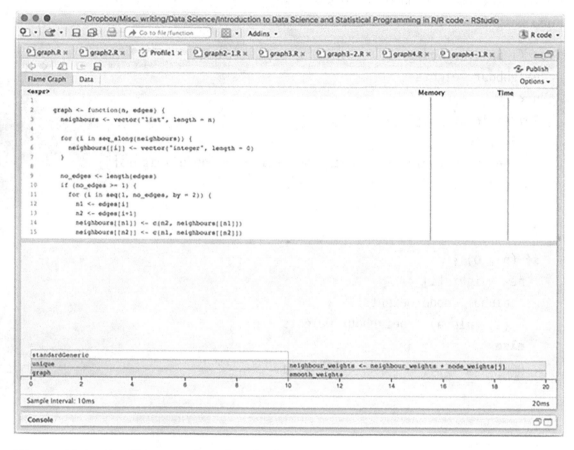

Figure 16-3. *Profiling results after the first change*

It should be said here, though, that the profiler works by sampling what code is executing at certain time points. It doesn't have an infinite resolution; it samples every 10ms as it says at the bottom left, so, in fact, it has only sampled twice in this run. The result we see is just because the samples happened to hit those two places in the graph construction and the smoothing, respectively. We are not actually seeing fine details here.

To get more details, and get closer to the size the actual input is expected to be, we can try increasing the size of the graph to 10,000 nodes and 600 edges:

```
profvis::profvis({
  n <- 10000
  nodes <- 1:n

  edges <- sample(nodes, 1200, replace = TRUE)
  weights <- rnorm(n)
  g <- graph(n, edges)
  smooth_weights(g, weights, 0.8)
})
```

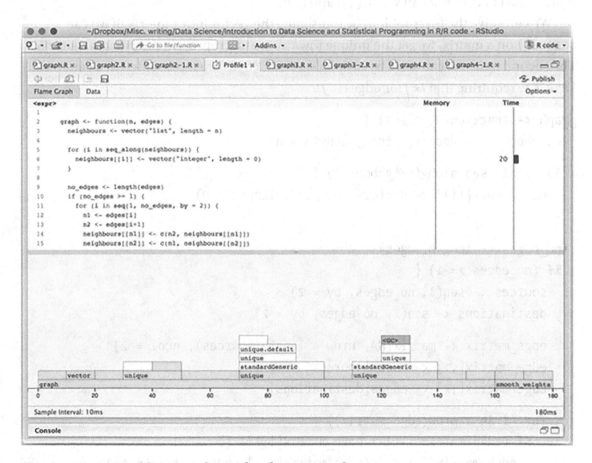

Figure 16-4. *Profiling results with a larger graph*

The result of this profiling is shown in Figure 16-4.

To our surprise, we see that for the larger graph we are actually spending more time constructing the graph than smoothing it. We also see that this time is spent calling the unique() function.

Now, these calls are necessary to avoid duplicated edges, but they are not necessarily going to be something we often see—in the random graph, they will be less likely, at least—so most of these calls are not really doing anything.

If we could remove all the duplicated edges in a single call to unique(), we should save some time. We can do this, but it requires a little more work in the construction function.

We want to make the edges unique, and there are two issues here. One is that we don't actually represent them as pairs we can call unique() on, and calling unique() on the edges vector is certainly not a solution. The other issue is that the same edge can be represented in two different ways: (i, j) and (j, i).

We can solve the first problem by translating the vector into a matrix. If we call unique() on a matrix, we get the unique rows, so we just represent the pairs in that way. The second issue we can solve by making sure that edges are represented in a canonical form, say requiring that $i < j$ for edges (i, j).

```
graph <- function(n, edges) {
  neighbours <- vector("list", length = n)

  for (i in seq_along(neighbours)) {
    neighbours[[i]] <- vector("integer", length = 0)
  }

  no_edges <- length(edges)
  if (no_edges >= 1) {
    sources <- seq(1, no_edges, by = 2)
    destinations <- seq(2, no_edges, by = 2)

    edge_matrix <- matrix(NA, nrow = length(sources), ncol = 2)
    edge_matrix[,1] <- edges[sources]
    edge_matrix[,2] <- edges[destinations]

    for (i in 1:nrow(edge_matrix)) {
      if (edge_matrix[i,1] > edge_matrix[i,2]) {
        edge_matrix[i,] <- c(edge_matrix[i,2], edge_matrix[i,1])
      }
    }
```

```
  edge_matrix <- unique(edge_matrix)

  for (i in seq(1, nrow(edge_matrix))) {
    n1 <- edge_matrix[i, 1]
    n2 <- edge_matrix[i, 2]
    neighbours[[n1]] <- c(n2, neighbours[[n1]])
    neighbours[[n2]] <- c(n1, neighbours[[n2]])
  }
}

  structure(neighbours, class = "graph")
}
```

Try profiling this code and see what results you get.

When I profiled, I found that the running time is cut in half, and relatively less time is spent constructing the graph compared to before. The time spent in executing the code is also so short again that we cannot be too certain about the profiling samples to say much more.

The graph size is not quite at the expected size for the application I had in mind when I wrote this code. We can boost it up to the full size of around 20,000 nodes and 50,000 edges[1] and profile for that size. Results are shown in Figure 16-5.

On a full-size graph, we still spend most of the time in constructing the graph and not in smoothing it—and about half of the constructing time in the unique() function—but this is a little misleading. We don't expect to call the smoothing function just once on a graph. Each call to the smoothing function will smooth the weights out a little more, and we might expect to run it around ten times, say, in the real application.

We can rename the function to flow_weights_iteration() and then write a smooth_weights() function that runs it for a number of iterations.

---

[1] There are more edges than nodes, but it is still a sparse graph. A complete graph would have the number of nodes squared or half that since we don't allow both (i, j) and (j, i) as edges, so a full dense graph would have up to 200 million edges. Compared to that, 50,000 edges is not much.

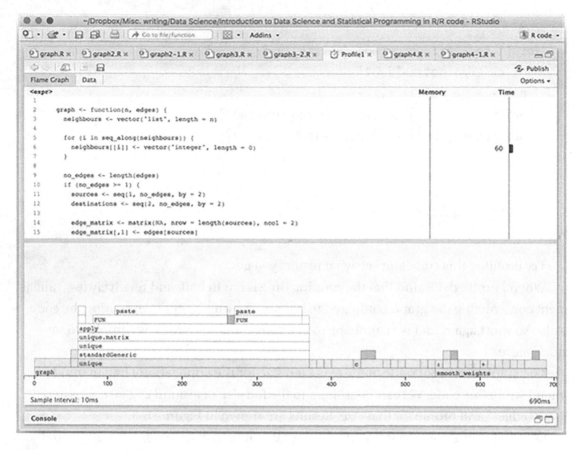

Figure 16-5. *Profiling results on a full time graph*

```r
flow_weights_iteration <- function(graph, node_weights, alpha) {
  if (length(node_weights) != length(graph))
    stop("Incorrect number of nodes")

  no_nodes <- length(node_weights)
  new_weights <- vector("numeric", n)

  for (i in 1:no_nodes) {
    neighbour_weights <- 0
    n <- 0
    for (j in graph[[i]]) {
      if (i != j) {
        neighbour_weights <- neighbour_weights + node_weights[j]
```

```
      n <- n + 1
    }
  }

  if (n > 0) {
    new_weights[i] <-
      alpha * node_weights[i] + (1 - alpha) * neighbour_weights / n
  } else {
    new_weights[i] <- node_weights[i]
  }

}
  new_weights
}

smooth_weights <- function(graph, node_weights, alpha, no_iterations) {
  new_weights <- node_weights
  replicate(no_iterations, {
      new_weights <- flow_weights_iteration(graph, new_weights, alpha)
  })
  new_weights
}
```

We can then profile with ten iterations:

```
profvis::profvis({
  n <- 20000
  nodes <- 1:n
  edges <- sample(nodes, 100000, replace = TRUE)
  weights <- rnorm(n)
  g <- graph(n, edges)
  flow_weights(g, weights, 0.8, 10)
})
```

The results are shown in Figure 16-6. Obviously, if we run the smoothing function more times, the smoothing is going to take up more of the total time, so there are no real surprises here. There aren't really any obvious hotspots any longer to dig into. I used the `replicate()` function for the iterations, and it does have a little overhead because it does more than just loop—it creates a vector of the results—and I can gain a few more milliseconds by replacing it with an explicit loop:

```
smooth_weights <- function(graph, node_weights,
                                  alpha, no_iterations) {
  new_weights <- node_weights
  for (i in 1:no_iterations) {
    new_weights <-
      smooth_weights_iteration(graph, new_weights, alpha)
  }
  new_weights
}
```

I haven't shown the results, so you will have to trust me on that. There is nothing major to attack any longer, however.

If you are in that situation where there is nothing more obvious to try to speed up, you have to consider if any more optimization is really necessary. From this point onward, unless you can come up with a better algorithm, which might be hard, further optimizations are going to be very hard and unlikely to be worth the effort. You are probably better off spending your time on something else while the computations run than wasting days on trying to squeeze a little more performance out of it.

Of course, in some cases, you really have to improve performance more to do your analysis in reasonable time, and there are some last resorts you can go to such as parallelizing your code or moving time-critical parts of it to C++. But for now, we can analyze full graphs in less than two seconds, so we definitely should not spend more time on optimizing this particular code.

Speeding Up Your Code

If you really do have a performance problem, what do you do? I will assume that you are not working on a problem that other people have already solved—if there is already a package available you could have used, then you should have used it instead of writing your own code, of course. But there might be similar problems you can adapt to your needs, so before you do anything else, do a little bit of research to find out if anyone else has solved a similar problem and, if so, how they did it. There are very few really unique problems in life, and it is silly not to learn from others' experiences.

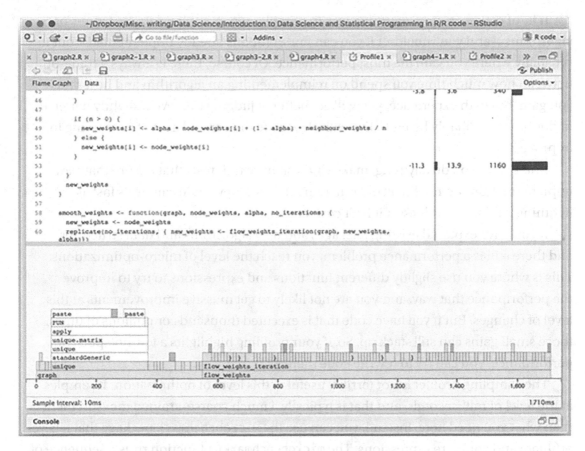

Figure 16-6. *Profiling results with multiple smoothing iterations*

It can take a little time to figure out what to search for, though, since similar problems pop up in very different fields. There might be a solution out there that you just don't know how to search for because it is described in terms entirely different from your own field. It might help to ask on mailing lists or Stack Overflow, but don't burn your karma by asking help with every little thing you should be able to figure out yourself with a bit of work.

If you really cannot find an existing solution you can adapt, the next step is to start thinking about algorithms and data structures. Improving these usually has much more of an impact on performance than micro-optimizations ever can. Your first attempts at any optimization should be to figure out if you could use better data structures or better algorithms.

It is, of course, a more daunting task to reimplement complex data structures or algorithms—and you shouldn't if you can find solutions already implemented—but it is usually where you gain the most performance. Of course, there is always a trade-off between how much time you spend on reimplementing an algorithm and how much you gain, but with experience, you will get better at judging this. Well, slightly better. If in doubt, it is probably better to live with slow code than spend a lot of time trying to improve it.

And before you do anything, make sure you have unit tests that ensure that new implementations do not break old functionality! Your new code can be as fast as lightning, and it is worthless if it isn't correct.

If you have explored existing packages and new algorithms and data structures and there still is a performance problem, you reach the level of micro-optimizations. This is where you use slightly different functions and expressions to try to improve the performance that way, and you are not likely to get massive improvements at this level of changes. But if you have code that is executed thousands or millions of times, those small gains can still stack up. So, if your profiling highlights a few hotspots for performance, you can try to rewrite code there.

The sampling profiler is not terribly useful at this level of optimization. It samples at the level of milliseconds, and that is typically a much coarse-grained measurement than what you need here. Instead, you can use the `microbenchmark` package that lets you evaluate and compare expressions. The `microbenchmark()` function runs a sequence of expressions several times and computes statistics on the execution time in units down to nanoseconds. If you want to gain some performance through micro-optimization, you can use it to evaluate different alternatives to your computations.

For example, we can use it to compare an R implementation of `sum()` against the built-in `sum()` function:

```
library(microbenchmark)

mysum <- function(sequence) {
  s <- 0
  for (x in sequence) s <- s + x
  s
}
```

```
microbenchmark(
  sum(1:10),
  mysum(1:10)
)

## Unit: nanoseconds
##          expr min     lq      mean median  uq
##    sum(1:10) 264 273.0    374.74  278.5 288
##  mysum(1:10) 755 768.5 26743.61  784.5 881
##       max neval cld
##      6409   100  a
##   2590600   100  a
```

The first column in the output is the expressions evaluated, then you have the minimum, lower quarter, mean, median, upper quarter, and maximum time observed when evaluating it, and then the number of evaluations used. The last column ranks the performance, here showing that sum() is a and mysum() is b so the first is faster. This ranking takes the variation in evaluation time into account and does not just rank by the mean.

There are a few rules of thumb for speeding up the code in micro-optimization, but you should always measure. Intuition is often a quite bad substitute for measurement.

One rule of thumb is to use built-in functions when you can. Functions such as sum() are actually implemented in C and highly optimized, so your own implementation will have a hard time competing with it, as we saw earlier.

Another rule of thumb is to use the simplest functions that get the work done. More general functions introduce various overheads that simpler functions avoid.

You can add together all numbers in a sequence using Reduce(), but using such a general function is going to be relatively slow compared to specialized functions:

```
microbenchmark(
  sum(1:10),
  mysum(1:10),
  Reduce(`+`, 1:10, 0)
)
```

```
## Unit: nanoseconds
##                     expr  min      lq    mean median
##              sum(1:10)  262   280.0  369.34  302.5
##            mysum(1:10)  767   792.5  987.13  830.0
##   Reduce(`+`, 1:10, 0) 5211 5373.5 5946.64 5504.0
##       uq   max neval cld
##    333.0  3661   100 a
##    891.0  5945   100  b
##   5649.5 19411   100   c
```

We use such general functions for programming convenience. They give us abstract building blocks. We rarely get performance boosts out of them, and sometimes they can slow things down substantially.

Thirdly, do as little as you can get away with. Many functions in R have more functionality than we necessarily think about. A function such as read.table() not only reads in data, it also figures out what type each column should have. If you tell it what the types of each column are using the colClasses argument, it gets much faster because it doesn't have to figure it out itself. For factor(), you can give it the allowed categories using the levels argument so it doesn't have to work it out itself:

```
x <- sample(LETTERS, 1000, replace = TRUE)
microbenchmark(
  factor(x, levels = LETTERS),
  factor(x)
)
```

```
## Unit: microseconds
##                         expr    min      lq
##   factor(x, levels = LETTERS) 13.915 15.5025
##                     factor(x) 52.322 59.4495
##      mean  median     uq     max neval cld
##   18.83221 16.1160 18.743  44.608   100   a
##   73.80394 66.4465 82.080 188.268   100    b
```

This is not just in effect when providing input, to help functions avoid spending time on figuring something out. Functions often also return more than you are necessarily interested in. Functions like unlist(), for instance, will preserve the names of a list into the resulting vector. Unless you really need those names, you should get rid of them since it is expensive dragging the names along with you. If you are just interested in a numerical vector, you should use `use.names = FALSE`:

```
x <- rnorm(1000)
names(x) <- paste("n", 1:1000)
microbenchmark(
  unlist(Map(function(x) x**2, x), use.names = FALSE),
  unlist(Map(function(x) x**2, x))
)
```

```
## Unit: microseconds
##                                                        expr
##  unlist(Map(function(x) x^2, x), use.names = FALSE)
##                     unlist(Map(function(x) x^2, x))
##     min        lq      mean    median        uq
##  497.608  539.9505  624.1242  577.4815  659.0905
##  533.906  572.9820  653.8293  600.9775  666.7030
##       max neval cld
##  1812.271    100    a
##  2059.874    100    a
```

Fourthly, when you can, use vector expressions instead of loops, not just because this makes the code easier to read but because the implicit loop in vector expressions is handled much faster by the runtime system of R than your explicit loops will.

Most importantly, though, is to always measure when you try to improve performance and only replace simple code with more complex code if there is a substantial improvement that makes this worthwhile.

Parallel Execution

Sometimes, you can speed things up, not by doing them faster, but by doing many things in parallel. Most computers today have more than one core, which means that you should be able to run more computations in parallel.

These are usually based on some variation of lapply() or Map() or similar; see, for example, package parallel but also check the package foreach that provides a higher-level looping construct that can also be used to run code in parallel.

If we consider our graph smoothing, we could think that since each node is an independent computation, we should be able to speed the function up by running these calculations in parallel. If we move the inner loop into a local function, we can replace the outer look with a call to Map():

```r
smooth_weights_iteration_map <- function(graph, node_weights, alpha) {
  if (length(node_weights) != length(graph))
    stop("Incorrect number of nodes")

  handle_i <- function(i) {
    neighbour_weights <- 0
    n <- 0
    for (j in graph[[i]]) {
      if (i != j) {
        neighbour_weights <- neighbour_weights + node_weights[j]
        n <- n + 1
      }
    }

    if (n > 0) {
      alpha * node_weights[i] + (1 - alpha) * neighbour_weights / n
    } else {
      node_weights[i]
    }
  }

  unlist(Map(handle_i, 1:length(node_weights)))
}
```

This is not likely to speed anything up—the extra overhead in the high-level Map() function will do the opposite, if anything—but it lets us replace Map() with one of the functions from parallel, for example, clusterMap():

```r
unlist(clusterMap(cl, inner_loop, 1:length(node_weights)))
```

Here, cl is the "cluster" that just consists of two cores I have on my old laptop:

```
cl <- makeCluster(2, type = "FORK")
```

```
microbenchmark(
  original_smooth(),
  using_map(),
  using_cluster_map(),
  times = 5
)
```

The changes gave me the results below. On my two-core laptop, we could expect the parallel version to run up to two times faster. In fact, it runs several orders of magnitude slower:

```
Unit: milliseconds
                 expr         min           lq
    original_smooth()    33.58665     33.73139

using_map()              33.12904     34.84107
using_cluster_map() 14261.97728 14442.85032
       mean       median            uq         max
   35.88307     34.25118      36.62977     41.21634
   38.31528     40.50315      41.28726     41.81587
15198.55138 14556.09176   14913.24566 17818.59187
neval cld
    5    a
    5    a
    5     b
```

I am not entirely sure what the problem we are seeing here is, but most likely the individual tasks are very short, and the communication overhead between threads (which are actually processes here) ends up taking much more time than the actual computation. At least my profiling seems to indicate that. With really lightweight threads, some of the communication could be avoided, but that is not what we have here.

Parallelization of this works better when each task runs longer so the threads don't have to communicate so often.[2]

---

[2] Parallelization on GPUs is a different case, but we won't go there in this book.

For an example where parallelization works better, we can consider fitting a model on training data and testing its accuracy on test data. We can use the cars data we have looked at before and the partition() function from Chapter 6.

We write a function that evaluates a single train/test partition and then call it n times, either sequentially or in parallel:

```
test_rmse <- function(data) {
  data$train %>% lm(dist ~ speed, data = .)

  model <- data$training %>% lm(dist ~ speed, data = .)
  predictions <- data$test %>% predict(model, data = .)
  rmse(data$test$dist, predictions)
}

sample_rmse <- function (n) {
  random_cars <- cars %>%
    partition(n, c(training = 0.5, test = 0.5))
  unlist(Map(test_rmse, random_cars))
}

sample_rmse_parallel <- function (n) {
  random_cars <- cars %>%
    partition(n, c(training = 0.5, test = 0.5))
  unlist(clusterMap(cl, test_rmse, random_cars))
}
```

When I do this for ten training/test partitions, the two functions take roughly the same time. Maybe the parallel version is a little slower, but it is not much overhead this time:

```
microbenchmark(
  sample_rmse(10),
  sample_rmse_parallel(10),
  times = 5
)
```

```
Unit: milliseconds
                        expr      min        lq
          sample_rmse(10) 28.72092 29.62857
 sample_rmse_parallel(10) 26.08682 27.15047
    mean    median      uq      max neval cld
 31.57316 33.05759 33.21979 33.23894     5   a
 34.75991 28.17528 29.37144 63.01556     5   a
```

If I create 1000 train/test partitions instead, however, the parallel version starts running faster than the sequential version:

```
microbenchmark(
  sample_rmse(1000),
  sample_rmse_parallel(1000),
  times = 5
)
```

```
Unit: seconds
                          expr     min       lq
          sample_rmse(1000) 3.229113 3.277292
 sample_rmse_parallel(1000) 2.570328 2.595402
    mean    median      uq      max neval cld
 3.459333 3.508533 3.536792 3.744934     5   b
 2.921574 2.721095 3.185070 3.535977     5   a
```

Since my laptop only has two cores, it will never be able to run more than twice as fast, and in general reaching the possible optimal speed-up from parallelization is rarely possible. The communication overhead between threads adds to the time used for the parallel version, and there are parts of the code that just have to be sequential such as preparing data that all threads should work on.

If you have a machine with many cores, and you can split your analysis into reasonably large independent chunks, though, there is often something to be gained.

Switching to C++

This is a drastic step, but by switching to a language such as C or C++, you have more fine-grained control over the computer, just because you can program at a much lower level, and you do not have the overhead from the runtime system that R does. Of course, this also means that you don't have many of the features that R does either, so you don't want to program an entire analysis in C++, but you might want to translate the time-critical code to C++.

Luckily, the Rcpp package makes integrating R and C++ pretty straightforward, assuming that you can program in both languages, of course. The only thing to really be careful about is that C++ index from zero and R from one. Rcpp takes care of converting this, so a one-indexed vector from R can be accessed as a zero-indexed vector in C++, but when translating code, you have to keep it in mind.

A full description of this framework for communicating between C++ and R is far beyond the scope of this book. For that, I will refer you to the excellent book *Seamless R and C++ Integration with Rcpp* by Dirk Eddelbuettel. Here, I will just give you a taste of how Rcpp can be used to speed up a function.

We will focus on the smoothing function again. It is a relatively simple function that is not using any of R's advanced features, so it is ideal to translate into C++. We can do so almost verbatim, just remembering that we should index from zero instead of one:

```
NumericVector
smooth_weights_iteration_cpp(List g,
                             NumericVector node_weights,
                             double alpha)
{
  NumericVector new_weights(g.length());

  for (int i = 0; i < g.length(); ++i) {

    IntegerVector neighbours = g[i];
    double neighbour_sum = 0.0;
    int n = 0;

    for (int j = 0; j < neighbours.length(); ++j) {
      neighbour_sum += node_weights[j];
      ++n;
    }
```

```
if (n > 0) {
  new_weights[i] = alpha * node_weights[i] +
    (1-alpha) * (neighbour_sum / n);
} else {
  new_weights[i] = node_weights[i];
}
}

return new_weights;
}
```

The types List, NumericVector, and IntegerVector correspond to the R types, and except for how we create the new_weights vector, the code very closely follows the R code.

There are several ways you can compile this function and wrap it into an R function, but the easiest is just to put it in a string and give it to the function cppFunction():

```
cppFunction("
NumericVector
smooth_weights_iteration_cpp(List g,
                             NumericVector node_weights,
                             double alpha)
{
  NumericVector new_weights(g.length());

  for (int i = 0; i < g.length(); ++i) {

    // The body here is just the C++ code
    // shown above...

  }

  return new_weights;
}
")
```

That creates a function, with the same name as the C++ function, that can be called directly from R, and Rcpp will take care of converting types as needed:

```
smooth_weights_cpp <- function(graph, node_weights,
                               alpha, no_iterations) {
  new_weights <- node_weights

  for (i in 1:no_iterations) {
    new_weights <-
      smooth_weights_iteration_cpp(graph, new_weights, alpha)
  }
  new_weights
}
```

If we compare the R and C++ functions, we see that we get a substantial performance boost from this:

```
microbenchmark(
  smooth_weights(g, weights, 0.8, 10),
  smooth_weights_cpp(g, weights, 0.8, 10),
  times = 5
)

Unit: milliseconds
                                    expr
     smooth_weights(g, weights, 0.8, 10)
 smooth_weights_cpp(g, weights, 0.8, 10)
       min          lq        mean    median
 1561.78635 1570.23346 1629.12784 1572.3979
   32.77344    33.38822    35.57017   36.5103
        uq         max neval cld
 1694.31571 1746.90573     5   b
   37.29083    37.88803     5   a
```

To translate a function into C++, you are not necessarily prevented from using R's more advanced features. You can call R functions from C++ just as easily as you can call C++ functions from R. Using R types translated into C++ can in many cases be used with vector expressions just like in R. Be aware, though, that the runtime overhead of using advanced features is the same when you use them in C++ as in R. You will likely not gain much performance from translating such functions. Translating low-level code like loops often gives you substantial performance boosts, though. If you have a few performance hotspots in your code that are relatively simple, just very time-consuming because they do a lot of work, it is worth considering translating these to C++, and Rcpp makes it easy.

Don't go overboard, though. It is harder to profile and debug code in C++, and it is harder to refactor your code if it is a mix of C++ and R. Use it, but use it only when you really need it.

Exercises

Find some code you have written and try to profile it. If there are performance hotspots you can identify, then try to optimize them. First, think algorithmic changes, then changes to the R expressions—checked using `microbenchmark()`—and if everything else fails, try parallelizing or implementing them in C++.

## CHAPTER 17

Project 2: Bayesian Linear Regression

The project for this chapter is building an R package for Bayesian linear regression. The model we will work with is somewhat a toy example of what we could imagine we could build an R package for, and the goal is not to develop all the bells and whistles of Bayesian linear regression. We will just build enough to see the various aspects that go into building a real R package.

Bayesian Linear Regression

In linear regression, we assume that we have predictor variables x and target variables y where $y = w_0 + w_1 x + \epsilon$ where $\epsilon \sim N(0, \sigma^2)$. That is, we have a line with intercept w_0 and incline w_1 such that the target variables are normally distributed around the point given by the line. We sometimes write σ^2 as $1/\beta$ and call β the precision. I will do this here and assume that β is a known quantity; we are going to consider a Bayesian approach to estimating the weights $w^T = (w_0, w_1)$.

Since we assume that we know the precision parameter β, if we knew the true weights of the model, then whenever we had an x value, we would know the distribution of y values: $y \sim N(w_0 + w_1 x, 1/\beta)$.

For notational purposes, I am going to define a function that maps x to a vector: $\phi : x \mapsto (1, x)^T$. Then we have $w^T \phi(x) = w_0 + w_1 x$ and $y \sim N(w^T \phi(x), 1/\beta)$.

Of course, we do not know the values of the weights but have to estimate them. In a Bayesian approach, we do not consider the weights as fixed but unknown values; we consider them as random variables from some distribution we have partial knowledge about. Learning about the weights means estimating the posterior distribution for the vector w conditional on observed x and y values.

© Thomas Mailund 2022
T. Mailund, *Beginning Data Science in R 4*, https://doi.org/10.1007/978-1-4842-8155-0_17

We will assume some prior distribution for w, $p(w)$. If we observe a sequence of matching x and y values $x^T = (x_1, x_2, \dots, x_n)$ and $y^T = (y_1, y_2, \dots, y_n)$, we want to update this prior distribution for the weights w to the posterior distribution $p(w \mid x, y)$.

We will assume that the prior distribution of w is a normal distribution with mean zero and diagonal covariance matrix with some (known hyperparameter) precision α, that is:

$$p(w|\alpha) = N(0, \alpha^{-1}I).$$

For reasons that I don't have time or space to go into here, this is a good choice of prior for a linear model since it means that the posterior will also be a normal distribution. It also means that, given x and y, we can compute the mean and covariance matrix for the posterior with some simple matrix arithmetic.

But first, we need to define our model matrix. This is a matrix that captures that the linear model we are trying to find is a line, that is, that $y = w_0 + w_1 x$. We define the model matrix for the observed vector x as such:

$$x = \begin{bmatrix} 1 & x_1 \\ 1 & x_2 \\ 1 & x_3 \\ \vdots & \vdots \\ 1 & x_n \end{bmatrix}.$$

In general, we would have a row per observation with the various features of the observation we want to include in the model, but for a simple line it is the incline and intercept, so for observation i it is 1 and x_i.

The posterior distribution for the weights, $p(w \mid x, y, \alpha, \beta)$, is then given by

$$p(w|x, y, \alpha, \beta) = N(m_{x,y}, S_{x,y})$$

where

$$m_{x,y} = \beta S_{x,y} x^T y$$

and

$$S_{x,y}^{-1} = \alpha I + \beta x^T x.$$

Exercises: Priors and Posteriors

Sample from a Multivariate Normal Distribution

If you want to sample from a multivariate normal distribution, the function `mvrnorm` from the package MASS is what you want:

```
library(MASS)

mvrnorm(n = 5, mu = c(0,0), Sigma = diag(1, nrow = 2))

##                [,1]        [,2]
## [1,] -0.1664955 -0.4859753
## [2,] -1.4915224 -1.0527432
## [3,] -0.2284911  0.4313458
## [4,]  0.4177218 -1.1576628
## [5,] -1.3341254 -0.2136770
```

You need to provide it with a mean vector, `mu`, and a covariance matrix, `Sigma`.

The prior distribution for our weight vectors is $N(0, S_0)$ with

$$0 = \begin{pmatrix} 0 \\ 0 \end{pmatrix}$$

and

$$S_0 = \frac{1}{\alpha} I = \begin{pmatrix} 1/\alpha & 0 \\ 0 & 1/\alpha \end{pmatrix}.$$

You can use the `diag` function to create the diagonal covariance matrix.

Write a function `make_prior(alpha)` that constructs this prior distribution and another function `sample_from_prior(n, alpha)` that samples n weight vectors wi from it. My version returns the samples as a data frame with a column for the w_1 samples and another for the corresponding w_0 samples. You can return the samples in any form that is convenient for you.

If you can sample from the prior, you can plot the results, both as points in w space and as lines in (x, y) space (a point (w_0, w_1) describes the line $y = w_0 + w_1 x$). See Figures 17-1 and 17-2 for samples from the prior, with the two alternative ways of looking at them.

Computing the Posterior Distribution

If we fix the parameters of the model, β and w $= (w_0, w_1)^T$, we can simulate (x, y) values. We can pick some random x values first and then simulate corresponding y values; see Figure 17-3, where the red line is the known line (given by (w_0, w_1)) and the black dots randomly sampled points given the model.

```
w0 <- 0.3 ; w1 <- 1.1 ; beta <- 1.3
x <- rnorm(50)
y <- rnorm(50, w1 * x + w0, 1/beta)
```

Of course, we do not know the true line in a setting such as this. What we have is a prior distribution of weights (w_0, w_1) and then observed (x, y) values. Think of the situation as Figure 17-3 but without the red line. If we have the points, and the prior, though, we have information about what the underlying line might be, and that information manifests as a posterior distribution for the weights—the distribution the weights have given that we have observed the points. When both our prior and our model say that the data points are normally distributed, the posterior is particularly simple, as we saw earlier. We can get the posterior distribution using the formula we saw before.

Write a function, `make_posterior(x, y, alpha, beta)`, that computes the posterior distribution (mean and covariance matrix) for the weights and a function `sample_from_posterior` that lets you sample from the posterior.

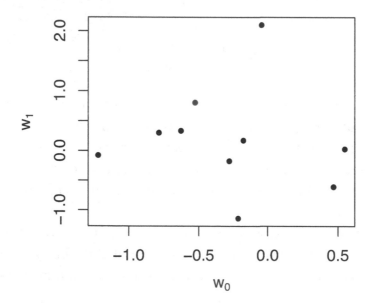

Figure 17-1. *Weight vectors sampled from the prior distribution*

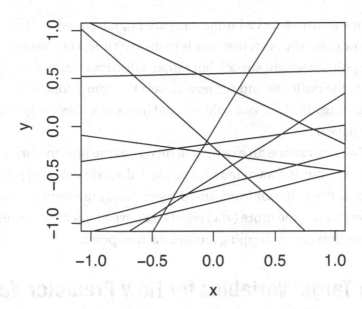

Figure 17-2. *Weight vectors sampled from the prior distribution represented as lines*

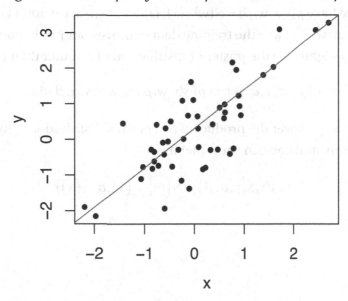

Figure 17-3. *Randomly sampled (x, y) values. The true linear model is shown in red*

If we again pretend that we can sample from the true distribution of (x, y) points, every time we get a bunch of points we can update the posterior knowledge we have. Naturally, in the real world, we don't have the true distribution to do this sampling, but the observations we have make it out for the data points, and we can still update our posterior each time we get new information.

475

Imagine that we have observed some n points, (x_1, y_1), (x_2, y_2), ... , (x_n, y_n), then we can ask ourselves how close our posterior now is to the true line. Our posterior, of course, isn't a line (or a point in weight space), but rather a distribution, so one way to do this is to sample from the posterior and see how closely they lump together—telling us how certain we are getting about the real values—and how close they lump around the true point, when we have that.

In Figure 17-4, you can see an example of this. The true line, the (w_0, w_1) the data points are sampled from, is shown in red. The black dots are the (x, y) points we have available to update the posterior from, and the gray lines are samples from the updated posterior. As you can see, the more (x, y) points we have to infer the posterior from, the closer the distribution gets to lumping around the true point.

Predicting Target Variables for New Predictor Values

Given a new value x, we want to make predictions about the corresponding y. For a fixed w, again, we have $p(y| x, \mathrm{w}, \beta) = N(\mathrm{w}^T \phi(x), 1/\beta)$, but since we don't know w, we have to integrate over all w. The way the training data improves our prediction is that we integrate over w weighted by the posterior distribution of w rather than the prior:

$$p\left(y|x, \mathrm{x}, y, \alpha, \beta\right) = \int p\left(y|x, \mathrm{w}, \beta\right) p\left(\mathrm{w}|\mathrm{x}, y, \alpha, \beta\right) d\mathrm{w}.$$

This kind of integral over the product of two normal distributions gives us another normal distribution, and one can show that it is

$$p\left(y|x, \mathrm{x}, y, \alpha, \beta\right) = N\left(\mathrm{m}_{\mathrm{x},\mathrm{y}}^T \phi\left(x\right), \sigma_{\mathrm{x},\mathrm{y}}^2\left(x\right)\right)$$

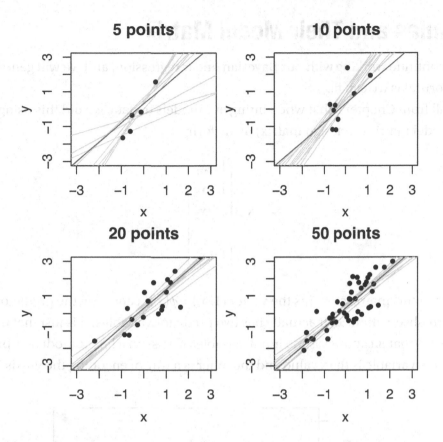

Figure 17-4. *Lines drawn from the posterior. The true line is shown in red*

where $m_{x,y}$ is the mean from the posterior distribution of w and where

$$\sigma^2_{x,y}(x) = \frac{1}{\beta} + \phi(x)^T S_{x,y} \phi(x)$$

where $S_{x,y}$ is the covariance matrix from the posterior distribution of w.

With the full distribution for the target value, *y*, given the predictor value, *x*, we can, of course, make predictions. The point prediction for *y* is, of course, the mean of this normal distribution, so $m^T_{x,y} \phi(x)$. But we can do more than just predict the most likely value; we can of course also get confidence values because we know the distribution.

Write a function that predicts the most likely *y* value for a given *x* value. Use it to plot the inferred model against (*x*, *y*) points. See Figure 17-5 for an example.

Use the fact that you also have the predicted distribution for *y* to write a function that gives you quantiles for this distribution and uses it to plot 95% intervals around the predictions; see Figure 17-6.

Formulas and Their Model Matrix

We will continue working with our Bayesian linear regression, and we will generalize the kind of formulas we can fit.

Recall from Chapter 6 that when fitting our models to data, we did this using a so-called model matrix (or design matrix) of the form

$$\mathbf{x} = \begin{bmatrix} 1 & x_1 \\ 1 & x_2 \\ 1 & x_3 \\ \vdots & \vdots \\ 1 & x_n \end{bmatrix}.$$

Row i in this matrix contains the vector $(1, x_i) = \phi(x_i)^T$ capturing the predictor variable for the i'th observation, xi. It actually has two predictor variables; it is just that one is a constant 1. What it captures is the input variables we use in a linear model for predicting targets. One variable is the x value and the other is a way of encoding the y-axis intercept.

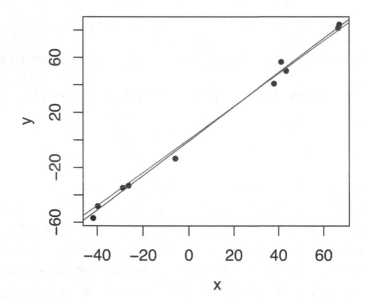

Figure 17-5. *True linear model in red and predicted values in blue*

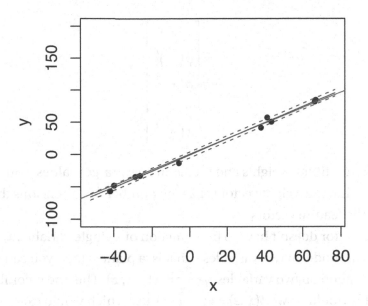

Figure 17-6. *Prediction with 95% support interval*

To predict the target, we use the inner product of this row and the weights of our fitted model:

$$y_i = \mathbf{w}^T \phi\left(x_i\right) + \dot{o} = w_0 \cdot 1 + w_1 \cdot x_i + \dot{o}.$$

We call $\phi(x)$ the feature vector for a point x, and the model matrix contains a row for each data point we want to fit or make predictions on, such that row i is $\phi(x_i)^T$. With the feature vector on the form we have used here, $\phi(x)^T = (1, x)$, we are fitting a line, but the feature doesn't have to have this form. We can make more complicated feature vectors.

If we instead used the feature vector $\phi(x)^T = (1, x, x^2)$ and added another weight to w so it now had three dimensions, (w_0, w_1, w_2), we could be predicting the target in the same way, $y = \mathbf{w}^T \phi(x_i) + \epsilon$, except now of course $\mathbf{w}^T \phi(x_i) = w_0 + w_1 x + w_2 x^2$, so we would be fitting a quadratic polynomial.

If you are thinking now "hey, that is no longer a linear model!", you are wrong. The linearity in the model was never actually related to the linearity in x. It is a linearity in w that makes the model linear, and as long as we are getting the predicted value as the inner product of a weight vector like this and some feature vector, it is a linear model we are working with. You can make the feature vector $\phi(x)$ as crazy as you like.

If you construct the model matrix the same way

$$\mathbf{x} = \begin{bmatrix} \phi(x_1)^T \\ \phi(x_2)^T \\ \phi(x_3)^T \\ \vdots \\ \phi(x_n)^T \end{bmatrix}$$

the mathematics for fitting weights and predicting new target values works the same, except of course that the weight vector w has the number of dimensions that reflects the dimensions in the feature vectors.

The feature vector doesn't have to be a function of a single variable, x, either. If you want to fit a linear model in two variables—that is, a plane—then you can just let the feature vector depend on two variables: $\phi(x, z)^T = (1, x, z)$. The linear combination with the weight vector would be $w^T \phi(x, z) = w_0 + w_1 x + w_2 z$ which would exactly be a linear model in two variables.

Working with Model Matrices in R

The way we specify both feature vectors and model matrices in R is a formula. A formula is created as an expression containing the tilde symbol, ~, and the target variable should be put to the left and the explanatory variables on the right.

R has quite a rich syntax for specifying formula, and if you are interested, you should read the documentation by writing

```
?formula
```

in the R shell.

For the linear model, we would write y ~ x. The intercept variable is implicitly there; you don't need to tell R that you want the feature vector to include the "-1", instead, you would have to remove it explicitly. You can also specify polynomial feature vectors, but R interprets multiplication, *, as involving interaction between variables.[1] To specify that you want the second-order polynomial of x, you need to write y ~ I(x^2) + x. The function I is the identity function, and using it here makes R interpret the x^2 as squaring the number x instead of trying to interpret it as part of the formula

---

[1] In formulas, x*z means x + z + x:z where x:z is the interaction between x and z—in practice, the product of their numbers—so y ~ x*z means $\phi(x, z) = (1, x, z, x.z)$.

specification. If you only want to fit the square of x, you would just write y ~ I(x^2). For a general n-degree polynomial, you can use y ~ poly(x,n, raw=TRUE).

To fit our linear model, we need data for two things. In the model we have already implemented, we had vectors x and y, but in the general case, the prediction variable x should be replaced with the model matrix. From the matrix and y we can fit the model.

R has functions for getting both from a formula and data. It isn't quite straightforward, though, because of scoping rules. If you write a formula somewhere in your code, you want the variables in the formula to refer to the variables in the scope where you are, not somewhere else where the code might look at the formula. So the formula needs to capture the current scope—similar to how a closure captures the scope around it. On the other hand, you also want to be able to provide data directly to models via data frames. Quite often, the data you want to fit is found as columns in a data frame, not as individual variables in a scope. Sometimes, it is even a mix.

The function model.frame lets you capture what you need for collecting data relevant for a formula. It will know about the scope of the formula, but you can add data through a data frame as well. Think of it as a data.frame, just with a bit more information about the data that it gets from analyzing the formula.

We can see all of this in action in the following small example:

```
predictors <- data.frame(x = rnorm(5), z = rnorm(5))
y <- with(predictors, rnorm(5, mean = 3*x + 5*z + 2))

model <- y ~ x + z

model.frame(model, data = predictors)

##            y          x           z
## 1   5.7166648  1.2943469  0.07347801
## 2  -7.3801586  0.7210624 -2.12012399
## 3   0.4894598 -0.6444302  0.15265587
## 4   5.5552442  1.6107988  0.01996477
## 5   7.8151347  0.2048122  0.98655386
```

Here, we have two predictor variables, x and z, in a data frame, and we simulated the response variable, y, in the global scope. We create the model using the formula y ~ x + z (which means $\phi(x, z)^T = (1, x, z)$), and we construct a model frame from this that contains the data for all the variables used in the formula.

The way the model frame gets created, R first looks in the data frame it gets for a variable, and if it is there, it uses that data; if it is not, it uses the data it can find in the scope of the formula. If it cannot find it at all, it will, of course, report an error.

The data frame is also used to construct expressions from variables. In the scope, you might have the variable x but not the variable x^2 where the latter is needed for constructing a model matrix. The model.frame function will construct it for you:

```
x <- runif(10)
model.frame(~ x + I(x^2))
```

```
##                 x         I(x^2)
## 1   0.9257926 0.857091....
## 2   0.2048174 0.041950....
## 3   0.3320932 0.110285....
## 4   0.5124721 0.262627....
## 5   0.4848565 0.235085....
## 6   0.1284884 0.016509....
## 7   0.9344755 0.873244....
## 8   0.8077187 0.652409....
## 9   0.7667746 0.587943....
## 10 0.9961101 0.992235....
```

In this example, we don't have a response variable for the formula; you don't necessarily need one. You need it to be able to extract the vector y of course, so we do need one for our linear model fitting, but R doesn't necessarily need one.

Once you have a model frame, you can get the model matrix using the function model.matrix. It needs to know the formula and the model frame (the former to know the feature function ϕ and the latter to know the data we are fitting).

In the following, we build two models, one where we fit a line that goes through $y = 0$ and the second where we allow the line to intersect the y-axis at an arbitrary point.

Notice how the data frames are the same—the variables used in both models are the same—but the model matrices differ:

```
x <- runif(10)
y <- rnorm(10, mean=x)

model.no.intercept <- y ~ x + 0
(frame.no.intercept <- model.frame (model.no.intercept))
```

```
##                y         x
## 1     0.05218288 0.7020026
## 2     0.78985212 0.2376200
## 3    -0.81501744 0.4099537
## 4    -0.71807916 0.5834204
## 5     0.40083214 0.5577035
## 6    -0.82443129 0.2163221
## 7    -0.98024996 0.1702824
## 8    -0.47115407 0.3512283
## 9     0.35959763 0.2272345
## 10   -0.47588279 0.9553629
```

```
model.matrix(model.no.intercept, frame.no.intercept)
```

```
##             x
## 1   0.7020026
## 2   0.2376200
## 3   0.4099537
## 4   0.5834204
## 5   0.5577035
## 6   0.2163221
## 7   0.1702824
## 8   0.3512283
## 9   0.2272345
## 10  0.9553629
## attr(,"assign")
## [1] 1
```

```
model.with.intercept <- y ~ x
(frame.with.intercept <- model.frame (model.with.intercept))
```

```
##                y         x
## 1     0.05218288 0.7020026
## 2     0.78985212 0.2376200
## 3    -0.81501744 0.4099537
## 4    -0.71807916 0.5834204
## 5     0.40083214 0.5577035
```

```
## 6   -0.82443129 0.2163221
## 7   -0.98024996 0.1702824
## 8   -0.47115407 0.3512283
## 9    0.35959763 0.2272345
## 10  -0.47588279 0.9553629
```

```
model.matrix(model.with.intercept, frame.with.intercept)
```

```
##    (Intercept)          x
## 1            1 0.7020026
## 2            1 0.2376200
## 3            1 0.4099537
## 4            1 0.5834204
## 5            1 0.5577035
## 6            1 0.2163221
## 7            1 0.1702824
## 8            1 0.3512283
## 9            1 0.2272345
## 10           1 0.9553629
## attr(,"assign")
## [1] 0 1
```

The target vector, or response variable, y, can be extracted from the data frame as well. You don't need the formula this time because the data frame remembers which variable is the response variable. You can get it from the model frame using the function model.response:

```
model.response(frame.with.intercept)
```

```
##              1            2            3            4
## 0.05218288   0.78985212 -0.81501744 -0.71807916
##              5            6            7            8
## 0.40083214 -0.82443129 -0.98024996 -0.47115407
##              9           10
## 0.35959763 -0.47588279
```

Exercises

Building Model Matrices

Build a function that takes as input a formula and optionally, through the ... variable, a data frame and build the model matrix from the formula and optional data.

Fitting General Models

Extend the function you wrote earlier for fitting lines to a function that can fit any formula.

Model Matrices Without Response Variables

Building model matrices this way is all good and well when you have all the variables needed for the model frame, but what happens when you don't have the target value? You need the target value to fit the parameters of your model, of course, but later on, you want to predict targets for new data points where you do not know the target, so how do you build the model matrix then?

With some obviously fake data, the situation could look like this:

```
training.data <- data.frame(x = runif(5), y = runif(5))
frame <- model.frame(y ~ x, training.data)
model.matrix(y ~ x, frame)
```

```
##   (Intercept)         x
## 1           1 0.1935202
## 2           1 0.4235126
## 3           1 0.8715640
## 4           1 0.0687407
## 5           1 0.7034587
## attr(,"assign")
## [1] 0 1
```

```
predict.data <- data.frame(x = runif(5))
```

but now, if we want to build a frame for the predicted data

```
frame <- model.frame(y ~ x, predict.data)
```

we would get an error. The formula tells `model.frame` that it needs the variable y, but `predict.data` doesn't have it; it only have x. So leave out the response side of the formula:

```
frame <- model.frame(~ x, predict.data)
model.matrix(~ x, frame)
```

```
##    (Intercept)          x
## 1            1 0.5181431
## 2            1 0.6967051
## 3            1 0.4965555
## 4            1 0.0729577
## 5            1 0.7235315
## attr(,"assign")
## [1] 0 1
```

This is not quite as easy if you don't know the formula, when it is input to your model. You cannot simply replace a formula you don't know with another that might not be related to the first. You can, however, remove the response variable from any formula using the `delete.response` function.

You cannot call `delete.response` directly on a formula; that is not the type of objects it works on. But you can combine it with the function `terms` to get a formula without the response variable that you can then use to build a model matrix for data where you don't know the target values:

```
# assume this is a parameter you don't know
unknown <- y ~ x
```

```
# get the formula without the response
```

```
responseless_formula <- delete.response(terms(unknown))
# and then you can use it with model.frame
frame <- model.frame(responseless_formula, predict.data) model.
matrix(responseless_formula, frame)
```

```
##    (Intercept)          x
## 1            1 0.5181431
## 2            1 0.6967051
## 3            1 0.4965555
```

```
## 4             1 0.0729577
## 5             1 0.7235315
## attr(,"assign")
## [1] 0 1
```

Exercises

Model Matrices for New Data

Write a function that takes as input a formula and a data frame as input that does not contain the response variable and build the model matrix for that.

Predicting New Targets

Update the function you wrote earlier for predicting the values for new variables to work on models fitted to general formula. If it doesn't already permit this, you should also extend it so it can take more than one such data point. Make the input for new data points come in the form of a data frame.

Interface to a `blm` Class

By now, we have an implementation of Bayesian linear regression but not necessarily in a form that makes it easy to reuse. Wrapping the data relevant for a fitted model into a class and providing various methods to access it is what makes it easy to reuse a model/class.

Generally, you want to access objects through functions as much as you can. If you know which $fields the class has, it is easy to write code that just accesses this, but that makes it hard to change the implementation of the class later. A lot of code that makes assumptions about how objects look like will break. It will also make it hard at some later point to change the model/class in an analysis because different classes generally do not look the same in their internals.

To make it easier for others—and your future self—to use the Bayesian linear regression model, we will make a class for it and provide functions for working with it.

This involves both writing functions specific to your own class and writing polymorphic functions that people, in general, expect a fitted model to implement. It is the latter that will make it possible to replace another fitted model with your `blm` class.

487

How you go about designing your class and implementing the functions—and choosing which functions to implement, in general—is up to you, except, of course, when you implement blm-specific versions of already existing polymorphic functions; in that case, you need to obey the existing interface.

How you choose to represent objects of your class and which functions you choose to implement for it is generally up to you. There is a general convention in R, though, that you create objects of a given class by calling a function with the same name as the class. So I suggest that you write a constructor called blm.

There aren't really any obvious classes to inherit from, so the class of blm objects should probably only be "blm" and not a vector of classes. If you want to make a class hierarchy in your implementation or implement more than one class to deal with different aspects of your model interface, you should knock yourself out.

Constructor

A constructor is what we call a function that creates an object of a given class. In some programming languages, there is a distinction between creating and initializing an object. This is mostly relevant when you have to worry about memory management and such and can get quite complicated, and it is not something we worry about in R. It is the reason, though, that in Python the constructor is called init—it is actually the initialization it handles. The same is the case for Java—which enforces the rule that the constructor must have the same name as the class, where for R it is just a convention. In Java, you have a special syntax for creating new objects: new ClassName(). In Python, you have to use the name of the class to create the object—ClassName()—but the syntax looks just like a function call. In R, it is only a convention that says that the class name and the constructor should be the same. The syntax for creating an object looks like a function call because it is a function call, and nothing special is going on in that function except that it returns an object where we have set the class attribute.

So you should write a function called blm that returns an object where you have set the class attribute to "blm". You can do this with the `class<-` replacement function or the structure function when you create the object. The object is a list—that is the only way you have of storing complex data, after all—and what you put in it depends on what you need for the functions that will be the interface of your class. You might have to go back and change what data is stored in the object from time to time as you develop the interface for your function. That is okay. Try to use functions to access the internals of the object as much as you can, though, since that tends to minimize how much code you need to rewrite when you change the data stored in the object.

Updating Distributions: An Example Interface

Let's consider a case of something we could have as an interface to Bayesian linear models. This is not something you have to implement, but it is a good exercise to try.

The thing we do when we fit models in Bayesian statistics is that we take a prior distribution of our model parameters, $P(\theta)$, and update them to a posterior distribution, $P(\theta \mid D)$, when observing data D. Think of it this way: the prior distribution is what we just know about the parameters. Okay, typically we just make the prior up based on mathematical convenience, but you should think about it as what we know about the parameters from our understanding of how the universe works and what prior experience has taught us. Then when you observe more, you add information about the world which changes the conditional probability of how the parameters look given the observations you have made.

There is nothing really magical about what we call prior and posterior here. Both are just distributions for our model parameters. If the prior is based on previous experience, then it is really a posterior for those experiences. We just haven't modelled it that way.

Let's say we have observed data D_1 and obtained a posterior $P(\theta \mid D_1)$. If we then later observe more data, D_2, we obtain even more information about our parameters and can update the distribution for them to $P(\theta \mid D_1, D_2)$.

We can of course always compute this distribution by taking all the old data and all the new and push it through our fitting code. But if we have chosen the prior distribution carefully with respect to the likelihood of the model—and by carefully I mean that we have a so-called conjugate prior—then we can just fit the new data but with a different prior: the old posterior.

A conjugate prior is a prior distribution that is chosen such that both prior and posterior are from the same class of distributions (just with different parameters). In our Bayesian linear model, both prior and posterior are normal distributions, so we have a conjugate prior. This means that we can, in principle, update our fitted model with more observations just by using the same fitting code but with a different prior.

I hinted a bit at this in the exercises earlier, but now you can deal with it more formally. You need a way of representing multivariate normal distributions—but you need this anyway to represent your blm objects—and a way of getting to a fitted one inside your blm objects to extract a posterior.

There are many ways to implement this feature, so you have something to experiment with. You can have an `update` function that takes a prior and new observations as parameters and outputs the (updated) posterior. Here, you need to include the formula as well somehow to build the model matrix. Or you can let `update` take a fitted object together with new data and get the formula and prior from the fitted object. Of course, if you do this, you need to treat the prior without any observations as a special case—and that prior will not know anything about formulas or model matrices.

We can try with an interface like this:

```
update <- function(model, data, prior) { ... }
```

where `model` is the formula, `data` a new data set, and `prior` the prior to use for fitting. This is roughly the interface you have for the constructor, except there you don't necessarily have `data` as an explicit parameter (you want to be able to fit models without data in a data frame, after all), and you don't have `prior` as a parameter at all.

Thinking about it a few seconds and realizing that whatever model fitting we put in here is going to be exactly the same as in `blm`, we can change the interface to get rid of the explicit `data` parameter. If we let that parameter go through ... instead, we can use exactly the same code as in `blm` (and later remove the code from `blm` by calling `update` there instead):

```
update <- function(model, prior, ...) { ... }
blm <- function(model, ...) {
    # some code here...
    prior <- make_a_prior_distribution_somehow()
    posterior <- update(model, prior, ...)
    # some code that returns an object here...
}
```

To get this version of `blm` to work, you need to get the prior in a form you can pass along to `update`, but if you did the exercises earlier, you should already have a function that does this (although you might want to create a class for these distributions and return them as such so you can manipulate them through an interface if you want to take it a bit further).

Of course, instead of getting rid of the model fitting code in the body of `blm`, you could also get rid of `update` and put that functionality in `blm` by letting that function take a prior parameter. If you do that, though, you want to give it a default so you can use the original one if it isn't specified:

```
blm <- function(model, prior = NULL, ...) {
    # some code here...

    if (is.null(prior)) {
        prior <- make_a_prior_distribution_somehow()
    }
    posterior <- update(model, prior, ...)
    # some code that returns an object here...
}
```

Let us stick with having update for now, though. How would we use update with a fitted model?

```
fit1 <- blm(y ~ x)
fit2 <- update(y ~ x, new_data, fit1)
```

This doesn't work because fit1 is a blm object and not a normal distribution. You need to extract the distribution from the fitted model.

If you have stored the distribution in the object—and you should because otherwise, you cannot use the object for anything since the fit is the posterior distribution—you should be able to get at it. What you don't want to do, however, is access the posterior directly from the object as fit1$posterior or something like that. It would work, yes, but accessing the internals of the object makes it harder to change the representation later. I know I am repeating myself here, but it bears repeating. You don't want to access the internals of an object more than you have to because it makes it harder to change the representation.

Instead, write a function posterior that gives you the posterior:

```
posterior <- function(fit) fit$posterior
```

This function has to access the internals—eventually, you will have to get the information, after all—but if this is the only function that does it, and every other function uses this function, then you only need to change this one function if you change the representation of the object.

With that function in hand, you can do this:

```
fit2 <- update(y ~ x, new_data, posterior(fit1))
```

You can also write update such that it can take both fitted models and distributions as its input. Then you just need a way of getting to the prior object (that might be a distribution or might be a fitted model's posterior distribution) that works either way.

One approach is to test the class of the prior parameter directly:

```
update <- function(model, prior, ...) {
    if (class(prior) == "blm") {
        prior <- posterior(prior)
    }
    # fitting code here
}
```

This is a terrible solution, though, for various reasons. First of all, it only works if you either get a prior distribution or an object with class "blm". What if someone, later on, writes a class that extends your blm? Their class attribute might be c("myblm","blm") which is different from "blm", and so this test will fail—and so will the following code because there you assume that you have a distribution but what you have is an object of a very different class.

To get around that problem, you can use the function inherits. It tests if a given class name is in the class attribute, so it would work if someone gives your update function a class that specializes your blm class:

```
update <- function(model, prior, ...) {
    if (inherits(prior, "blm")) {
        prior <- posterior(prior)
    }
    # fitting code here
}
```

This is a decent solution—and one you will see in a lot of code if you start reading object-oriented code—but it still has some drawbacks. It assumes that the only object that can provide a distribution you can use as a prior is either the way you have implemented priors by default (and you are not testing that earlier) or an object of class "blm" (or specializations thereof).

You could, of course, make a test for whether the prior, if it isn't a fitted object, is of a class you define for your distributions, which would solve the first problem, but how do you deal with other kinds of objects that might also be able to give you a prior/posterior distribution?

Whenever you write such a class that can provide it, you can also update your update function, but other people cannot provide a distribution for you this way (unless they change your code). Explicitly testing for the type of an object in this way is not a good code design. The solution to fixing it is the same as for accessing object internals: you access stuff through functions.

If we require that any object we give to update as the prior parameter can give us a distribution if we ask for it, we can update the code to be just

```
update <- function(model, prior, ...) {
    prior <- posterior(prior)
    # fitting code here
}
```

This requires that we make a polymorphic function for posterior and possibly that we write a version for distribution objects as well. I will take a shortcut here and make the default implementation the identity function:

```
posterior <- function(x) UseMethod("posterior")
posterior.default <- function(x) x
posterior.blm <- function(x) x$posterior
```

The only annoyance now is that we call it posterior. It is the posterior distribution when we have a fitted object, but it isn't really otherwise. Let us change it to distribution:

```
distribution <- function(x) UseMethod("distribution")
distribution.default <- function(x) x
distribution.blm <- function(x) x$posterior
```

and update update accordingly:

```
update <- function(model, prior, ...) {
    prior <- distribution(prior)
    # fitting code here
}
```

This way, it even looks nicer in the update function.

Designing Your `blm` Class

As you play around with implementing your `blm` class, think about the interface you are creating, how various functions fit together, and how you think other people will be able to reuse your model. Keep in mind that "future you" is also "other people," so you are helping yourself when you do this.

The `update` function we developed earlier is an example of what functionality we could put in the class design and how we made it reusable. You should think about other functions for accessing your objects and design them.

One example could be extracting the distribution for a given input point. You implemented a function for predicting the response variable from predictor variables already, and later you will do it in the `predict` function again, but if you want to gain the full benefits of having a distribution for the response at a given input, you want to have the distribution. How would you provide that to users? How could you use this functionality in your own functions?

Play around with it as you develop your class. Whenever you change something, think about whether this could make other functions simpler or if things could be generalized to make your code more reusable.

Model Methods

There are some polymorphic functions that are generally provided by classes that represent fitted models. Not all models implement all of them, but the more you implement, the more existing code can manipulate your new class, another reason for providing interfaces to objects through functions only.

The following is a list of functions that I think your `blm` class should implement. The functions are listed in alphabetical order, but many of them are easier to implement by using one or more of the others. So read through the list before you start programming. If you think that one function can be implemented simpler by calling one of the others, then implement it that way.

In all cases, read the R documentation for the generic function first. You need the documentation to implement the right interface for each function anyway, so you might at least read the whole thing. The description in this note is just an overview of what the functions should do.

coefficients

This function should return fitted parameters of the model. It is not entirely straightforward to interpret what that means with our Bayesian models where a fitted model is a distribution and not a single point parameter. We could let the function return the fitted distribution, but the way this function is typically used that would make it useless for existing code. Existing code expects that it can get parameters of a fitted model using this function, so it is probably better to return the point estimates of the parameters which would be the mean of the posterior you compute when fitting.

Return the result as a numeric vector with the parameters named. That would fit what you get from lm.

confint

The function confint gives you confidence intervals for the fitted parameters. Here, we have the same issue as with coefficients: we infer an entire distribution and not a parameter (and in any case, our parameters do not have confidence intervals; they have a joint distribution). Nevertheless, we can compute the analogue to confidence intervals from the distribution we have inferred.

If our posterior is distributed as w ~ $N(m, S)$, then component i of the weight vector is distributed as $w_i \sim N(m_i, S_{i,i})$. From this, and the desired fraction of density you want, you can pull out the thresholds that match the quantiles you need.

You take the level parameter of the function and get the threshold quantiles by exploiting that a normal distribution is symmetric. So you want the quantiles to be c(level/2, 1-level/2). From that, you can get the thresholds using the function qnorm.

deviance

This function just computes the sum of squared distances from the predicted response variables to the observed. This should be easy enough to compute if you could get the squared distances or even if you only had the distances and had to square them yourself. Perhaps there is a function that gives you that?

fitted

This function should give you the fitted response variables. This is not the response variables in the data you fitted the model to, but instead the predictions that the model makes.

plot

This function plots your model. You are pretty free to decide how you want to plot it, but I could imagine that it would be useful to see an x-y plot with a line going through it for the fit. If there is more than one predictor variable, though, I am not sure what would be a good way to visualize the fitted model. There are no explicit rules for what the plot function should do, except for plotting something, so you can use your imagination.

predict

This function should make predictions based on the fitted model. Its interface is

```
predict(object, ...)
```

but the convention is that you give it new data in a variable `newdata`. If you do not provide new data, it instead gives you the predictions on the data used to fit the model.

print

This function is what gets called if you explicitly print an object or if you just write an expression that evaluates to an object of the class in the R terminal. Typically, it prints a very short description of the object.

For fitted objects, it customarily prints how the fitting function was called and perhaps what the fitted coefficients were or how good the fit was. You can check out how `lm` objects are printed to see an example.

If you want to print how the fitting function was called, you need to get that from when you fit the object in the `blm` constructor. It is how the constructor was called that is of interest, after all. Inside that function, you can get the way it was called by using the function `sys.call`.

residuals

This function returns the residuals of the fit. That is the difference between predicted values and observed values for the response variable.

summary

This function is usually used as a longer version of print. It gives you more information about the fitted model.

It does more than this, however. It returns an object with summary information. What that actually means is up to the model implementation, so do what you like here.

Building an R Package for blm

We have most of the pieces put together now for our Bayesian linear regression software, and it is the time we collect it in an R package. That is the next step in our project.

You already have an implementation of Bayesian linear regression with a class, blm, and various functions for accessing objects of this type. Now it is time to collect these functions in a package.

Deciding on the Package Interface

When you designed your class functionality and interface, you had to decide on what functionality should be available for objects of your class and how all your functions would fit together to make the code easy to extend and use. There is a similar process of design involved with making a package.

Everything you did for designing the class, of course, is the same for a package, but for the package, you have to decide on which functions should be exported and which should be kept internal.

Only exported functions can be used by someone else who loads your package, so you might be tempted to export everything you can. This, however, is a poor choice. The interface of your package is the exported functions, and if you export too much, you have a huge interface that you need to maintain. If you make changes to the interface of a package, then everyone using your package will have to update their code to adapt to the changing interface. You want to keep changes to the package interface at a minimum.

You should figure out which functionality you consider essential parts of the package functionality and what you consider internal helper functions and only export the functions that are part of the package interface.

Organization of Source Files

R doesn't really care how many files you use to have your source code in or how the source code is organized, but you might. At some point in the future, you will need to be able to find relevant functions to fix bugs or extend the functionality of your package.

Decide how you want to organize your source code. Do you want one function per file? Is there instead some logical way of splitting the functionality of your code into categories where you can have a file per category?

Document Your Package Interface Well

At the very least, the functions you export from your package should be documented. Without documentation, a user (and that could be you in the future) won't know how a function is supposed to be used.

This documentation is mostly useful for online help—the kind of help you get using ?—so it shouldn't be too long but should give the reader a good idea of how a function is supposed to be used.

To give an overall description of the entire package and how various functions fit together and how they should be used, you can write documentation for the package as a whole.

Like with package data, there isn't a place for doing this, really, but you can use the same trick as for data. Put the documentation in a source code file in the R/ directory.

Here is my documentation for the admixturegraph package:

```
#' admixturegraph: Visualising and analysing admixture graphs.
#'
#' The package provides functionality to analyse and test
#' admixture graphs against the \eqn{f} statistics described
#' in the paper
#' \href{http://tinyurl.com/o5a4kr4}{Ancient Admixture in Human History},
#' Patternson \emph{et al.}, Genetics, Vol. 192, 1065--1093, 2012.
#'
```

```
#' The \eqn{f} statistics, \eqn{f_2}, \eqn{f_3}, and \eqn{f_4},
#' extract information about correlations between gene frequencies
#" in different populations (or single diploid genome samples),
#' which can be informative about patterns of gene flow between
#' <<more description here>>
#'
#' @docType package
#' @name admixturegraph
NULL
```

It is the @docType and @name tags that tell Roxygen that I am writing documentation for the entire package.

Adding README and NEWS Files to Your Package

It is customary to also have a README and a NEWS file in your package. The README file describes what your package does and how and can be thought of as a short advertisement for the package, while the NEWS file describes which changes you have made to your package over time.

Many developers prefer to use "markdown" as the format for these files—in which case they are typically named README.md and NEWS.md—and especially if you put your package on GitHub,[2] it is a good idea to have the README.md file since it will be prominently displayed when people go to the package home page on GitHub.

README

What you write in your README file is up to you, but it is customary to have it briefly describe what the package does and maybe give an example or two on how it is used.

If you write it in markdown—in a file called README.md—it will be the home page if you put your package on GitHub.

You might want to write it in R Markdown instead to get all the benefits of knitr to go with the file. In that case, you should just name the file README.Rmd and put this in the header:

---

[2] We return to git and GitHub in a later session.

```
---
output:
  md_document:
    variant: markdown_github
---
```

This tells `knitr` that it should make a markdown file as output—it will be called `README.md`.

NEWS

This file should simply contain a list of changes you have made to your package over time. To make it easier for people to see which changes go with which versions of the package, you can split it into sections with each section corresponding to a version.

Testing

In the package, we should now make sure that all of our functions are tested by at least one unit test and that our package can make it through a package test.

GitHub

Sign up to GitHub and create a repository for the project. Move the code there.

Conclusions

Well, this is the end of the book but hopefully not the end of your data science career. I have said all I wanted to say in this book. There are many things I have left out—text processing, for instance. R is not my favorite language for processing text, so I don't use it, but it does have functionality for it. It just goes beyond the kind of data we have looked at here. If you want to process text, like genomes or natural languages, you need different tools than what I have covered in this book. I have assumed that you are just working on data frames. It made the book easier to write, which matched my laziness well. But it doesn't cover all that data science is about. For more specialized data analysis, you will need to look elsewhere. There are many good books you can consult. It just wasn't within the scope of this book.

It is the end of this book, but I would like to leave you with some pointers for learning more about data science and R. There are different directions you might want to go in depending on whether you are more interested in analyzing data or more about developing methods. R is a good choice for either. In the long run, you probably will want to do both. The books listed in the following will get you started in the direction you want to go.

Data Science

- *The Art of Data Science* by Roger Peng and Elizabeth Matsui

This is a general overview of the steps and philosophies underlying data science. It describes the various stages a project goes through—exploratory analysis, fitting models, etc.—and while it doesn't cover any technical details, it is a good overview.

Machine Learning

- *Pattern Recognition and Machine Learning* by Christopher Bishop

© Thomas Mailund 2022
T. Mailund, *Beginning Data Science in R 4*, https://doi.org/10.1007/978-1-4842-8155-0

This is a book I have been using to teach a machine learning class for many years now. It covers a lot of different algorithms for both supervised and unsupervised learning—also types of analysis not covered in this book. It is rather mathematical and focused on methods, but if you are interested in the underlying machine learning, it is an excellent introduction.

Data Analysis

- *Linear Models in R* by Julian J. Faraway

- *Extending the Linear Model with R: Generalized Linear, Mixed Effects and Nonparametric Regression Models* by Julian J. Faraway

Linear models and generalized linear models are the first things I try. Pretty much always. These are great books for seeing how those models are used in R.

- *R Graphics* by Paul Murrell

- *ggplot2: Elegant Graphics for Data Analysis* by Hadley Wickham

The first book describes the basic `graphics` package and the `grid` system that underlies `ggplot2`. The second book, obviously, is the go-to book for learning more about `ggplot2`.

R Programming

- *Advanced R* by Hadley Wickham

- *R Packages* by Hadley Wickham

These are great books if you want to learn more about advanced R programming and package development.

- *Seamless R and C++ Integration with Rcpp* by Dirk Eddelbuettel

If you are interested in integrating C++ and R, then Rcpp is the way to go, and this is an excellent introduction to Rcpp.

The End

This is where I leave you. I hope you have found the book useful, and if you want to leave me any comments and criticism, please do. It will help me improve it for future versions. If you think things should be added, let me know, and I will add a chapter or two to cover it. And definitely, let me know if you find any mistakes in the book. I will be particularly grateful if you spot any errors in the code included in the book.

Index

A

Automating testing, 411

B

Bare repositories, 425, 426
Bayesian linear model fitting, 374–376
Bayesian linear regression, 471, 472
  blm class
    building R package, 497
    designing, 494
    interface to, 487, 488
  building model matrices, 485
  constructor, 488, 489
  deviance, 495
  fitting general models, 485
  GitHub, 500
  model matrices without response
    variables, 485, 486
  model methods, 494
    coefficients, 495
    confint, 495
    fitted response variables, 496
    plots, 496
    predictions, 496
    print, 496
    residuals, 497
    summary, 497
  new data, model matrices
    for, 487
  news, 500
  organization of source files, 498

package interface, deciding on,
    497, 498
package interface well, document,
    498, 499
predicting new targets, 487
priors and posteriors
  formulas and model
    matrix, 478–480
  multivariate normal distribution,
    sample from, 473
  new predictor values/predicting
    target variables, 476, 477
  posterior distribution,
    computing, 474–476
README and NEWS file, 499
R, working with model matrices
    in, 480–484
testing, 500
updating distributions, 489–493
BugReports, 396
.Buildignore, 393

C

C++, 466–469
Classes, 376–379
Class hierarchies, object-oriented
    programming, 382, 383
Cloning repositories, 425, 426
Code, speeding up, 456–461
Confint, 495
Conjugate prior, 173, 489

Y, Z

Printed in the United States
by Baker & Taylor Publisher Services